Environmental Modeling with Stakeholders

Steven Gray • Michael Paolisso
Rebecca Jordan • Stefan Gray
Editors

Environmental Modeling with Stakeholders

Theory, Methods, and Applications

Springer

Editors
Steven Gray
Department of Community Sustainability
Michigan State University
East Lansing, MI, USA

Michael Paolisso
Department of Anthropology
University of Maryland
College Park, MD, USA

Rebecca Jordan
Department of Human Ecology
Rutgers University
New Brunswick, NJ, USA

Stefan Gray
Senior Adviser, Parliamentary Commissioner for the Environment
Wellington, New Zealand

ISBN 978-3-319-25051-9 ISBN 978-3-319-25053-3 (eBook)
DOI 10.1007/978-3-319-25053-3

Library of Congress Control Number: 2016951336

Printed on acid-free paper

This Springer imprint is published by Springer Nature
The registered company is Springer International Publishing AG Switzerland
The registered company address is: Gewerbestrasse 11, 6330 Cham, Switzerland

To my parents and role modelers, Gary and Diane, for their unwavering support.

– Steven Gray

To my students and the many environmental stakeholders who have taught me so much about the richness and utility of participatory approaches.

– Michael Paolisso

To Dave, Shelby, Lucas, Richard, Nancy, Dan, Chris, Cara, Delphine, Amanda, and all the other modelers who have collaboratively generated explanations supported with evidence.

– Rebecca Jordan

To Jess, Jackson, and Joel, your wonderful and mostly baffling nonlinear dynamics give rise to constantly entertaining emergent properties. Thanks crew.

– Stefan Gray

Finally, all of the editors are extremely thankful for Paula Szocik's outstanding editing and coordination abilities. Thanks for being our cheerleader and coach wrapped up into one! We also want to thank Alison Singer for her jokes (and editing capabilities).

Foreword

This book demonstrates the "coming of age" of participatory modeling (PM). We see applications of PM that vary: from the support of livelihoods and critical rangeland resources for the Maasai people in Kenya to regulating recreational boating in Australia and from understanding the tacit knowledge of fishermen in the Chesapeake area, USA, to managing the Ria Formosa Natural Park in Portugal. There is also a remarkable collection of modeling tools described in the book: ranging from fuzzy cognitive mapping (FCM) to system dynamics, from influence diagrams to Bayesian networks, and from structured surveys to decision trees. The editors have surveyed the field and attracted a very diverse company of authors from three continents—North America, Europe, and Australia—with various backgrounds and disciplinary training all unified by theoretical or applied interest in PM.

What I particularly like about the book is how it stimulates the reader to think about the most exciting part of PM, which is the interaction of individuals' informal mental models with the more formal models that are built using equations, computers, and software. On the one hand, we have the humans involved in the modeling process, with their qualitative, conceptual, cultural, and mental models. These models are largely driven by intuitive, implicit, and tacit knowledge, powered by Kahneman's "system one" or fast type of thinking. As humans or communities themselves constantly evolve, acquiring new knowledge and personal experience, changing priorities, values, party affiliations, judgments, decisions, or sympathies, so do their mental models and ideas about how systems work. On the other hand, we have the quantitative knowledge, based on data collected in other case studies and formalized in terms of system models and software code that is powered by the analytical and computational ability of the computer. These models are the embodiment of Kahneman's "system two," or slow type of thinking, driven by data, logic, and expertise. These quantitative models can also be changed through the acquisition of new information, but their alterations require more formal and intentional model revisions and are well explained by data and logic.

The PM experience is the marriage of the two types of models and two types of thinking that inform one another and that, when combined, can lead to a better understanding of our world. However, as is clear from the book chapters when considered collectively, there is still a gap between the operations of the two systems and a lack of integration of these knowledge systems. For example, as presented in one of the book chapters, there is a well-developed practice of developing surveys and questionnaires to explore human perceptions and values and to understand their roles in the overall system performance. However, we should remember that by even administering these surveys, we might be already changing the system. By just asking a question, we may be changing the answer. In physics this is known as the "observer effect": the act of observation may have an impact on the phenomenon being observed. Clearly with surveys we find some very similar effects. Consider asking the question about the value of a certain ecosystem function. By only presenting the question, we are likely to already increase this value, since some people may have never even thought about this function as bearing value, but by being asked about it, they may give it a higher rating. People are driven by many biases and heuristics, which have an impact on human responses, judgments, and actions.

The world of computer models is more structured, formalized, and static. This is not to say that there is less uncertainty in the observations we make and in model results we generate. But this is a different type of uncertainty. In ecology, hydrology, or biology, there are more "known unknowns," whereas in social science we are routinely dealing with "unknown unknowns."

The other problem we face when synching the mental and computer models is that the latter are developed, tested, and applied, but rarely well documented and prepared for reuse and further integration with other models. While there is growing interest in developing tools for model integration, to synch mental and computer models, there is still a long way to go before navigation in this space of models will become simple and intuitive enough.

In PM we launch the stakeholders into this space of models, while at the same time enriching the existing models and algorithms with the knowledge and information provided by the stakeholders. So far there are not many tools that can be offered to assist them. We still see that at some point the stakeholder discussions are stopped and the modelers go behind the scenes and do their magic, translating the results of workshop deliberations into some model codes, which are now no longer understood by the workshop participants. It is not until the computer spews out the model results in the form of maps, graphs, or tables that the results can be brought back to the stakeholders.

Just like in integrated modeling, where we are trying to connect various models as components and make them work in concert, exchanging information and updating each other, here we face the challenge of integrating computer models with mental models. The only integration tool that exists so far for this task is the model interface, though it is rarely built with this purpose in mind. This is certainly worth looking at and further developing.

These limitations notwithstanding, the field of PM has exciting new implications in this information age to increase our scientific understanding of current socio-environmental dynamics, improve decision-making, and incite collective action to address shared problems.

The field of PM is certainly not immune to contradictions, which is also demonstrated in the book. Given the wide range of researchers and practitioners from multiple fields working in PM, it is not surprising that the language used to describe different modeling techniques can be confusing, if not altogether contradictory. For example, one chapter refers to FCM approaches as "dynamic" and draws analogies with system dynamics, forgetting that FCM by definition can deal only with concepts, with no time per se (or in fact any other units) involved in the analyses. Further, another chapter correctly recognizes that quantification in FCM is performed based solely on the relationships between the concepts in a model, highlighting that the outcomes can only be compared within the system and that concepts, their relationships, and assigned weights are subjective, hardly leaving space for actual validation of the model. In yet another chapter, we again find quite misleading claims that FCMs can provide the modeler with insights about dynamics in the system and understanding of the stability of the output, as well as suggestions about actual calibration and validation of FCMs. Such contradictions are unfortunate but part and parcel of scientific progress, and I suspect that more robust knowledge of FCM, and indeed of other approaches, will emerge as these methods become more commonplace. I also see them as even useful for the readers who can conduct their own analyses and draw their own conclusions about this and similar issues so that PM can continue to move forward with help from the scientific community.

Third, we also see in this book that PM continues to be a very attractive area with many attempts to carve a particular niche and apply a specific brand for the specific type of PM that is involved. As was noted in my earlier paper with Francios Bousquet, there are numerous versions of PM that are available, such as "group model building," "community modeling," "mediated modeling," "shared vision planning," etc. which essentially are very similar in terms of the basic ideas and goals, yet they are presented as different "trademarks" or "brands," making them look quite special. We see that this trend continues with the introduction of the "structured decision-making" (SDM) or "level of sustainable activity" (LSA) frameworks, which seem to still be versions of PM, though rebranded for specific application domains and purposes. The proliferation of various brands may become somewhat confusing, but this phenomenon is probably inevitable, when different groups and developers need to have a special name or trademark for what they are doing. In most cases, what they do is still good old participatory modeling, but with a specific modeling.

Despite these current limitations and knowledge gaps, the "coming of age" of participatory modeling is well demonstrated in this book. Indeed, "coming of age" implies a certain grappling with the challenges of growth. Next we need to bring the field to maturity by further developing appropriate tools and methods that are aligned with the technologically capable times in which we now live. The new era of social media, smart phones, and personal devices and distributed and cloud computing certainly bears the promise of making PM available on a broader scale and operational for even bigger problems and challenges. My expectation is that with modern social media tools, the ease of access, unlimited connection time, breadth of sharing tools, etc., PM will become operational not just for local or regional problems where we can get all the stakeholders in one room but also for large-scale regional, national, or even global problems. "If it works for Facebook, why can't it work for PM?"

University of Twente
Enschede, The Netherlands

Alexey Voinov

Preface

The popularity of environmental modeling with stakeholders has grown considerably in recent years. This proliferation has been spurred by the assumption that the inclusion of stakeholders and a wide variety of scientific perspectives is required to improve our understanding of social-ecological systems and current environmental problems. As Voinov and Bousquet (2010) point out in their seminal paper, "Stakeholder engagement, collaboration, or participation, shared learning or fact-finding, have become buzz words and hardly any environmental assessment or modeling effort today can be presented without some kind of reference to stakeholders and their involvement in the process" (p. 1268). They go on to identify two major objectives that drive environmental modeling with stakeholders: to increase and share knowledge and understanding of a system and its dynamics under various conditions and to identify and clarify the impacts of solutions to a given problem. Currently, a wide range of stakeholder-centered modeling programs and practices exist, which all essentially aim to provide decision support and facilitation in participatory planning contexts. However, as the field of "participatory modeling" becomes more mainstream with new technological advances, important questions such as "what constitutes quality participation?," "how is the public included in modeling efforts?," "what exactly do we mean by the term model?," and "what tools and processes actually lead to improved understanding or a change in environmental policy?" remain largely unanswered.

In this edited volume, we begin to address these questions by bringing together, in a central text, the work of more than 50 scholars working at the intersection of modeling, the natural and social sciences, and public participation. Contributors to this book include a wide range of natural scientists, social scientists, and interdisciplinary researchers interested in methodological and technological approaches to including the public in environmental decision-making via different forms of modeling. Therefore, not surprisingly, the answers to these questions show a considerable degree of variation.

In semi-structured interviews conducted by the editors with a majority of chapter authors, considerations for *how and why stakeholders were involved* and what constituted "participation" varied by context and methodological approach. Chapter authors mentioned that the value of modeling with stakeholders came in many forms, including:

- A source of data
- Model calibrators
- Model validators
- A way to integrate opinions
- Defining attitudes and preferences for outcomes
- A way to refine and revise models
- A feasible way to bridge the qualitative/quantitative divide
- A way to promote buy-in of decision-making

Additionally, defining the *purpose of* "modeling" also varied by context and included:

- Moving from conceptual to empirical understanding of the world
- Providing simple contextual understanding of environmental problems through knowledge sharing
- A way to evaluate competing hypotheses
- A way to understand public "thinking"
- A way to define the desirability of competing scenarios
- To guide project design and implementation
- A way to ensure the right questions were being asked and the right data was used to answer such questions

Even questions about what chapter authors meant by *model* varied and included:

- An abstraction of reality
- A formal and logical set of interactions
- A mathematical representation
- An informal mental model representation
- A tool for reasoning and a representation of a hypothesis

In our view, such diversity in chapter author responses to these questions indicates two things. First, environmental modeling with stakeholders is an extremely versatile tool that can meet a variety of contemporary environmental decision-making objectives that are not exclusive to one another. Second, the field is still maturing, and the terms and approaches associated with participatory modeling are, to date, largely unstandardized if not altogether ambiguous. This variation in the field is exciting, but also daunting because we are far from generating norms and conceptual agreement, although we hope that the perspectives in this book can be used to move us closer to a shared understanding and language.

Structure of the Book

To move these discussions forward, we divide the book into two sections. The first part of the book covers basic considerations for including stakeholders in the *modeling process* and its intersection with the theory and practice of public participation in environmental decision-making. The second part of the book is devoted to specific *applications and products* of the various methods available through case study examination. This second part of the book also provides insight from several international experts currently working in the field about their approaches, types of interactions with stakeholders, models produced, and the challenges they perceived based on their practical experiences.

Book Chapters

In terms of the process of environmental modeling with stakeholders, in Chap. 1, Paolisso and Trombley begin the book by disentangling the interactions between cognitive, material, and technological dimensions of participatory environmental modeling. The authors argue that there is a need to recognize the implicit value orientations and cognitive biases that lead to using computational and technologically supported approaches to understanding social-ecological system dynamics. Further, these authors highlight the way in which these tools influence interpretations of meaning and shape understanding of environmental issues. In Chap. 2, Tuler et al. present a sociocultural approach to understanding participatory modeling. Using qualitative data collected from participants in their process-oriented program called VCAPS, the authors describe the complex social and cultural environment in which the practice of participatory modeling often takes place and highlight that the modeling process should be designed to promote learning during the model building process.

Following these discussions of how models are situated within cultural contexts, in Chap. 3, Voinov and Gaddis provide a review of the popularity of participatory modeling and highlight the lessons learned from their experiences, calling attention to the fact that values, in addition to knowledge, should be considered part and parcel of the inclusive modeling process. The authors then provide clear suggestions on design considerations with specific attention to the "best practices" of adopting this type of approach. These first three chapters provide an important conceptual framework, and specific guidelines, for how the processes of participatory modeling should be designed, and further, the types of questions that should be asked by researchers before modeling tools are selected.

After these more theoretical concerns are reviewed, Chaps. 4–6 move in a more practical direction and highlight specific methods that can be used to operationalize some of the concepts identified in earlier chapters. In Chap. 4, Nelitz and Beardmore demonstrate how structured survey methods are not only tenable but highly effective in the effort to balance trade-offs when considering decision alternatives. The authors call attention to the value that structured surveys can play in capturing the judgments, priorities, and values that different stakeholders bring to collaborative decision-making contexts. These values can be used to statistically parametrize a range of computational modeling approaches. These authors provide three cases where structured survey methods enabled reliable decisions when other group engagement methods were not practical. In Chap. 5, Robinson and Fuller present a comprehensive review of structured decision-making (SDM), focusing on the value of earning trust and multiple perspectives through stakeholder participation. As individuals come together during the modeling process, they are able to evaluate multiple alternatives in the resource management process. Because stakeholder values are discussed during the objective setting phase, there is an expectation of greater fulfillment on the part of all participants. The first section of the book concludes with Wolfe et al. in Chap. 6 which addresses "participation" from a slightly different lens than the previous two chapters and includes a case study of using a Delphi method to integrate disciplinary perspectives, including different scientific studies within a region. Their approach ultimately facilitated negotiation and defined shared research priorities among expert groups working on land management issues. These authors call attention to the different worldviews and practices associated with different experts and practitioners working in the same area.

In the second section of the book, we focus on understanding how and when different tools and approaches have been used, how stakeholders were involved in the modeling process, and what decision-making affordances emerged. This section begins with Chap. 7's overview of fuzzy cognitive mapping (FCM), where Malek provides a useful primer and history of this modeling approach. Of particular interest is the attention given to FCM's ability to represent feedbacks and explore hypothetical scenarios of change, which are common goals of many participatory modeling approaches. In Chap. 8, van Vliet et al. continue the discussion of FCM and demonstrate how models developed by regional experts can be compared with those developed using more formal modeling techniques. Doing so identifies commonalities between informal and stakeholder-driven descriptions of a system and those offered up by more formal and quantitative modeling practices. Penn et al., in Chap. 9, conclude the FCM subsection and present an extension of FCM beyond representations of stakeholder understanding of a system and promote the idea of including discussions of human agency that the authors term "controllability," which extends the applications previously discussed as a starting point for collaboratively developing effective policy options.

Chapters 10 and 11 focus on the participatory application of agent-based modeling (ABM) and present two very different case studies that illustrate how this approach can be applied to understand the dynamic behavior and emergent properties of coupled human natural systems. In Chap. 10, Mwangi demonstrates how ABM can be used to understand the influence of climate change on interactions between environmental, social, and economic conditions affecting the livelihoods of Maasai pastoralists, providing an interesting case of how local knowledge can be used to parametrize and predict future social and environmental states. In Chap. 11,

Itami et al. employ ABM in a different context and present a framework called "level of sustainable activity" that integrates information from several stakeholder communities. These authors show interactions between boating activity, business activity, and management activity in a process that creates simulations that are useful to a range of decision-makers in Australia. Both of these ABM applications highlight the nature of understanding environmental and social change from the user or stakeholder perspective.

Chapters 12–15 provide case studies of using various system dynamics modeling approaches with stakeholders. In Chap. 12, Videira et al. introduce a case study in Portugal that begins with a discussion of generic causal loop and stock-and-flow diagrams. These authors extend their review to include the implications of these approaches within multi-criteria decision-making and visioning exercises to improve stakeholder participation in sustainability decision-making. In Chap. 13, Webler et al. outline the detailed use of their VCAPS process (see Tuler et al. in Chap. 2) that couples dialog-based group concept mapping and system dynamics modeling to understand lobstermen's, community members', and scientists' modeling practices formed around collective concerns with regard to the lobster fishery and climate change. In Chap. 14, Inouye et al. discuss the coproduction of knowledge using system dynamics approaches to move from informal and qualitative stakeholder understanding to more formal and quantitative understanding of local-scale dynamics of water usage. By creating a collaborative knowledge to action network (KTAN), the authors demonstrate how a network of diverse individuals can conceptualize and visualize the impacts of water scarcity under alternative scenarios of climate and population change as the group collectively learns about uncertainty in model-based projections. Finally, to finish this subsection, in Chap. 15, Santoso and Halog outline how extracted information from the public in previously published literature can be used to understand integrated coastal zone management and establish policy scenarios for managing complex coastal areas.

We conclude the book with two chapters that demonstrate more integrated approaches to participatory modeling. In Chap. 16, Mokrech and colleagues use a "meta-modeling" approach to develop a continental (European)-scale integrated assessment methodology, allowing dynamic and cross-sectoral simulations of flood impacts and wetland change to be developed under varying social and environmental conditions. The value of this chapter comes in the form of technical modeling practices and their goal of creating decision support for a range of stakeholders in the CLIMSAVE Integrated Assessment Platform across several stakeholder groups in Europe. Finally, in Chap. 17, we see another example of integrated modeling that employs both a Bayesian and an ABM approach to simulate the dynamics of ranchers in Mexico. In this final chapter, Pope and Gimblett evaluate how modeling with stakeholders can be used to understand conflict for water between human use and ecological function. They evaluate models under shifting environmental and economic conditions using information about rancher decision-making to build "Bayesian cognitive maps" that create a probability of the likelihood a decision will be made. Further, the authors use these maps to parametrize an ABM that incorporates temporal dynamics to understand the complexity and dynamic nature of the issue.

Concluding Remarks

We are pleased to have the opportunity to present such diverse perspectives in this book and are very thankful for the wide range of contributions we received and are able to share. Additionally, we were honored that Alexey Voinov agreed to write a preface for this volume so that his perspective could be included along with the perspectives of our chapter authors. We hope the sharing of these many views on public participation in stakeholder-driven modeling research from across the globe will provide a new foundation from which further progress can be made, allowing new frontiers to be identified and new technological advances to be informed by high-quality social science and environmental research.

East Lansing, MI, USA — Steven Gray
College Park, MD, USA — Michael Paolisso
New Brunswick, NJ, USA — Rebecca Jordan
Wellington, NZ — Stefan Gray

Contents

Contributors

Paula Antunes CENSE – Center for Environmental and Sustainability Research, Faculdade de Ciências e Tecnologia, Universidade Nova de Lisboa, Caparica, Portugal

Taryn Arthur Environmental Sciences Division, Oak Ridge National Laboratory, Oak Ridge, TN, USA

Fox Run Dr., Plainsboro, NJ, USA

I. Bärlund Department Aquatic Ecosystem Analysis (ASAM), Helmholtz Centre for Environmental Research – UFZ, Magdeburg, Germany

Latha Baskaran Environmental Sciences Division, Oak Ridge National Laboratory, Oak Ridge, TN, USA

Ben Beardmore Bureau of Science Services, Wisconsin Department of Natural Resources, Madison, WI, USA

John P. Bolte Climate Impacts Research Consortium, Oregon State University, Corvallis, OR, USA

E.H. Çakmak Department of Economics, TED University, Ankara, Turkey

Georgios Chalkias Department of Sociology, Evolution and Resilience of Industrial Ecosystems Project, University of Surrey, Guildford, UK

D. D'Agostino Mediterranean Agronomic Institute of Bari, Valenzano, Italy

Virginia H. Dale Environmental Sciences Division, Oak Ridge National Laboratory, Oak Ridge, TN, USA

Kirstin Dow Department of Geography, University of South Carolina, Columbia, SC, USA

H. Dudu Institute for Prospective and Technological Studies, EC-JRC, Seville, Spain

P. Esteve Department of Agricultural Economics, Statistics and Management, Universidad Politécnica de Madrid, Madrid, Spain

M. Flörke Center for Environmental Systems Research, University of Kassel, Kassel, Germany

A.K. Fuller Department of Natural Resources, US Geological Survey, New York Cooperative Fish and Wildlife Research Unit, Cornell University, Ithaca, NY, USA

Erica Brown Gaddis Division of Water Quality, Utah Department of Environmental Quality, Salt Lake City, UT, USA

Randy Gimblett School of Natural Resources and Environment, University of Arizona, Tucson, AZ, USA

Rob Goble Environmental Science and Policy Program, Clark University, Worcester, MA, USA

Anthony Halog School of Geography, Planning, and Environmental Management (GPEM), The University of Queensland, Brisbane, QLD, Australia

Allison M. Inouye Climate Impacts Research Consortium, Oregon State University, Corvallis, OR, USA

Robert M. Itami GeoDimensions Pty Ltd, Sorrento, VIC, Australia

A.S. Kebede Faculty of Engineering and the Environmental, Tyndall Centre for Climate Change Research, University of Southampton, Highfield, Southampton, UK

R. Khadra Mediterranean Agronomic Institute of Bari, Valenzano, Italy

Christopher J.K. Knight Evolution and Resilience of Industrial Ecosystems Project, Department of Mathematics, University of Surrey, Guildford, UK

Jennifer Koch Department of Geography and Environmental Sustainability, The University of Oklahoma, Norman, OK, USA

K. Kok Soil Geography and Landscape Group, Wageningen University, Wageningen, The Netherlands

Denise H. Lach Climate Impacts Research Consortium, Oregon State University, Corvallis, OR, USA

David J.B. Lloyd Evolution and Resilience of Industrial Ecosystems Project, Department of Mathematics, University of Surrey, Guildford, UK

Žiga Malek Environmental Geography Group, Faculty of Earth and Life Sciences, Vrije Universiteit Amsterdam, Amsterdam, The Netherlands

Risk, Policy and Vulnerability Program, International Institute for Applied Systems Analysis (IIASA), Laxenburg, Austria

M. Mokrech School of Science and Computer Engineering, University of Houston Clear Lake, Houston, TX, USA

Margaret Mwangi Department of Geography & African Studies Program, The Pennsylvania State University, University Park, PA, USA

Marc A. Nelitz ESSA Technologies Ltd, Vancouver, BC, Canada

R.J. Nicholls Faculty of Engineering and the Environmental, Tyndall Centre for Climate Change Research, University of Southampton, Highfield, Southampton, UK

Michael Paolisso Department of Anthropology, University of Maryland, College Park, MD, USA

Alexandra S. Penn Evolution and Resilience of Industrial Ecosystems Project, Department of Sociology, Centre for Environmental Strategy, University of Surrey, Guildford, UK

Aaron Poe Aleutian and Bering Sea Islands LCC, U.S. Fish and Wildlife Service, Anchorage, AK, USA

Aloah Pope School of Natural Resources and the Environment, University of Arizona, Tucson, AZ, USA

K.F. Robinson Quantitative Fisheries Center, Department of Fisheries and Wildlife, Michigan State University, East Lansing, MI, USA

Rui Santos CENSE – Center for Environmental and Sustainability Research, Faculdade de Ciências e Tecnologia, Universidade Nova de Lisboa, Caparica, Portugal

Purwono Budi Santoso Directorate General of Marine and Fisheries Resources Surveillance, The Ministry of Marine Affairs and Fisheries of The Republic of Indonesia, Jakarta, Indonesia

Esperanza Stancioff University of Maine Cooperative Extension and Maine Sea Grant, ME, USA

John R. Stevenson Climate Impacts Research Consortium, Oregon State University, Corvallis, OR, USA

Jeremy Trombley Department of Anthropology, University of Maryland, College Park, MD, USA

Seth P. Tuler Interdisciplinary and Global Studies Division, Worcester Polytechnic Institute, Worcester, MA, USA

C. Varela-Ortega Department of Agricultural Economics, Statistics and Management, Universidad Politécnica de Madrid, Madrid, Spain

Anne P.M. Velenturf Evolution and Resilience of Industrial Ecosystems Project, Centre for Environmental Strategy, University of Surrey, Guildford, UK

Nuno Videira CENSE – Center for Environmental and Sustainability Research, Faculdade de Ciências e Tecnologia, Universidade Nova de Lisboa, Caparica, Portugal

M. van Vliet Public Administration and Policy Group, Wageningen University, Wageningen, The Netherlands

Water Consultancy Group, Movares B.V., Utrecht, The Netherlands

Soil Geography and Landscape Group, Wageningen University, Wageningen, The Netherlands

Alexey Voinov University of Twente, Faculty of Geo-information Science and Earth Observation, Enschede, The Netherlands

Thomas Webler Department of Environmental Studies, Keene State College, Keene, NH, USA

Jessica Whitehead North Carolina Sea Grant, North Carolina State University, Raleigh, NC, USA

Amy K. Wolfe Environmental Sciences Division, Oak Ridge National Laboratory, Oak Ridge, TN, USA

Part I
The *Process* of Environmental Modeling with Stakeholders

Chapter 1
Cognitive, Material and Technological Considerations in Participatory Environmental Modeling

Michael Paolisso and Jeremy Trombley

1.1 Introduction

Participatory environmental modeling is a relatively recent practice that involves scientists and members of the public working together to develop conceptual and dynamic models to address environmental issues (cf. Voinov and Gaddis 2008; Sandker et al. 2010; Gaddis et al. 2010; Voinov and Bousquet 2010; Whatmore 2009). Scientists involved in the process lend their expertise in creating computational models and help to keep the model scientifically sound while public stakeholders contribute first-hand knowledge of environmental systems and concerns about environmental quality and human dimensions (Voinov and Bousquet 2010). Additionally, since stakeholders are included in the modeling process, it may be assumed that more attention will be paid to their needs and concerns during the development of solutions to the environmental problems the models are designed to address (Voinov and Gaddis 2008; Korfmacher 2001).

As the many contributions in this volume attest, integrating the perspectives of those affected by decisions based on contemporary scientific modeling practices is an attempt to improve the democratization of science and the degree of public participation. Often environmental stakeholders possess knowledge of the complex socio-ecological dimensions and processes that environmental models attempt to capture and simplify. Stakeholders often have extensive experience with the socio-ecological dynamics that are at the heart of what modelers are attempting to identify and explain (Berkes 1999). Many stakeholders also may have engaged in their own study of relevant environmental issues in an attempt to understand complex socio-ecological dynamics that threaten environmental well-being and, in some cases, their livelihoods.

M. Paolisso • J. Trombley (✉)
Department of Anthropology, University of Maryland, College Park, MD, USA
e-mail: jmtrom@umd.edu

S. Gray et al. (eds.), *Environmental Modeling with Stakeholders*,
DOI 10.1007/978-3-319-25053-3_1

However, stakeholders bring more than their expertise to the modeling process. They are participating because the approaches and results of environmental modeling influence public policy and decision-making. Participants would like a "seat at the table" in order to understand and make sure their voices are heard, and even more proactive roles in terms of shaping the modeling's form and processes. Modelers are also beginning to recognize the research, policy, and political dimensions of participatory environmental modeling, as they expand their conceptualizations of how and when to integrate "participation" into the modeling process (Paolisso et al. 2015; Weller et al. 2013).

Thus, participatory environmental modeling is, in most cases, driven by good intent from both stakeholders and modelers; at best participation is recognized as needed, and at worst it is accepted as a broader social or policy requirement. But, participation by itself is a complex process, whose form and process varies by environmental topic and broader socio-ecological context (Callon 1999). Participation is not a simple or linear set of procedures for integrating stakeholder knowledge. Rather, it is a process by which complex knowledge and values are made to fit new conceptual frameworks (models) that in turn are dynamic depending on social, material, technological, and environmental conditions. Furthermore, the resulting participatory environmental model becomes a lens or integrated knowledge system that helps to define and influence broader socio-ecological systems. These larger considerations are often beyond the immediate participatory environmental modeling process, although their presence influences modeling participants and modeling outcomes.

Related to this, an important assumption of participatory environmental modeling is that the success or failure of the effort is primarily the result of quality, process, and content of the participatory process itself: for example, identifying representative stakeholders and a clear problem, selecting appropriate modeling tools, engaging stakeholders in discussions about uncertainty, and transparency (Korfmacher 2001; Voinov and Gaddis 2008). Implicit in the above assumption is the belief that a successful participatory process overcomes previous cultural differences in understanding and promotes learning, at least to a point sufficient for the modeling process to be seen as successful by participants. Ultimately, a successful participatory process should lead toward building consensus among the participants.

Yet, the participatory process is very limited in time and scope. It may have very specific objectives and goals or significant constraints or parameters, such as management decisions that need to be made or regulations that must be implemented. Some form of an agenda will be present to structure group meetings, which will be limited in number and time. Practically speaking, because an environmental problem or set of problems is the explicit reason for groups coming together, the problem(s) also constrains the breadth and range of knowledge, values, and experiences exchanged. Thus, although participatory environmental modeling is an improvement over modeling practices that exclude stakeholders, its implementation *de facto* limits the range of knowledge, values, and experiences that can be included often by attempting to constrain understanding of a problem by the tools available and common to scientific practices.

It is impractical and probably unproductive to develop participatory modeling processes that are without a problem focus or some specified facilitation process. Realistically, stakeholders and scientists are being asked to collaborate to address specific problems; they would not be meeting, in many cases, unless there was a problem to address. And, they may be in conflict over that problem, which can lead to focused and sustained attention directed at key conflict issues. The question is not whether we should or should not focus participatory modeling on a particular topic, but rather what is the range and content of knowledge, values, and practices that we need to include in the participatory process.

How much and what type of knowledge and values needs to be included in participatory modeling is certainly an empirical question. During the modeling process, discoveries about what is important and the topics for which there is agreement and disagreement can be discovered. However, it is also true that the specific knowledge or values that help promote success or limit the effectiveness of participatory modeling processes are often not understood or explicitly represented in the modeling process by the modelers, stakeholders, or facilitators. The knowledge, values, and practices are present but not readily identifiable in terms of their effects on the modeling process. They are implicit and tacit, yet may be active in shaping perceptions and influencing behaviors in the participatory modeling process. What might this knowledge look like? How can it be understood and incorporated into participatory environmental modeling? How should the process be structured so that these dimensions can be integrated into the process?

In this chapter, we draw on philosophical and social science critiques of environmental modeling and our own cognitive and environmental anthropology research focused on the Chesapeake Bay region to argue that research and practice focused on participatory environmental modeling also needs to consider that there are: (1) unrecognized implicit, cognitive knowledge and values embedded in the participatory modeling process that affect outcomes (models and decisions derived from models); (2) social, material, and technological dynamics in the participatory modeling process that create meaning and shape understanding, and (3) participatory environmental modeling leads to changes in conceptualizations and the potential dynamics of the broader socio-ecological system of which the models and processes are a part.

Consequently, it is useful to focus critically on participation as a site for the meeting of multiple types of knowledge and the construction of new knowledge about the broader socio-ecological system that is being modeled. The perspectives and experiences included and the activities that stakeholders undertake as they participate in modeling activities can be seen as a lens through which to critically analyze how modeling frames and generates environmental knowledge and behavior. Moreover, with a better understanding of the role of participation in the generation of cultural meanings and values among stakeholders, it is important to step back and investigate the significance of participatory models for how we understand and interact with broader socio-ecological systems.

We begin with an analysis of the various roles that cultural knowledge can play in environmental controversy and decision-making. We draw on research in cognitive environmental anthropology to demonstrate the importance of understanding how different "cultural models" can shape beliefs, values, behaviors, and interactions between different stakeholder groups. We next discuss the way that technological and material aspects of computational environmental modeling can significantly shape and structure knowledge and values. We then shift the discussion to focus on a few possible ways that these cognitive, material, and technological factors converge to shape the broader socio-ecological systems in which they are embedded: the participatory environmental modeling itself recombines these factors to re-shape our definitions and understandings of broader socio-ecological systems. We conclude with some observations on how this chapter's philosophical and theoretical musings can be grounded in the practice of collaborative learning, drawing upon our ongoing research in the Chesapeake Bay watershed.

1.2 Cognitive Environmental Knowledge and Values

How and where does one begin to look for the knowledge and values that affect the success of participatory environmental modeling? What guiding theories and methods can be deployed across diverse environmental problems, participatory dynamics, and cultural configurations of both scientists and stakeholders? What conceptual shifts must practitioners of participatory modeling make to help ensure that the relevant knowledge, values, and practices are identified? While there are many social science approaches to the study of environmental knowledge, values, and practices, one useful approach is derived from research in cognitive anthropology that emphasizes how shared, mental processes organize and define environmental knowledge. Fundamentally, cognitive anthropologists are interested in how cultural knowledge is a function of shared human cognition (Holland and Quinn 1987; D'Andrade 1995).

Environmental anthropologists have productively used cognitive approaches to study a wide range of environmental issues for diverse groups (Atran 2005; Paolisso 2002; Paolisso 2007; Kempton et al. 1996; Ross 2004). At first look, a cognitive approach to environmental research may appear to be very similar to other systematic qualitative or quantitative studies of environmental beliefs, values, and behaviors. While a cognitive anthropological approach to environmental issues does share fundamental orientations and interests with standard ethnographic research, its core research interests make it different from and particularly useful to the study of participatory environmental modeling.

Cognitive environmental studies seek to understand the *implicit* or *tacit* knowledge that helps explain explicitly stated environmental beliefs, values, and related behaviors (Paolisso 2007; Ross 2004). Broadly stated, cognitive anthropologists are interested in understanding how the interaction between mental and cultural processes creates cognitive frames or lenses that individuals "see with" as they attempt to understand, value, and behaviorally respond (Kronenfeld et al. 2011; D'Andrade 1995). In addition, cognitive environmental anthropology also has a strong tradition of applying its theories and methods in support of environmental policymaking and decision making (cf. Kempton et. al. 1996; Ross 2004; Paolisso 2007).

More specifically, cognitive environmental anthropologists are interested in how knowledge and values are actually organized in the mind (D'Andrade 1995). There are a wealth of theories and methods within cognitive anthropology (Kronenfeld et al. 2011), and many have been applied to environmental issues. One approach, known as cultural models, has been particularly useful for the study of environmental knowledge and values. Cultural model research is being used to study the complex interaction of attitudes, values, modes of understanding, and discourses surrounding an array of environmental and conservation issues, including global climate change (Kempton et al. 1996), toxic phytoplankton impacts (Kempton and Falk 2000; Paolisso and Chambers 2001; Paolisso and Maloney 2000), protected areas management (Pfeffer et al. 2001), landscape conservation (Dailey 1999), and coastal planning (Christel et al. 2001).

Cultural models are cognitive frameworks used by individuals to process and organize information, make decisions, and guide behavior (Holland 1987). They are shared, implicit, tacit, and—most of the time—unquestioned understandings about how the world works. They are core, defining knowledge that individuals "take for granted," foundational beliefs (what is known or believed), and values (moral, ethical, aesthetic, culturally "correct") that individuals acquire as members of social groups through enculturation, social interactions, language, and institutions (e.g., religion) (Holland 1987). Cultural models are the knowledge and values that individuals, based on repeated, confirmatory and shared experiences, perceive as so true and fundamental that group members do little questioning or critiquing of this knowledge. Cultural models do change as social groups evolve and respond to new situations, creating new forms of shared and implicit knowledge and values.

Technically, cultural models are nested cognitive hierarchies in the mind, and are composed of schemas, which are interconnected building blocks of knowledge and values. Schemas conceptually frame the external world for individuals. They are heuristics that are cognitively stored and available to individuals to assist in understanding or interpreting a physical or socio-cultural situation. Schemas accomplish this through the use of default values or open slots that can be variously filled with appropriate specifics (D'Andrade 1995). Schemas may consist of images (e.g., examples of pollution) or knowledge stored as short propositions (e.g., humans need to be environmental stewards; nature can be managed; nature is unpredictable). Cultural models can be visualized as schemas unfolding, in a simplified fashion, to account for the core features of an event or process (e.g., $CO2$ emissions leading to global warming leading to drought leading to wildfires leading to land-use policies).

Cultural models can link a very wide range of cultural beliefs and values to explicit statements/propositions about the environment—in essence, situating and contextualizing this environmental knowledge. Thus, in discussing environmental issues, stakeholders may be situating explicit environmental information in broader socio-cultural contexts, all without conscious effort or cognizance. In fact, what is environmental or cultural becomes blurred and individuals draw upon both cultural and environmental knowledge and experiences to construct cultural schemas and models. Also, understanding conflicting cultural models can improve dialog among stakeholders and create policies and environmental solutions that benefit from a combination of lay and expert knowledge (Paolisso 2002). Findings from cultural model research can be effectively used to promote collaboration and learning among stakeholders and to increase public participation in decision-making (Morgan 2002; Paolisso 2002; Paolisso and Chambers 2001; Paolisso and Maloney 2000; Kempton et al. 1996).

Focusing more on the application of cultural modeling to participatory environmental modeling highlights even further a number of additional fundamental assumptions associated with cultural models. First, cultural models and schemas are particularly relevant and useful when individuals encounter new forms of environmental knowledge for which they have had little exposure. Immediately, individuals need to make sense of this new information (e.g., changes in climate as computer-generated models with relationships between variables quantitatively defined; changes in coastal marshes as computer displays of increased flooding and inundation due to sea level rise and subsidence) and the associated new interactions (e.g., listening to scientists with little local knowledge or experience but extensive comparative data from other regions and times). The most readily available information for stakeholders is cognitively stored cultural model and schema knowledge that can be "fit" to these new knowledge and social interactions. This fitting is not just one model, but a series of cultural models that are rapidly drawn upon and applied and discarded as conversations and interactions advance, though some initial cultural models that are highly salient and emotive may be foundational and actually become more reified with time and more information and interactions. Initially, at least from a cognitive perspective, the research focus is not on whether the cultural models or schemas applied are right or wrong in any objective and scientific sense. Rather, the focus is on what shared cognitive knowledge is deployed by individuals to assist them in understanding, evaluating, and valuing the new knowledge and interactions. Later, of course, in applied work the question surfaces of how and why cultural models of different stakeholder groups for the same environmental phenomena are similar or different (Paolisso 2007). Again, this need to use simplified cognitive frameworks is necessary because individuals cannot attend to or comprehend all of the information presented due to its complexity, volume, or form of presentation. Consequently, individuals must use simplified, cognitive models to reason with or calculate by mentally manipulating the parts of the model to solve problems or interpret situations or events (D'Andrade 1995). In the absence of this information, stakeholders will draw upon other knowledge frameworks to understand the modeled outputs, which in turn will form a basis for agreement or disagreement and either support or resistance.

A closely related second assumption is that cultural models need to contain essential or primary cultural knowledge that forms or reinforces core cultural beliefs and values among a group who shares that cultural model. In exchanges of information among scientists/modelers and public stakeholders, it may be equally important to pay attention to the broader, tacit understandings that are present, rather than the details of particular modeling parameters or calculations, such as debates about what information is knowable with some degree of certainty, and how the modeling process generates shared understanding of what can be known. For example, research has shown that Chesapeake Bay watermen (commercial fishers) believe that scientists studying the blue crab and oyster fisheries know a lot about crabs and oysters, what watermen refer to as "book knowledge." But, they do not accept the scientists' stock assessment results because they tacitly and implicitly do not believe that "you can predict nature with numbers" ("You cannot understand a crab by counting.") (Paolisso 2002). Thus, in focusing on the details, such as specific assumptions inherent in calculations of maximum sustainable yield of the crab fishery, instead of equally paying attention to the broader human-environment cognitive frame, scientists and managers create a situation opposite of what their sincere efforts aim to accomplish: Watermen begin to question why the scientist is working so hard to try to convince them of something that they just know, without question, to be impossible and not true (Paolisso 2002). As watermen seek answers to this question, a new cultural model emerges (e.g., that the scientist has a job and needs to produce results that his boss will find satisfactory enough to keep the scientist employed) (Paolisso 2002). Interestingly, watermen relate to this latter model because of their own beliefs and values that one must work and support his family, so the scientist is practicing this responsibility.

Third, the knowledge and values encapsulated with cultural models motivate and explain behavior both within groups that share the model and toward groups that operate using different cultural models (Holland 1987). Cultural models are shared among a group of individuals—to be determined empirically—and that shared, implicit knowledge is a reference for individuals that they are a group. Thus, when information does not fit with a group's cultural model, they can collectively develop their alternative explanation and engage in activities to resist what may come to be external and imposed knowledge and behavior. For example, in research on cultural models of land conservation among farmers on the Eastern Shore of the Chesapeake Bay, it was found that the models of land conservation organizations prioritized the placement of land with high ecological value into permanent conservation easements, *de facto* removing the land from further human agency. Implicit in this model of land conservation is the belief that nature is best served when separated and protected from humans, who do not have a very good track record of protecting and managing the environment for its ecological services and benefits. Relatedly, nature is seen as separate from humanity, and the former's future is dependent on removing humans from key areas of high ecological value (Paolisso et al. 2013).

Farmers, on the other hand, do not share widely in the land trust or environmental organization model of land conservation. For farmers, their wider, implicit and tacit cultural model understandings of land conservation give humans a more active role in sustaining natural resources and valuable ecosystems. Farmers' modeling of land conservation emphasizes making agriculture economically profitable. While on a small scale they recognize the importance of putting land that is not good for agriculture into easements, if they have high ecological value, on the larger scale they argue that land conservation can only be achieved if agriculture is profitable. A profitable agriculture reduces the extent to which farmers sell to developers; motivates farmers to practice soil conservation and best management practices on lands best suited for agriculture; supports secondary businesses in the region that are the backbone for place-based, local communities; and creates an ethos of environmental stewardship (Paolisso et al. 2013). Thus, in any participatory modeling project involving land trust and environmental organizations and farmers, it is critically important to focus on the broader cultural and social knowledge that lies implicitly behind the details of easements and land-management practices. For example, Eastern Shore farmers do value land conservation easements, but as part of an economic model for a profitable farm, not as a mechanism to "lock land away," which, if that land is good farmland, could not be further from farmer's cultural model of land conservation.

Finally, cultural models are used by individuals when confronted with incomplete and contradictory information that requires a decision or viewpoint. In fact, cultural models may be used to fill in missing information. Most of the information used in environmental modeling is incomplete, contains uncertainty, and is subject to unpredictable change. A very relevant example of this is modeling that attempts to predict the type and extent of local impacts to global environmental processes, such as climate change. Such downscaling from higher-scale data to local or regional impacts is very difficult and requires many assumptions about model parameters and size and direction of changes in affected socio-ecological systems (Edwards 2010). Consequently, there are ample opportunities for the need to estimate the size and direction of change in key model variables, along with how such changes are accommodated by model assumptions and parameters. In our ongoing research with Chesapeake Bay farmers, it is clear that they are very familiar with and experienced in adapting their agricultural practices to changing climate and weather. In fact, being able to adapt to climate trends and seasonal weather dynamics is one of the characteristics of a successful farmer. As farmers incorporate the scientific interests and activities focused on climate change for the Chesapeake Bay region into their knowledge system and practices, our cognitive environmental research revealed that they use two cultural models, which are able to handle the range of predictions about impacts and the significant amount of uncertainty in our estimates of impacts. First, farmers maintain a cultural model of climate change as a process in nature driven by dynamics that are largely beyond human influence. This cultural model of *climate change as natural change* it is not something that is knowable or predictable to any significant degree, which is not to say that farmers do not develop strategies to adapt to it. Given this model, farmers do not see much use or value in the scientific estimates of climate change impacts for the region. Not only do these scientific efforts not fit their cultural model of *climate change as natural change,* but farmers also have developed a second cultural model that better helps them explain and accommodate the scientific efforts at modeling climate change. This cultural model focuses on how humans have integrated more of nature in their own environmental management domains. This cultural model of *climate change as environmental change* is applied to situations where human institutions (e.g., universities, environmental organizations, management agencies) seek to understand and regulate human and environment interactions. For farmers, this cultural model of climate change captures their understanding that science and management are drawing natural climate change into their human-environmental frameworks. In doing so, it is not surprising that natural climate change does not fit well, in terms of knowing future climate change impacts with great certainty. Also, the cultural model fits farmers' beliefs that environmentalists, scientists, and managers are trying to control and manage more of the natural world, often with negative impacts on farm operations.

In summary, a major advantage in using a cultural model approach is its ability to link wide ranges of cultural beliefs and values to explicit statements/propositions on the environmental issues that surface during participatory environmental modeling. Thus, unless they are overtly identified, stakeholders will be situating explicit information on the environmental and modeling issues into broader cultural contexts without being aware that these broader cultural frames are present.

1.3 Material and Technological Dynamics

In addition to an awareness of differences in cultural knowledge and values present in the participatory environmental modeling process, it is also important to recognize the ways in which the material, conceptual, and technological qualities of the models, and the modeling process, enable and constrain certain understandings and interactions. These qualities can play a significant role in shaping the processes and outcomes of participatory modeling projects, and can affect the way individuals in a participatory modeling project interact and engage with one another.

The field of science and technology studies (STS) offers a way to understand the social dimensions of scientific knowledge production, and the influence of material, technological, and conceptual factors in participatory environmental modeling. Conceptually, STS is a broad field of research with a wide range of approaches (Latour and Woolgar 1986; Fischer 2007; Haraway 1988; Barnes et al. 1996). However, there is general consensus on the goal of STS research, which is to situate scientific practices—whether successful or unsuccessful—within their social contexts (Latour and Woolgar 1986; Barnes et al. 1996; Haraway 1988). STS researchers ask what social processes contribute to the production of scientific knowledge and how scientific knowledge in turn shapes social processes.

Actor-Network Theory (ANT) is one STS conceptual framework that allows researchers to focus on the role that material or non-human "actors" play in these social processes, and in the production of scientific knowledge. From this perspective, "facts" come to be seen as assemblages of material and discursive "actors" (e.g., plants, animals, microbes, molecules, microscopes, mass spectrometers, publications, people, laboratories, institutions, and so on) that combine to produce an effective structure of knowledge—one that is capable of influencing broader social practices while holding up against critiques (Latour 1988; Law 1992; Pickering 1995; Haraway 1990). It is by producing such "well-constructed" assemblages or networks that science is able to position itself as a powerful method for understanding and acting in the world (Latour 1988; Pickering 1995). Latour's most influential example is that of the vacuum pump, which was used by Robert Boyle to demonstrate that space is not filled with ether (Latour 1993). The pump serves as an actor that convinced scientific observers of the mid-1600s of the nature of the vacuum. It is by assembling these actors—the pump, the feather, the laboratory, the scientific observers, and so on—that Boyle was able to construct the "fact" of the vacuum. This "fact" then propagated outward from its initial assemblage to shape various fields such as engineering, transportation, and space travel (Latour 1993).

Like Boyle's vacuum pump, models play a significant—and increasingly essential—role in understanding and responding to complex environmental problems (Paolisso et al. 2015). They too can be seen as "actors" within the networks that develop around resolving these environmental problems. As a result, situating modeling within broader social processes, and understanding the role that models play in those processes is important to understanding and undertaking participatory environmental modeling projects. However, models are themselves heterogeneous assemblages. They are produced through a confluence of conceptualization, computational technologies, and coding structures so that it is necessary to explore the various effects of all of these features when examining the role of models in environmental management. Below, we provide a brief discussion of each of these dimensions as they relate to the social processes and the knowledge generated by the model-building process.

1.3.1 The Conceptual Limitations of Models

Models are, by definition, simplifications of complex systems (Oreskes 1998). They isolate key processes within the system and represent the effects different activities might have on those processes. Therefore, it is always necessary to discriminate those aspects of a system that are relevant to the particular issue at hand (Oreskes 1998; Munk 2013). Although at first glance these choices are merely rational determinations—choosing those factors that have causal significance within a mechanistic system—embedded within these choices are many assumptions about what factors are and are not relevant. It would be unreasonable to suggest that the model should attempt to encompass everything that might be relevant, since this would make the model unnecessarily complex and probably unusable. Nevertheless, it is important to be conscious and critical of the assumptions that underlie these discriminations and the ways they might be viewed by others, especially during the participatory environmental modeling process.

These simplifications shape the way knowledge of the environment is integrated and applied by defining the nature of interactions. Generally, what are modeled in a computational environmental model are the physical relationships between organisms and the geophysical landscape. Conceptually, this sets up a situation where humans are external to the environmental system. For example, Munk (2013) describes his experience developing a "perceptual model" of flooding. Not a computer model, but a conceptual model sketched out with pen and paper; this practice allows the modeler to think through all of the different factors that might come into play in a flood scenario: evapotranspiration, rainfall, soil, vegetation, human settlement, etc. Nature is defined as a distinct domain, for which the model speaks, but separate from the politics that are ever-present in the causes, consequences, and management of flooding. In other words, the modeler is disciplined by the constraints of the model to fill a particular role (informing society about nature) within a system that is organized in part by the model itself (nature/society).

The questions, then, are: What gets left out, and how are those decisions made in a participatory environmental modeling project? In many cases, the important factors for participants will not be the same as those for the modelers. It may even be the case that the knowledge and aspects of the system that participants find important simply cannot fit within a computational modeling framework. As a result, it is important to be conscious and critical of the selection of factors that are incorporated into the model, and the effect those choices might have on the participants and project outcomes.

1.3.2 The Materiality of Computational Modeling

Participatory environmental modeling depends on several material components of computation and the energy to run them: mainframes, personal interfaces, wires and cables connecting everything together, and the energy to power those systems. Institutional circumstances determine the physical and technological systems to which different modeling projects have access. Some modelers will have access to powerful supercomputers that require significant amounts of energy to power and human labor to run and maintain. Others will only have access to personal computers, or even just pens and papers. Each project group will have to work with whatever systems they have at their disposal. However, the relationship between institutional circumstances and computational systems is not as simple as certain institutions being able to access certain kinds of computational systems. Rather, a reciprocal relationship exists where the material dimensions of computational systems can come to shape the institutions in which they are embedded. As a result, it is important not only to consider what kinds of computational systems are accessible to a given project, but also the way that these material components are to be managed and the way this management shapes institutional and relational structures.

Edwards (2010) uses the term "friction" to indicate "the expenditures of energy and limited resources in the processing of numbers" (112). His definition incorporates the processes of overcoming discursive barriers such as translating data from one format to another, and communicating scientific findings to the public. However, many of the frictions he describes are purely material. For example, prior to the increase in computer technologies, human labor was required to perform complex calculations. Organizing and managing this human labor required complex institutional structures. With increased use of computers, fewer people are needed to perform the same calculations, but the systems still require maintenance and management in order to run properly and perform the appropriate calculations

(Edwards 2010). Questions arise such as where to house the physical computer systems, how to maintain them effectively, who will program them to perform those computations, what kind of interfaces will be used, how the energy to power the computers will be procured, how to cool the computer systems if they overheat, and so on. Addressing these questions and concerns may have significant effects on the kinds of relationships that are ultimately built around these systems. Edwards (2010) demonstrates this with the image of the "forecast factory" imagined by Lewis Fry Richardson. Richardson devised a way to use Bjerknes's equations[1] to analyze and predict weather patterns. In the early twentieth century, people did the computations by hand because digital computers were not available. The amount of data and manpower required to make the calculations on a global scale would have required a global forecast factory where individuals would access data and make calculations for a particular point on a grid and these would be used to inform the calculations of other individuals working on adjacent portions of the grid (Edwards 2010). Ultimately, it was impossible to implement such a massive undertaking, and the advent of computing made the idea of a climate factory unnecessary, but the vision of such a project highlights the way that material factors can shape institutional and relational dynamics. The shift from human computation to electronic computation meant that the forecast factory was unnecessary, and made possible the meteorological institutions we have today.

In a participatory environmental modeling project, these material and technological dimensions of computational models could play a significant role in shaping the outcome. Different computing systems will place different demands on the project and potentially may require additional institutional arrangements that could influence how the participants interact with one another, with the model, and with the environmental systems in question. For example, the Chesapeake Bay Modeling System (CBMS) is a large, complex model that requires a lot of computing power to run. It is situated within the Chesapeake Bay Program (CBP), which is a partnership between several state governments and the Environmental Protection Agency (EPA) (Paolisso et al. 2015). It is doubtful that the model could exist outside this institutional structure because few others are equipped to manage its demands, such as the costs of operating and maintaining the powerful computers required to run the model. As a result, using the CBMS in a participatory project or attempting to add a collaborative component to the CBMS framework would require changing the model, and, in the process, changing the institutional structure of the CBP. That is not to say that such an endeavor could not or should not be undertaken, but it is important to recognize this aspect of the problem. Participation, in other words, is not simply about transforming the model; it is about changing the very structure of the models and the computing systems on which they operate.

[1] Vilhelm Bjerknes developed a set of nonlinear differential equations that provided a link between fluid dynamics and thermodynamics. These equations were later used by Lewis Fry Richardson and others to approximate global atmospheric flows in order to make weather and climate predictions. These equations are at the core of the general circulation models used to model climate change (Edwards 2010).

1.3.3 Code Structures

Closely linked to the material dimensions of modeling is the problem of the structure of the computer code itself. As any software developer knows, code is not a neutral or malleable language capable of producing any effects we desire (Kelty 2005). Code is, in fact, a complex linguistic system that enables and constrains the kinds of processes that can be modeled, the way those processes are represented, and the efficiency and usefulness of the model for analysis and prediction (Kelty 2005; Sundberg 2009).

Sundberg (2009) found that there is a division within the meteorological and climate sciences between those who seek *practical modeling*, and those who seek *more accurate models*. In order to be practical, a model must be coded efficiently so that it can be run quickly and at little cost. These models tend to run on parameters that work whether or not they accurately represent the physical processes involved. In other words, if a parameter produces the expected effects, then that is sufficient for it to be incorporated into the model (Sundberg 2009). On the other hand, there are those who seek to model the physical processes more directly, generally for research or theoretical reasons. For these modelers, the economy of the code is not as important as the accuracy of the model in simulating physical processes. In other words, the structure of modeling code affects not only the accuracy of the representation, but also the uses to which the model can be put.

Landstrom et al. (2013) encountered issues with the limitations of code in their participatory project modeling flooding in rural UK. Data was abundant as scientists provided accumulated data on flooding in the region, and local participants provided information and data from their own experiences with the hydrologic system. However, the model that was initially chosen by the researchers was unable to show bank overflow because it had not been designed for this kind of research (Landström et al. 2013). As a result, they had to abandon that model and develop a simple model of their own that allowed them to both model stream overflow and represent it through a Graphical User Interface (GUI) that allowed local participants to visualize and provide feedback on the model as it progressed. By shifting from one modeling system to another, the researchers were able not only to represent the system better, but also to engage the participants more effectively. It is clear, then, that the decision of what modeling system and what kind of code language to use is an important consideration for any modeling project, but especially for participatory projects.

1.3.4 Broader Socio-Ecological Systems

Finally, it is important to understand how the models themselves are situated in relation to the ecological systems they represent. A socio-ecological systems perspective recognizes that social and ecological systems are intertwined with one another such that change in one inevitably effects change in the other as well (Berkes 1999).

But a socio-ecological perspective cannot be complete unless the very tools and methods used to understand the two interacting systems are themselves included. The question is: How do these tools and methods affect the way the overall socio-ecological system is structured? And, how can these tools and methods reshape the system in beneficial ways?

White (1996) illustrates the link between computational environmental modeling and ecological systems with an example from the Columbia River:

> In the virtual Columbia, electronic fish swim past electronic dams on video terminals. Change the electronic river and the fate of the electronic fish is graphically displayed… The virtual river influences events in the actual Columbia. How electronic fish behave will lead to decisions on how fish in the actual Columbia—the organic machine—will be managed. That the various virtual Columbias depend on the actual Columbia for some of their own electrical power only compounds the ironies and connections (White 1996, p. 116).

This is a rather obvious case where there is a direct link between the "virtual Columbias" and the actual Columbia by way of hydroelectric power. However, it is the relationship between the behavior of "electronic fish" and management policy that has the most significant impact upon the socio-ecological system as a whole.

An environmental modeling project—participatory or otherwise—becomes a microcosm where concern for the broader socio-ecological system can play out in relation to the models, and these dynamics can "spill out" into the broader socio-ecological system producing different outcomes and interests. These effects can range from cognitive and cultural changes in our knowledge and understanding of the socio-ecological system and the challenges it faces, to social and material changes in the relationships between people, organizations, and environmental systems. As a result, the development and deployment of participatory environmental modeling should not be concerned only with improving the quality of the science or efforts in public outreach. When considering the structure of the participatory environmental modeling project, it is also important to consider its effects upon the broader socio-ecological system.

Because environmental models typically focus on modeling only environmental processes and human activities are, at best, depicted as impacts, they tend to result in a "humans versus nature" mentality in which the problems natural systems face are inherently intractable—humans must lose in order for nature to win. Through a participatory engagement that is self-conscious of its position within the socio-ecological system, the problems may still be difficult to overcome, but a different kind of mentality can emerge and the outcome—an improved socio-ecological system—can be seen as beneficial to everyone involved. Furthermore, such participatory environmental modeling projects can, theoretically, through their practice and performance, generate the cultural, social, and material changes that might ultimately overcome seemingly intractable problems.

How might such changes take place? Looking back at the factors discussed previously, it is possible to speculate on some of the ways that a participatory environmental modeling project might feed back into the broader socio-ecological system in beneficial ways. For example, we recently collaborated in an interdisciplinary effort to evaluate the research and management opportunities and challenges in using multiple models for key Chesapeake Bay ecosystems, rather than relying on one model, the Chesapeake Bay Modeling System (CBMS) (Weller et al. 2013). Among the many findings and outcomes of this effort was the realization of how the requirements of the modeling system helped to structure how participants viewed the larger socio-ecological system. For the modelers present, it was clear that the actual practice of modeling was not the place for stakeholder involvement. Rather, modeling was meant to inform political debates, but not be subject to political interests and concerns. Instead, the place for political debate was thought to be after the modeling has been done. This can be done through the legislative process, where models are used as tools to inform policy making, but where the public and special interests can negotiate the best solutions. Thus, the view of participatory or collaborative modeling that we have seen is one where the *results* of models are used collaboratively to make decisions. Another mode is where modelers collaborate with decision makers and managers to produce a model that will fit their needs. When modelers work with stakeholders in the actual modeling process, the intent is to improve "buy in" to the model results. Thus the results are assumed from the beginning, and the process is meant as a way to get the public to accept the results more readily.

In addition, because of the way coding and computer systems, to some degree, end up structuring environmental management programs, a participatory environmental modeling project could produce new stakeholder arrangements and networks to address environmental challenges. For example, a participatory environmental modeling project that uses simpler models operating at different scales and able to run on smaller computer systems might encourage a more polycentric approach to addressing environmental concerns (Ostrom 2009). Greater interaction and negotiation between scientists and stakeholders might develop into ongoing relationships that foster novel approaches to environmental management, thereby again affecting how the broader socio-ecological system is understood and enacted (Whatmore 2009). Again, an example from our Chesapeake Bay research illustrates this point. Scientists working with local community stakeholders collaborated on developing strategies to reduce harmful algal blooms (HAB) of *Mycrosistis* in small lakes and ponds in rural, agricultural areas (Van Dolah et al. 2013). Scientists drawing initially on previous harmful algal bloom mediation proposed to use clay flocculation to control bloom outbreaks. However, local farmers had experience with using bales of hay to capture and bind the algae, some of which was *Mycrosistis*, an approach less familiar to scientists. In the end, the farmers' strategy was successfully used. Beyond the HAB mitigation outcome, it is relevant that this collaboration between farmers and scientists forced both groups to reevaluate their understanding of the broader socio-ecological system that included extensive and long-term agricultural nutrient runoff and recent aggressive policies by the state to implement formal farm-nutrient management plans, which did not prioritize local farmer knowledge and experience. Initially, farmers were seen as polluters to be regulated (Van Dolah et al. 2013). The result of the project was a changed perspective on the broader socio-ecological system that situated farmers as co-problem solvers.

1.4 Conclusion: Insights from the Chesapeake Bay

In the above, we have attempted to shift the intellectual focus beyond the immediate methods available to participatory modelers to the process of conceptualizing and participating in environmental modeling. In any participatory environmental modeling project, there will be implicit, tacit knowledge present that will affect understandings and valuing of the modeling process. Also present will be meanings and values embedded in material relations and the technology utilized to model environmental processes;both will shape the direction and scope of participatory activities. Certainly, simply being aware of or sensitive to these cognitive, material, and technological factors is a positive contribution (though one not easily measured) to efforts to develop and sustain participatory environmental modeling. Still, it would be useful to identify practical steps or procedures that would "pause" and "expand" the participatory modeling process just enough to create social relations and exchanges that could foster more integration of cognitive, material, and technological factors. Drawing on our Chesapeake Bay research, we share a few examples of how we have used the approach of collaborative learning (Daniels and Walker 2001; Feurt 2008) to foster the integration of these wider factors in participatory and collaborative research, some of which includes modeling.

Steven Daniels and Gregg Walker in their book, *Working Through Environmental Conflict: The Collaborative Learning Approach* (2001), provide a useful and flexible framework for approaching environmental issues that are contentious, characterized by strong differences of opinions, and generative of cultural and economic conflict. While not all participatory environmental modeling projects include extreme conflict and contention, there are the broader cognitive, material, and technological factors discussed above that lead to differences in understandings and priorities, and possibly enough conflict to derail the project. Fundamentally, collaborative learning is a set of concepts and tools that can help craft effective policy (cf. Feurt 2008). Effective policy is defined as an adaptive process that uses the most appropriate science and technology, is implementable, and has low transaction costs (ibid., p. 2). Additionally, effective policy needs social legitimacy: Decisions should be rational and technically sound; also, if people's lives are affected, they should have a voice in the process (ibid., p. 4). Finding ways to increase the quality of technical expertise, while simultaneously increasing the inclusivity of decision processes, is perhaps the fundamental challenge of effective policy formation (ibid., p. 6). Daniels and Walker argue that collaborative processes that are inclusive and sincere have the potential to achieve balance between technical competence and inclusive deliberation (ibid., p. 10). "Collaboration involves interdependent parties identifying issues of mutual interest, pooling their energy and resources, addressing their differences, charting a course for the future, and allocating implementation responsibility among the group" (ibid., p. 10).

A fundamental premise of collaborative learning is that controversy and conflict should be used to generate "social learning" among stakeholders (Daniels and Walker 2001, p. 6). The goal of this social learning is a shared, deeper understanding of a complex environmental situation. Thus, the role of natural resource managers and policymakers is to promote social learning, rather than to only make decisions on the public's behalf, and to help the public deliberate over the decisions that need to be made. Rather than viewing debate and controversy as managerial failure that makes policymaking and implementation more difficult, policymakers should see them as natural and desirable aspects of the formation of public values, contributing to society's self-understanding. The goal is to help stakeholders have genuine engagement, dialogue that takes into account a wide variety of factors, with increased emphasis on the normative, valuing, ethical sides, along with honest discussions of scientific and local-based knowledge and findings. The output is to have stakeholders generate a set of implementable improvements in a situation of mutual concern (ibid., p. 21).

In implementing collaborative learning among Chesapeake Bay stakeholders, we have found it useful to organize project activities and practices into three stages that progressively allow the incorporation of more specific cognitive, material, and technological factors.[2] During the first stage, meetings and informal workshops are used to allow participants to share their views and concerns and to listen to other perspectives and priorities. What is important during this first phase is to refrain from establishing rigorously-defined project outcomes and steps to achieve them. While there does need to be broad project goals and objectives, it is important during stage one to be inclusive and value all contributions and perspectives equally in the shaping of these goals and objectives, even if they are in conflict or appear to be "off the mark" or "extreme." Applied to participatory environmental modeling, this is the time to build stakeholder rapport: We all share the same goal of building the best environmental model possible that accomplishes our collective goals and objectives. It is also a time for participants to collaboratively learn about each other's views and experiences, all in a nonjudgmental and non-critical environment. Of relevance, an indicator of the success of this first stage is that participants are a little confused though satisfied. Given past experiences, many may expect and even need more explicit project goals and objectives, and expect that the workshops or meetings would identify and discuss the steps to accomplishing the project's objectives. However, when engaged in the collaborative learning activities, participants are able to voice more of their views and share more of their experiences, all of which are rewarding but do not immediately contribute to refining objectives and project tasks. Implicitly and tacitly, the collaborative learning process is opening social space and creating dialogue that allows a wider range of cultural knowledge, material conditions and technological dynamics to be brought into the discussion.[3] Consistent with collaborative learning theory, the focus of stage one in our projects has been to foster and facilitate stakeholders learning about each other, all in the context of some broad project goals, such as developing and implementing a participatory environmental model.

[2] A current example of this staged implementation of the collaborative learning is Deal Island Peninsula Project (www.dealislandpeninsulaproject.org).

[3] Excellent practical guidelines for collaborative learning activities can be found in Feurt 2008.

In the second stage of our collaborative learning research, we build on the stakeholder rapport and learning developed in stage one to implement actual collaborative research projects. In our current "Deal Island Peninsula Project," which is using a collaborative learning approach to increase marsh and community resilience to climate change (www.dealislandpeninsulaproject.org), we have organized close to 50 project stakeholders, including community residents, social and natural scientists, and local and state resource managers, into three Collaborative Research Projects (CRPs). The three CRPs focus on, respectively: Flooding and Coastal Erosion, Marsh Restoration, and Community Heritage. These themes emerged out of stage one discussions and dialogues as topics or areas that the project stakeholders wanted to focus on, and areas that they saw as broadly relevant to the project's goal of increasing marsh and community resilience to climate change impacts. Each CRP has co-leaders and membership representative of the range of stakeholders involved in the project. This second stage of the collaborative learning process emphasizes action and engagement, versus only discussion of views and experiences. It is more about putting those views and experiences into activities that each CRP believes are important to undertake. For example, recently our Marsh Restoration CRP had a marsh field day, where natural scientists, local community members, and state resource managers visited different sites in local marshes and discussed each site from their varying perspectives. Scientists talked about marsh subsidence and accretion process, local community members talked about hunting and fishing experiences and past settlements in the marshes, and resource managers talked about state and county efforts to enhance vegetation growth around a large water impoundment that had been built to attract wildfowl for hunters. The Marsh CRP is now taking the information collected from the various stakeholders and creating multi-perspective GIS maps of the marshes: sociocultural, ecological and management. Our two other CRPs are equally engaged in collaborative activities, focused on interviews to collect histories and heritage information to collecting information on flooding risks from community residents. Again, similar to stage one, the second stage also promotes the integration of a wide range of cognitive knowledge (e.g., the value of marshes; notions of flooding risks), materials (e.g., the structure of built and natural environments related to flooding and historical settlements in marshes) and technological factors (e.g., commercial fishing technology as community heritage and use of models to understand marsh dynamics and climate change impacts).

It is in the third stage that we build upon the inter-stakeholder relations and broad understandings established in stages one and two to move toward defining and specifying the project's objectives and outcomes. Our project stakeholders and CRPs are quite familiar with our project goal: to integrate the human and ecological dimensions to increase community and marsh resilience to climate change. We certainly could have defined some variables and set some parameters from the beginning of the project on the definition of socio-ecological resilience, and then enlisted project stakeholders to help us collect information from their various areas of expertise. However, that would have narrowed the breadth of cognitive, material, and technological information available to us as we collectively craft both a definition of socio-ecological resilience and a plan for achieving it. The Deal Island Peninsula Project is currently at stage three of the collaborative learning process. What is clear so far is that we have a wide range of committed project stakeholders and a very robust conceptual framework that includes socio-cultural, economic, ecological, institutional, and material/technological considerations. The collaborative learning approach has allowed us to integrate cognitive, material, and technological factors. Equally important, it has created a project-wide acceptance of different viewpoints and experiences and a sincere and genuine interest among all project stakeholders in learning about and participating with individuals who represent different interest groups. Implicitly and tacitly, it is our sense that this is the collaborative learning definition of socio-ecological resilience: We have created a new network of relationships that collectively are better situated to respond to future climate change impacts for communities and marshes.

We believe that the lessons learned from our use of collaborative learning apply to participatory environmental modeling. There are excellent sources available on how to develop and implement collaborative learning (Feurt 2008), and the approach is extremely flexible and adaptive. It can help participatory environmental modeling projects accomplish many goals related to the processes of participation, including, importantly, the goal of moving forward even when there are differences of views and experiences; and reaching some final decisions even when you have disagreement (Daniels and Walker 2001). Furthermore, using collaborative learning and integrating more cognitive, material, and technological factors will increase the resilience of participatory environmental modeling projects, and ultimately increase and expand definitions of what defines successful environmental models in general. It may not be too unrealistic to argue that the quality and form of participation will be a driving factor in the success of future environmental models, and that participation would be well-served by the inclusion of cognitive, material, and technological information.

References

Atran S (2005) The cultural mind: environmental decision making and cultural modeling within and across populations. Psychol Rev 112(4):744–776

Barnes B, Bloor D, Henry J (1996) Scientific knowledge: a sociological analysis. University of Chicago Press, Chicago

Berkes F (1999) Sacred ecology: traditional ecological knowledge and resource management, 1st edn. Routledge, New York

Callon M (1999) The role of lay people in the production and dissemination of scientific knowledge. Sci Technol Soc 4(1):81–94

Christel D, Kempton W, Harris J (2001) The effects of values and cultural models on policy: an anthropological approach to environmental policy in Tampa Bay. University of Delaware/College of Marine Studies, Newark

D'Andrade R (1995) The development of cognitive anthropology. Cambridge University Press, Cambridge/New York

Dailey M (1999) Cultural models of forests and ecological change on the Appalachian Plateau, 1750–1840. PhD Diss, University of Georgia, 1999

Daniels SE, Walker GB (2001) Working through environmental conflict: the collaborative learning approach. Praeger, Westport
Edwards PN (2010) A vast machine: computer models, climate data, and the politics of global warming. MIT Press, Cambridge
Feurt CB (2008) Collaborative learning guide for ecosystem management. Environmental Studies Faculty Publications. Paper 5
Fischer M (2007) Four geneologies for a recombinant anthropology of science and technology. Cult Anthropol 22(4):539–615
Gaddis EJB, Falk HH, Ginger C, Voinov A (2010) Effectiveness of a participatory modeling effort to identify and advance community water resource goals in St. Albans, Vermont. Environ Model Softw 25(11):1428–1438
Haraway D (1988) Situated Knowledges: The Science Question in Feminism and the Privilege of Partial Perspective. Fem Stud 14:575–599
Haraway D (1990) Primate visions: gender, race, and nature in the world of modern science. Reprint. Routledge, New York
Holland D, Quinn N, eds. (1987) Cultural models in language and thought. Cambridge University Press, Cambridge/New York
Kelty C (2005) Geeks, social imaginaries, and recursive publics. Cult Anthropol 20(2):185–214
Kempton W, Falk J (2000) Cultural models of Pfiesteria: toward cultivating more appropriate risk perceptions. Coast Manag 28:273–285
Kempton W, Boster J, Hartley J (1996) Environmental values in American culture. The MIT Press, Cambridge
Korfmacher KS (2001) The politics of participation in watershed modeling. Environ Manag 27(2):161–176
Kronenfeld D, Bennardo G, de Munck VC, Fischer MD (eds) (2011) A companion to cognitive anthropology. Wiley-Blackwell, Malden
Landström C, Whatmore SJ, Lane SN (2013) Learning through computer model improvisations. Sci Technol Hum Values 38(5):678–700
Latour B (1988) The Pasteurization of France. Harvard University Press, Cambridge
Latour B (1993) We have never been modern. Harvard University Press, Cambridge
Latour B, Woolgar S (1986) Laboratory life. Princeton University Press, Princeton
Law J (1992) Notes on the theory of the actor-network: ordering, strategy, and heterogeneity. Syst Pract Action Res 5(4):379–393
Morgan MG (2002) Risk communication: a mental models approach. Cambridge University Press, Cambridge
Munk AC (2013) Emancipating nature: what the flood apprentice learned from a modelling tutorial. In: Hastrup K, Skydstrup M (eds) The social life of climate change models: anticipating nature. Routledge, New York
Oreskes N (1998) Evaluation (not validation) of quantitative models. Environ Health Perspect 106(Suppl 6):1453
Ostrom E (2009) A polycentric approach for coping with climate change. Background Paper for the WDR 2010
Paolisso M (2002) Blue crabs and controversy on the Chesapeake Bay: a cultural model for understanding watermen's reasoning about blue crab management. Hum Organ 61(3):226
Paolisso M (2007) Taste the traditions: crabs, crab cakes, and the Chesapeake Bay blue crab fishery. Am Anthropol 109(4):654
Paolisso M, Chambers E (2001) Culture, politics, and toxic dinoflagellate blooms: the anthropology of Pfiesteria. Hum Resour Abstr 36(3):309–456
Paolisso M, Maloney R (2000) Recognizing farmer environmentalism: Nutrient runoff and toxic dinoflagellate blooms in the chesapeake bay region. Hum Organ 59(2):209–221
Paolisso M, Weeks P, Packard J (2013) A cultural and applied analysis of land conservation. Hum Organ 74(1):12–22

Paolisso M, Trombley J, Hood RR, Sellner KG (2015) Environmental models and public stakeholders in the Chesapeake Bay watershed. Estuaries Coast 38:97–113
Pfeffer MJ, Schelhas JW, Day LA (2001) Forest conservation, value conflict, and interest formation in a Honduran National Park. Rural Sociol 66(3):382–402
Pickering A (1995) The mangle of practice: time, agency, and science, 1st edn. University of Chicago Press, Chicago
Ross NO (2004) Culture and cognition: implications for theory and method. Sage Publications Inc, Thousand Oaks
Sandker M, Campbell BM, Ruiz-Pérez M, Sayer JA, Cowling R, Kassa H, Knight AT (2010) The role of participatory modeling in landscape approaches to reconcile conservation and development. Ecol Soc 15(2):1
Sundberg M (2009) The everyday world of simulation modeling the development of parameterizations in meteorology. Sci Technol Hum Values 34(2):162–181
Van Dolah E, Paolisso M, Sellner K, Place A (2013) Beyond the bloom: using a socioecological framework to investigate stakeholder response to HAB management in the Chesapeake Bay, USA. In: Proceedings of the international conference on harmful algae, Seoul, Korea
Voinov A, Bousquet F (2010) Modelling with stakeholders. Environ Model Softw 25(11):1268–1281
Voinov A, Gaddis EJB (2008) Lessons for successful participatory watershed modeling: a perspective from modeling practitioners. Ecol Model 216(2):197–207
Weller DE, Benham B, Friedrichs M, Najjar R, Paolisso M, Pascual P, Shenk G, Sellner K (2013) Multiple models for management in the Chesapeake Bay. STAC Publication Number 14-004, Edgewater
Whatmore SJ (2009) Mapping knowledge controversies: science, democracy and the redistribution of expertise. Prog Hum Geogr 33(5):587–598
White R (1996) The organic machine: the remaking of the Columbia river. Hill and Wang, New York

Chapter 2
Learning Through Participatory Modeling: Reflections on What It Means and How It Is Measured

Seth P. Tuler, Kirstin Dow, Thomas Webler, and Jessica Whitehead

2.1 Introduction

Participatory modeling has tremendous potential to promote learning in environmental decision-making processes. Participatory modeling (Jones et al. 2009; Mendoza and Prabhu 2005; Voinov and Gaddis 2008), also known as mediated modeling (van den Belt 2004), group model building (Rouwette et al. 2002), cooperative modeling (Tidwell and van den Brink 2008), or scenario building (Berkhout et al. 2002), is viewed as a way of bringing stakeholders together to organize information about complex systems into tools that are more useful for local decision-making than those designed by scientists and decision makers alone. Practitioners and researchers report that individual and group learning are both sought and achieved through the process of creating and using conceptual models (Gaddis et al. 2010; Jones et al. 2009; Knapp et al. 2011; Sieber 2006; Tidwell and van den Brink 2008; van den Belt 2004). The same is said about public participation in environmental and risk assessment more generally (National Research Council 2008; Muro and Jeffrey 2008; Pahl-Wostl and Hare 2004; Schusler et al. 2003; Webler et al. 1995).

S.P. Tuler (✉)
Interdisciplinary and Global Studies Division, Worcester Polytechnic Institute, 100 Institute Rd., Worcester, MA 01609, USA
e-mail: STuler@wpi.edu

K. Dow
Department of Geography, University of South Carolina, Columbia, SC 29208, USA

T. Webler
Department of Environmental Studies, Keene State College, 229 Main Street, Keene, NH 03435, USA

J. Whitehead
North Carolina Sea Grant, North Carolina State University, 1575 Varsity Dr., Campus Box 8605, Raleigh, NC 27695, USA

S. Gray et al. (eds.), *Environmental Modeling with Stakeholders*,
DOI 10.1007/978-3-319-25053-3_2

It is widely reported that individual and group learning are often among the important outcomes identified by participants in participatory modeling processes (Antunes et al. 2006; Gaddis et al. 2010; Webler et al. 2014). Models, and the technologies and activities to create and use them, facilitate thinking and group interactions. They facilitate learning by leading to the development of new knowledge and skills. For example, they can provide insights about system structure and dynamics and trade-offs of management strategies. Participants also can learn new methods of organizing, analyzing, and presenting information.

During the last several years we have been developing and evaluating a participatory modeling process intended as a systematic approach to integrate local and scientific knowledge on the topic of climate change vulnerability and adaptation. By design, our Vulnerability, Consequences, and Adaptation Planning Scenarios (VCAPS) process promotes learning. We emphasize learning through the dynamic process of developing capacities to think critically, creatively, and collaboratively in support of individual and group decision-making and planning to remedy a problem of common concern. The objective is to promote among participants a form of knowing how, rather than the internalization of isolated facts, such as future climate predictions or the relative costs of adaptation strategies; conceptual frameworks, such as representations of social-ecological systems and procedures for vulnerability assessment, and; skills, such as how to use a computer-based simulation tool. Furthermore, we believe it is vital to avoid conflating the idea of learning with specific desired outcomes, such as agreement, shared understandings, or enhanced trust; a distinction that is often not made sharply enough in the literature on social learning (Reed et al. 2010).

The distinction between learning via development of capacities that support activities of thinking and planning and learning via the transfer of knowledge prompted us to consider more deeply how we design participatory modeling processes and evaluate learning in them. In this endeavor, we turned, as others have, to socio-cultural theories that help illuminate human action, development, and learning (Berkhout et al. 2002; Cundill 2010; Muro and Jeffrey 2012; Pahl-Wostl and Hare 2004). We have primarily grounded our work in the concepts and methods of Wertsch (1991, 1998) and Rogoff (1991, 1994, 1995). Their work suggests an approach to promoting and evaluating how individual and group learning emerge through participatory modeling, characterized by relatively short term and episodic interactions among peers in a guided group activity.

This chapter is organized into three parts. First, we provide a brief overview of the participatory modeling process we developed and applied in the context of climate adaptation and hazard mitigation planning and how our experiences with this process have informed our interest in developing a broader approach to assess learning. Second, we interpret learning in our VCAPS process using theories and methods of socio-cultural analysis and a framework from developmental psychology. Third, we reflect upon how these insights informed the design of the VCAPS process. Fourth, we comment on the lessons learned for theories of learning in participatory modeling.

2.2 Our Approach to Participatory Modeling and Efforts to Assess Participant Learning

Since 2008 we have implemented the Vulnerability, Consequences, and Adaptation Planning Scenarios (VCAPS) process in 14 communities in seven states, with four efforts still ongoing (Kettle et al. 2014; Webler et al. 2014).[1] The VCAPS process is intended to support local vulnerability assessment and climate adaptation planning by providing a systematic approach to integrate local knowledge and scientific information through facilitated, deliberative learning-based activities. Our development of VCAPS draws on the intellectual history of hazard management (Clark et al. 1998; Kates et al. 1985), climate vulnerability assessment (Dow and Carbone 2007; Kasperson et al. 2005; Smit and Wandel 2006), and analytic-deliberation (National Research Council 1996, 2008; Webler and Tuler 2008).

As part of our research program we have endeavored to understand how VCAPS promotes systems-based thinking and how it empowers learning. We have found that participants learn about the socio-ecological system. In particular, they learn about changes in climate and weather, how these changes may impact individuals and communities, and how individuals and communities might respond to mitigate impacts. But people also told us that they learned concepts and skills that support thinking and planning, such as perspective taking, strategies for problem solving, methods for data analysis and simulation, procedures for collaborative decision making, and strategies for building trust. Consequently, our interest has broadened to understand how participants learn to engage in the collaborative activity of adaptation planning, by integrating various facts, concepts, and skills.

2.2.1 The VCAPS Process

VCAPS is implemented in three basic phases, as illustrated in Fig. 2.1: preparing, scenario building, and reporting. In the preparing phase we identify and recruit participants and collect background information relevant to understanding past planning, hazard events, and ongoing concerns within the community. As part of the interviews we also inquire about an appropriate design for VCAPS, including the number of meetings, timing of meetings, and number of participants. It is imperative that the process (specifically the scenario-building phase) be designed in a way that is responsive to the community's need and preferences. This helps to promote legitimacy, building of trust, motivation to participate, and accessibility, which we view as important factors mediating learning.

The second scenario-building phase involves holding meetings, during which exploration and learning about climate-change-related risks, vulnerabilities, and adaptation strategies and the development of scenarios take place. The number of

[1] For additional information see: www.vcapsforplanning.org.

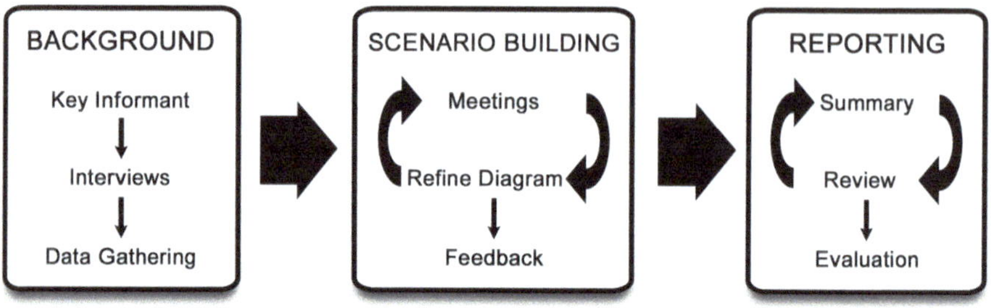

Fig. 2.1 Schematic of the three phases of a VCAPS process

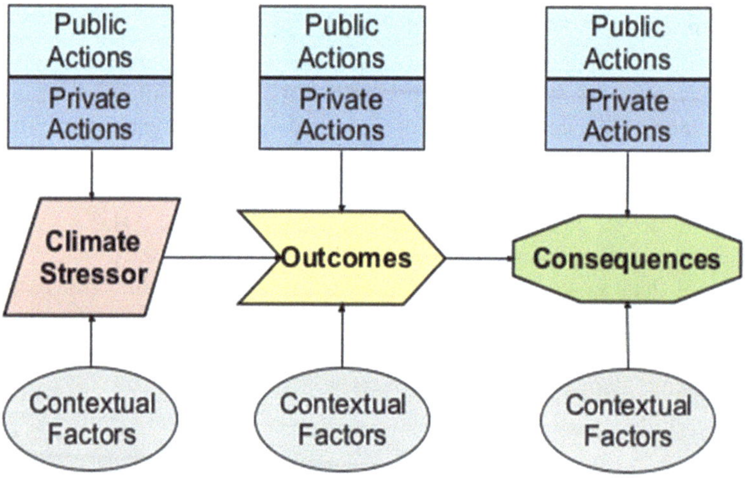

Fig. 2.2 Building blocks of VCAPS diagrams

meetings depends on the preferences of the participants. The meetings weave between opportunities to learn about relevant topics, such as climate change predictions for the area and discussions to elaborate scenarios that represent impacts, vulnerabilities, and management actions to address specific climate-related stressors. For example, we start the first meeting with a presentation by a local climate expert to discuss regional climate trends, projections, and potential impacts with the participants. The purpose is to help participants think about how climate variability and change may influence possible futures.

Subsequently, we provide a brief introduction to VCAPS, and specifically the diagramming of scenarios, and identify the climate stressors of most concern and interest. A small set of components is used to create a VCAPS diagram as shown in Fig. 2.2. We often begin diagrams with a *management concern*, which frames the diagram in the context of an issue the participants are examining in a decision-making context, such as stormwater management or coastal erosion. Management concerns may be known before beginning a VCAPS session, or may be decided upon during the meeting through group consensus after initial review of pertinent climate information. *Stressors*, which may include sea level rise, more intense storms, temperature extremes, or other climate stresses, initiate the hazard chain and are represented by trapezoids; in some cases we begin the diagram with a specific stressor. The decision about how to start a diagram depends on the interests of the participants. Climate stressors lead to intermediary *outcomes* in the coupled human-environment system, represented by block arrows. Finally, outcomes lead to *consequences*, represented as octagons. Consequences are distinguished from outcomes by the significance of their implications for the community. Distinguishing between outcomes and consequences is important because consequences signify why participants care about a particular hazard chain. In cases where individuals may differ on what is an outcome versus a consequence and participants are not able to reach an agreement quickly, it is useful to note the need for further discussion. The diagrams exclude feedback loops to maintain simplicity, but some aspects of feedback can still be captured in the more linear diagrams. For example, management actions can lead to new outcomes and consequences, including altering the character of stressors or other intermediate outcomes. Attached to the bottom of the climate stressors, intermediary outcomes, and consequences are *contextual factors*, represented by ellipses. Contextual factors contain vulnerability qualities of that component of the coupled human-environmental system that may increase or decrease exposure, sensitivity, or the ability to act. The vulnerability concept is retained, but the jargon is eliminated as it poses confusion and can lead to time-consuming discussions that do not add significant value to the diagram. The final elements in a diagram are two types of *management actions*, represented at the top of the diagram as boxes. Management actions can be public or private, and represent actions that can be taken to interrupt or alter the chain of the hazard event at that point. In some cases, management actions can lead to a new series of outcomes and consequences.

We continue with the scenario-building phase by facilitating discussions that are informed by the components of VCAPS diagrams. Diagramming of scenarios occurs in real time using a laptop and a projector. We create VCAPS diagrams using the freeware called VUE (available at http://vue.tufts.edu/). As causal chains become developed, the facilitator encourages participants to identify a broad range of management actions by public and private entities, including both "no regret" strategies, which offer immediate benefits whether or not projected storm and flooding events occur, and "low regret" strategies, which present greater resilience at limited cost. Identifying these management actions on the diagram allows participants to see both upstream and downstream actions that could be taken at each step in the pathway. For example, managers could implement "upstream" regulations to decrease the area covered by impervious surfaces to lessen surface runoff or increase the "downstream" treatment capacity of the wastewater treatment facility. It also enables participants to discuss trade-offs and local contextual features that enhance adaptive capacity and coping capacity. The facilitated discussion aims to promote systems-based thinking and learning, as well as to identify critical gaps in knowledge.

The resulting diagrams can be quite complex and large. Figure 2.3 illustrates a simplified example of pathways associated with sea level rise and storm surge on coastal erosion and near-shore shellfish aquaculture in one of the communities in

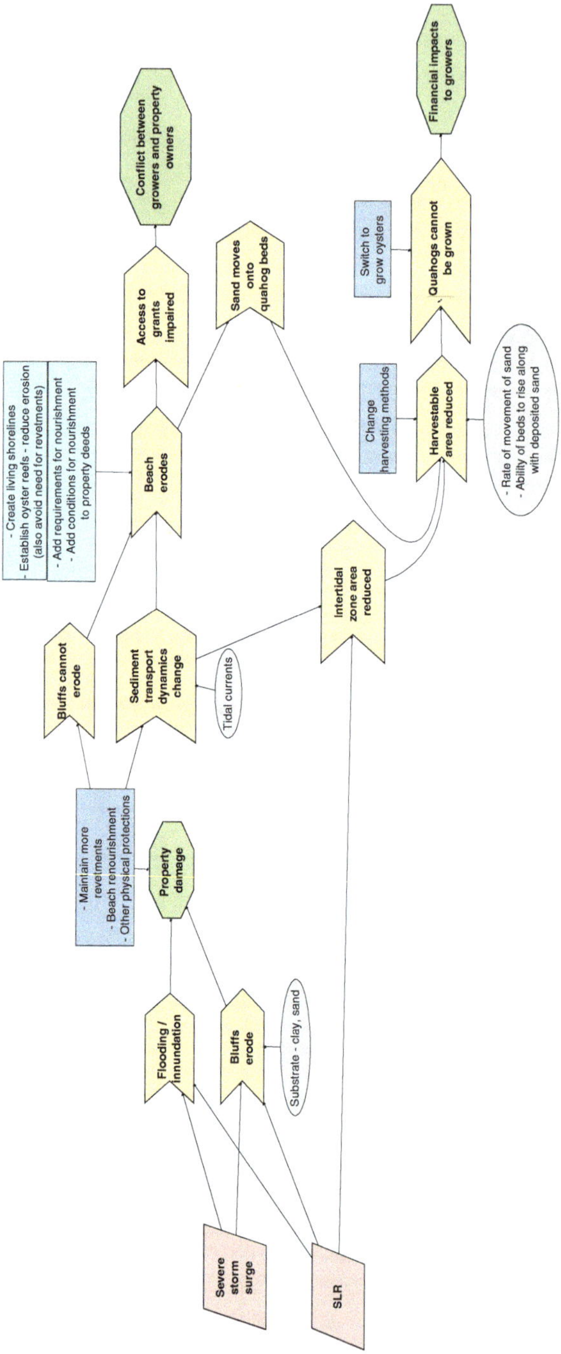

Fig. 2.3 Simplified VCAPS diagram

which we implemented VCAPS. Sea level rise can reduce the intertidal zone, where oyster and clam (quahog) aquaculture grants are located. This limits the ability of commercial growers to produce shellfish. Sea level rise and storm surge can also lead to property damage by eroding bluffs and increasing flooding. A strategy for reducing property damage from these outcomes is coastal armoring, represented by the box above the consequence of *property damage*. Coastal armoring, however, can lead to additional outcomes and consequences to commercial aquaculture businesses, by creating additional coastal erosion and changing sediment transport dynamics. While these outcomes have economic impacts on commercial growers, property owners' decisions to increase armoring can also produce social conflict.

In the final reporting phase, the team summarizes, reviews, and evaluates results from the meetings. The information is presented in ways that facilitate its integration into local planning, which may be associated with hazard-mitigation planning, comprehensive planning, and adaptation planning. Depending on participant preferences, the process may conclude with discussions about how to prioritize and schedule implementation of management actions. We use participant checking to validate results.

2.2.2 Challenges to Assessing Participant Learning in VCAPS

We have implemented VCAPS with communities that are interested in planning for climate change and variability, but the intended outcome has been to support planning rather than to produce a specific plan. Consequently, our consideration of learning has focused on processes of individual and group activities, such as identifying management strategies. We are less concerned with the achievement of particular outcomes, such as the choice of a specific low regret hazard mitigation management strategy. Similarly, our attention has focused on supporting the capacities of participants to engage with new information and other people, rather than to teach about particular facts or options.

We have evaluated VCAPS processes using questionnaires, close observation of discussions, and semi-structured interviews to inquire about the usefulness of the process for planning and decision making, what participants learned (if anything), and the individual and group processes associated with learning. Comments from participants highlighted challenges of how to assess learning in a group process such as VCAPS. These results inspired us to reflect on whether the questions we were asking were capturing the scope of learning and how we might improve our approach to evaluation.

First, participants' self-reflections about learning do not necessarily distinguish among a reflective and an unreflective acceptance of new information or another perspective. For, example, a participant in a process stated:

> "I would say that it brings to light—as a regulator—what I should be thinking about when looking at new proposed projects and how they should be designed with regard to more frequent storms and sea level rise."

The statement suggests that the participant is looking at projects in a new light. However, it does not clarify the source of the new awareness. It may have been proposed by a respected source or be the result of personal reflection and transformation. Similarly, a participant told us:

> "What I learned were new perspectives, different stories about how they are impacted, the community, and parts of the community."

This statement leaves ambiguous whether the differing perspectives were integrated, if at all, or maintained as separate viewpoints. Furthermore, we do not know if the new perspectives were adopted or rejected. The change associated with learning can, with good reason, reinforce prior beliefs and actions. Participants may remain skeptical of others' viewpoints, even while better understanding the values or perspectives of others, or they may even become more cynical with reasonable justification. There may also be failure to agree or share a perspective resulting from rejection on the basis of who is expressing the perspective. On the other hand, acceptance of a new perspective or idea can be based on passive deference to someone perceived as an expert.

Second, it is difficult to determine how shifts in thinking occur. For example, a participant stated:

> "I think the VCAPS process was an opportunity for everybody to see the same information displayed at the same time and have an opportunity to synergize our knowledge base."

The development of shared understandings is commonly cited as an important goal of participatory modeling and VCAPS participants cited it as an important outcome in all of our case studies. However, for the purposes of assessing learning, it would be useful to know if sharedness resulted from the overlap of individuals' prior knowledge or perspectives, the sharing of individual knowledge and perspectives with a group, or generative development of knowledge among individuals in the group that cannot be attributed to specific individuals.

Third, it is difficult to assess how conceptual frameworks and processes influence learning. We commonly heard that the causal structure of VCAPS diagrams (scenarios) was helpful, as illustrated by this quotation:

> "What I really like about them [the VCAPS diagrams] is that they are visual and they level the playing field for everyone at the table. People bring in very different backgrounds, very different sets of experiences all trying to communicate around what can be a very complex area, so it being very visual and going from one step to another, very cause/effect-oriented it levels the playing field for everyone there. That is its biggest value."

However, conceptual frameworks can hide as much as they reveal. To assess learning we need to understand the impact of frameworks and processes on, for example, individual thinking and group problem solving. When participants learn about a new conceptual framework or analytic tool, an important distinction is whether they also gain an associated understanding of its appropriateness in particular contexts. An alternative is that it becomes the singular approach in all future activities without consideration of its strengths and weaknesses in different contexts. In other words, we want to understand if they are adopting a new technique or mastering a new skill.

These challenges raised questions in our efforts at evaluating VCAPS about how to best make sense of changes to participants' ways of thinking and collaborating. These challenges are not unique to VCAPS. In our view, it is not enough to inquire about changes in individual knowledge or preferences, although these can be important kinds of learning. We also want to inquire about changes in how people engage in different mental actions such as perspective taking, problem solving, categorizing, and system conceptualizing. Furthermore, we want to understand the underlying character of changes in individual and group actions, such as deference to expertise, critical reflection, and inter-mingling of multiple perspectives. To approach these issues we turned to theories and methods of socio-cultural analysis and a framework from developmental psychology.

2.3 Characterizing Learning in Participatory Modeling Activities

A common starting point in theories of learning is that people learn by interacting with their environment and interacting in social activities. Participatory modeling does not usually emphasize learning from direct interaction with the environment. Scholars concerned with learning through participatory modeling have drawn on a variety of theoretical frameworks that emphasize the social character of learning, including those of Bandura (1971), Argyris and Schon (Argyris and Schon 1978; Argyris 1995), and Wenger (Lave and Wenger 1991; Wenger 1998). We draw primarily on the socio-cultural approaches to the study of mind developed by Wertsch (1991, 1998) and Rogoff (1991, 1994, 1995) to understand individual and group processes and outcomes of learning in collaborative activities like participatory modeling. We believe they suggest an approach to characterizing processes of learning through the kinds of interactions and contexts that are more typical of participatory modeling. These include relatively short-term and episodic interactions of peers to understand and manage complex systems, where problem solving and decision making are generally emphasized more than learning and apprenticeship. In order to characterize the kind of change associated with learning opportunities, we draw on a framework from developmental psychology (Kaplan and Crockett 1968; Werner and Kaplan 1963; Werner 1957). This framework provides a means for differentiating levels of knowledge and mastery in individual and group actions, including cognitive and communicative actions.

2.3.1 A Socio-Cultural Approach to Learning

A central claim of social-cultural approaches to the study of mind is that human action cannot be analyzed by reductive approaches that isolate people from the means by which they carry out an action. This claim applies to both individuals and groups. Rather, human action is best understood by considering both the person(s) and the tools employed in the action. Wertsch refers to human action as "mediated action." He describes the "tool" as a "mediational means" or "cultural tool." Tools can be cognitive, procedural, or physical. For example, mental models of a social-ecological system can be elicited using a series of procedural steps and represented using computer software, illustrating cognitive, procedural, and physical tools, respectively. Additional types of cultural tools relevant to participatory modeling are identified in Table 2.1. Furthermore, the practice of participatory modeling is itself a cultural tool.

Rogoff argues that three levels of analysis are useful for understanding human action and processes of development and learning in socio-cultural activities: community, interpersonal, and individual (Rogoff 1995). Each level of analysis can provide insights about the ways cultural tools are structured and transformed, how they are used by groups and by individuals, and how they are learned.

Analysis of socio-cultural activities at the community level refers to social, cultural, legal, and institutional contexts in which newcomers or novices develop their knowledge and skills through their participation with others. These contexts involve participation in varied activities, such as those between a scientist modeler and a decision maker and a teacher and a student. They can also involve small groups engaged in goal-oriented activities involving "peers who serve as resources and challenges for each other in exploring an activity, along with experts (who, like peers, are still developing skill and understanding in the process of engaging in activities with others of varying experience)" (Rogoff 1995, p. 143). Participatory modeling processes exemplify this kind of community context. Through interaction in social activities people are exposed to and learn new cultural tools. Wertsch refers to "cultural tools" to emphasize the idea that tools reflect social, cultural, and institutional forces in both their structure and use. Social, cultural, and institutional forces range from the concrete and immediate to the more distant and indirect: disciplinary languages and methods, organizational structures and norms of practice, ideological beliefs, and socio-political and economic systems. As part of our efforts to evaluate the participatory modeling process of VCAPS we consider how the design of the process affects interactions of participants and their opportunities for learning. While it is also possible to consider the institutional and social context in which VCAPS is embedded, to date we have focused less on how these issues affect opportunities for learning.

The interpersonal level of analysis involves analysis of how cultural tools are learned via socio-cultural activity and how cultural tools are used in socio-cultural activity. Analysis at this level can consider how the introduction of a new tool can alter the form and process of an activity. For example, the shift from pen and paper to word processors—both technical tools—can alter how people write, as well as how they remember and organize information. Similarly, shifting the framing of problems has been suggested as a means for resolving conflicts and finding better solutions; an action achieved by shifting the cognitive and procedural tools employed. Rogoff's concept of *guided participation* focuses our analysis on the ways varied types of guidance and

Table 2.1 Topics and tools participants can learn in VCAPS processes

Topics that participants can learn about		Examples of cultural tools learned and used
The system being modeled	Facts; theories; system structure and dynamics; uncertainties and knowledge gaps; implications of framing and boundary setting; implications of management actions.	Conceptual frameworks for representing social-ecological systems and their components. Ways of organizing information (e.g., taxonomies). Assessment frameworks for evaluating best practices for management.
The societal context	Values and norms held by participants; problem definitions and policy perspectives held by participants; opinions about roles of different stakeholders; relationships of different stakeholders; history of addressing problems within community; resources that may be mobilized for environmental management; legal frameworks; opportunities and barriers to implementing decisions.	Conceptual frameworks for moral and ethical reflection.
Forms of expression	Ability to understand and use technical terminology; ability to construct arguments (i.e., make claims, marshal evidence and theories); ability to articulate one's own view.	Disciplinary jargon used in speaking and writing. Skills of presentation. Skills of argumentation. Interpretive frames.
Ways to interact with others	How to listen; how to show empathy; skills of persuasion, negotiation, and compromise; methods for reaching closure.	Rules of facilitation and interaction. Skills of adversarial and collaborative discourse. Frameworks of negotiation (e.g., distinguish between interests and values). Culturally and socially appropriate cues for signaling active listening and empathy.
Methods of sense making	Analytic methods and tools for representing the social-ecological system (e.g., GIS, systems dynamics modeling); analytic methods for eliciting and representing individual preferences, values, beliefs; methods for choosing among conceptual and analytical frameworks; decision making strategies; ethical and moral frameworks; use of evidence and knowledge in reasoning.	Computers and computer software for modeling (e.g., STELLA, GIS, Bayesian networks, distributed computer-based dialogue systems, actor-networks). Analytic approaches to eliciting and weighing individual preferences (e.g., surveys, multi-attribute utility theory). Conceptual frameworks for moral and ethical reflection. Drawings, sketches, notes, etc. made with pen and paper.
Self and identity	Preferences and norms of self and others; roles and purposes of self and others; compatibility with group beliefs.	Mental models. Conceptual frameworks for moral and ethical reflection.

participation can affect learning. She uses this term to refer to the ways that people observe, communicate, and coordinate while participating in a joint activity. For example, interactions can be face-to-face or mediated by computer technologies. Guidance can take the form of experts teaching lay people or collaborative exploration within a group of individuals having diverse types of recognized expertise in a goal-oriented activity.

The personal level of analysis of socio-cultural activity focuses on how people change through participation in socio-cultural activity. Individual learning is often associated with the concept of internalization of something new, such as facts, conceptual frameworks, and skills. When applied to learning, the concept of internalization is problematic when it is taken to mean acquisition of something new by an individual from the external world such that static, invariant meanings are preserved (Rogoff 1995; Wertsch 1998). Rogoff and Wertsch share a theoretical perspective on human mental functioning and development. This perspective emphasizes the development, by individuals and groups, of *knowing how* to engage in mediated action in specific situations. This kind of knowing is developed through social activity, including contexts where multiple individuals with varying types and levels of expertise engage in purposive activity (Lave and Wenger 1991; Rogoff 1991, 1995; Wenger 1998; Wertsch 1998; Vygotsky 1986). Wertsch (1998) argues that there are two dimensions to knowing how to engage in mediated action. The first dimension is described as a form of *mastery* or transformation,[2] whereby the skill of knowing how to use a cultural tool with some degree of proficiency is developed by engaging in its use (i.e., mediated action). The second dimension is described as a form of *appropriation*, whereby a person chooses to embrace or reject a cultural tool. The concept of appropriation highlights that people are not "mindless, helpless consumers of the meditational means provided by their sociocultural settings" (Wertsch 1998, p. 55).

For example, consider the playing of a guitar, a mediated action. Through guided participation, a player can master the knowledge and skills central to playing compositions, including chords and scales. The learning of cultural tools such as chords and scales can be mastered to varying degrees; they are internalized. But, not every aspiring guitar player succeeds in the process of appropriation, the "taking something that belongs to others and making it one's own" (Wertsch 1998, p. 53). They may learn to play, via mimicry, some delta blues songs, but not develop the expertise to play the delta blues with deep feeling and employ their developing mastery in new, creative ways. Wertsch further notes "the appropriation of meditational means need not be related to their mastery in any simple way. In some cases, mastery and appropriation are correlated at high or low levels, but in others the use of cultural tools is characterized by a high level of mastery and a low level of appropriation. Thus these two forms of 'internalization' need to be differentiated…" (1998, p. 57).

[2] Rogoff uses the term *participatory appropriation*. We adopt the terminology of Wertsch (1998) because he makes the distinction between *mastery* and *appropriation*.

2.3.2 A Developmental Approach to Assessing Learning

If we are to study learning—and try to promote it in practice such that it supports environmental decision making in a constructive way—its assessment in processes such as participatory modeling cannot avoid making claim to some sort of goal. We believe it is important to avoid conflating learning with specific ideal outcomes desired either by organizers or participants, such as agreement, trust, or prevention of conflict. Instead, we focus our assessment of learning on *how* mediated action is accomplished.

Our approach is informed by a conceptual framework of developmental psychology represented by Hanz Werner (Werner and Kaplan 1963; Werner 1957) and is consistent with the interest in mastery of cultural tools and mediated action. From this perspective, mental development proceeds "from a state of relative globality and undifferentiatedness towards states of increasing differentiation and hierarchic integration" (Werner and Kaplan 1963, p. 7). Individuals exhibiting expertise are defined by their ability to employ varied cultural tools and their ability to consciously reflect on and choose among them depending on context of activity. Novices, on the other hand, are more limited in the set of tools they can draw on, as well as their mastery in using different cultural tools especially in different contexts (Kaplan and Crockett 1968). These claims are consistent with insights from research on expertise and organizational learning (Boreham and Morgan 2004; Engestrom et al. 1995; Engestrom 2001; Lave and Wenger 1991; Schon 1983).

Attention to the "differentiation, articulation, and hierarchic integration" of mediated action by individuals and groups can be applied to the characterization of learning and change among participants in participatory modeling processes. Involvement in a participatory modeling process might cause transformation of how an individual represents a social-ecological system, understands links between components, and construes "facts" about the world. Involvement might also confront individuals with different perspectives, ethical frameworks, and ways of speaking. For example, our assessments are building on efforts to understand if the VCAPS process involved participants:

- accepting information on the basis of "source" (e.g., passive deference to "experts")?
- applying new conceptual or analytical frameworks and identifying their strengths and weaknesses for a particular problem context?
- developing the ability to adopt different perspectives, without simply accepting one as more "correct" than others, in contexts of uncertainty and system complexity?
- developing the capacity to identify underlying points of agreement or disagreement among perspectives, rather than accepting or rejecting based on more global assessments, such as compatibility with prior beliefs?

This approach to characterizing change provides a mechanism for assessing how different process designs or modeling tools can affect the actions of participants and the mastery of cultural tools by participants. Our approach considers the three levels of analysis (community, interpersonal, and individual). In the following section we turn to the ways that this approach leads to consideration of different types of cultural tools that participants might learn and the ways that process design can influence what is learned. These are aspects of participatory modeling over which organizers have the most influence.

2.4 Designing Participatory Modeling Processes to Promote Learning

In this section, we apply a socio-cultural analysis and the developmental framework to our experiences with the VCAPS process. First, we identify the variety of possible learning outcomes and cultural tools that can be used to promote them. We have intentionally designed VCAPS to promote learning, via processes of mastery, of some of these tools. Second, a socio-cultural analysis suggests how processes may be designed to facilitating learning. We discuss examples from our implementations of VCAPS to illustrate a small set of process-design choices that can impact learning.

2.4.1 Cultural Tools That Participants Can Learn by Engagement in Participatory Modeling

In a participatory modeling process, participants may potentially learn a variety of cultural tools. While others distinguish among cognitive, moral, and relational learning (Muro and Jeffrey 2008; Webler et al. 1995), we constructed an elaborated set of learning categories based on a range of topics that participants in VCAPS processes have identified as part of their learning outcomes: the system being modeled, the societal context, forms and methods of expression, ways to interact with others, and methods of sense making. Their ability to learn about these topics is mediated by the cultural tools used. At the same time, VCAPS has provided an opportunity for participants to learn, with varying degrees of mastery and appropriation, some of these cultural tools. Table 2.1 illustrates the types of learning and associated cultural tools using examples from VCAPS processes.

2.4.2 Process Designers Can Promote Interactions and Activities to Promote Learning of Cultural Tools

Process design can create or limit opportunities for sharing knowledge and skills among participants in participatory modeling activities. We illustrate three aspects of process design that can impact dynamics of mastery and appropriation of cultural tools associated with representing the social-ecological system, characterizing the societal context, how to interact with others, and methods of sense making.

First, we intentionally frame the purpose of the causal models and scenarios as "*thinking devices*" (Lotman 1990; Wertsch 1998, p. 115) about how climate stressors may impact a community. We frame VCAPS as a means to inform ongoing and future climate adaptation and hazard management planning, rather than generating models that provide an accurate representation of the existing system or of future system states. To emphasize the role of scenarios as thinking devices the facilitator continuously poses "what if" questions. For example, when specific management actions are proposed, the facilitator can encourage reflection about a broad range of implications of the action. For example, if shoreline armoring or beach nourishment are proposed as ways to reduce flooding of coastal properties, we encourage discussion of how these actions may impact erosion and sediment transport (see Fig. 2.3 above).

Second, we consider how the number and duration of meetings can affect the ways that participants interact with each other, the depth and breadth of the discussions, and the opportunities for engaging with new cultural tools. For example, implementations of VCAPS have ranged from holding a single 7-hour meeting on 1 day, to two 3-hour meetings split over 2 days, to a series of five meetings spread over a year. In our experience, a single-day meeting limits participants' elaboration of complexities of causal diagrams and their ability to learn how to consider the implications of others' perspectives. Fewer meetings also limit the ability of participants to understand the analytic frameworks that provide structure to VCAPS diagrams and how other frameworks and methods (e.g., GIS mapping) can be integrated to provide different perspectives of the system being discussed. The pros and cons of the underlying framework are not made explicit through exploration of examples and participants may accept or reject outcomes on the basis of more superficial attributes, such as visual appeal.

On the other hand, in communities where a greater number of meetings have been held, some participants have reported developing new skills for elaborating and exploring scenarios. As part of the process to develop a Hazard Mitigation Plan revision for Boston, a VCAPS participant noted the difference between two cultural tools for planning: the FEMA hazard mitigation planning framework, which emphasizes a focus on mitigation actions, and the VCAPS framework, which emphasizes the causal structure of hazards:

> "If you focus on an action and it proves undoable for whatever reason, it comes full stop, but if you focus on the hazard and have a very clear understanding of the hazard, that simply means you go back to the diagram and identify another intervention point...if the action doesn't work it is not like the end of the road, but simply means you come back to your understanding of the hazard and look for a different approach."

The greatest impact of VCAPS on learning of cultural tools has been on the ability of participants to learn about new perspectives and the capacity for perspective-taking. For example, as part of a post-process interview a participant stated:

> "It brought some barriers to the forefront. I don't know that it had to do with understanding them. We kind of know what the barriers are but it did help to give me a little insight instead of looking at it always from a regulatory standpoint, to look at it from the other side, and what is important to the actual homeowner or property owner as far as mitigation is concerned. And sometimes their interest in mitigation and the interest from the regulatory standpoint are totally different."

However, like others we have found that participants may not always have the background knowledge or skills to engage with tools, such as GIS or causal modeling, in a productive way (Ghose et al. 2003); we have found this to be a barrier, for example, to encouraging local planners to continue using VCAPS after our projects have ended. The VCAPS process in Boston also provides insight into the difference between mastery and appropriation. In this case, the participant, a regional planner, began to develop mastery with the causal model concept and its applications to hazard management planning. However, a community level of analysis makes explicit the constraints to appropriation resulting from institutional requirements; FEMA requirements were viewed as making it hard to incorporate the results of the VCAPS process into the plan.

2.4.3 Process Designers Can Select Participants and Define Roles to Promote Learning of Cultural Tools

The ability to learn about the perspectives of others is impacted by the ability to come into contact and explore alternative perspectives, mental models, normative frameworks, and other cultural tools. The information and cultural tools that are available for "intermingling" in dialogue depend to a great extent on who is involved in a participatory modeling process and how they are involved. In the context of participatory modeling, participants can include people who own property (local or otherwise), use an area for recreation, live in the area, or are members of civic and environmental organizations. Scientists and decision makers or managers are also typically part of the process, and in some cases, those who organize or facilitate can play a role in the discussion. Hence, all are included in processes of learning.

Most of our applications of VCAPS have included relatively small numbers of local officials and staff, with some representation of other community stakeholders because we have been interested in working with people strongly engaged in local climate adaptation and hazard-management planning. In Boston, however, over 100 people participated over a series of five meetings. This had the effect of broadening the information available to share, but limited opportunities for actually sharing or discussing details.

In all of our cases the presence of people with differing perspectives was clearly important for the quality of learning that occurred. For example, participants in different VCAPS implementations stated in post-process interviews:

> "It also opened the door to the possibility that we could have found some places where infrastructure improvements that might not otherwise have been identified were identified because of the human services people there and their particular perspective. Or maybe changed how that mitigation action might have been advanced. Another example was having the historic preservation people there who wouldn't normally show up. Again another very different perspective."

> "The community throws things out from a different perspective that are really important that perhaps from a staff perspective we may have missed."

> "When you have different departments there in the room, what seems like a good idea when you have just the infrastructure-based people in the room becomes less of a good idea when you have social service providers who are actually talking about different people with different disabilities or health conditions, and so on and so forth, then interact with that process."

The types of roles given to different groups within a process can also influence learning in two ways. First, certain cultural tools may be privileged over others. For example, a particular ethical framework may be used, and left unquestioned, when advanced by those with authority or without someone being present to articulate an alternative. Second, the ways that people interact can be influenced by the roles given, implicitly or explicitly, to participants. How people interact can affect if and how cultural tools are mastered. In the case of scientists and decision makers, their roles may be, for example, as central designers of models, as resources for others developing and using models, or as the audience for model results. Similarly, in processes described as participatory the roles of the public and other stakeholders have included, for example: providing information to inform the design of a model or policy proposal, helping to define problems and models in collaboration with decision makers and disciplinary experts, and providing input about decision preferences. It is less likely that analytical frameworks, such as systems dynamics modeling, will be mastered by participants in a process when scientists are understood as the expert modelers and other participants viewed as providers of information. On the other hand, when scientists and others are viewed as co-constructors of models and facilitators emphasize methods as well as outcomes, then there are more opportunities for the mastery of relevant cultural tools. This is the approach we take with VCAPS. For example, we introduce participants, including expert scientists, to the rationale for causal modeling and the structural components of causal models of hazards. Together, we then construct causal diagrams, as a means of both producing relevant scenarios and transferring knowledge about how to construct scenarios. We have adopted the same approach in more recent efforts to develop systems dynamics models (see Webler et al. this volume).

2.4.4 *Process Designers Can Selectively Use Cultural Tools to Promote Learning*

Disciplinary frameworks, conceptual frameworks, and analytic tools always provide a certain viewpoint and influence what can be understood and shared. For example, what can be learned from a conceptual model depends on what the model represents. Similarly, GIS provides information about spatial relationships, but not about system dynamics. Boundaries separating endogenous and exogenous variables and connections focus attention.

As part of VCAPS we primarily emphasize the causal linkages between climate stressors, consequences, and management actions. At times we have supplemented discussions with GIS maps, to highlight spatial relationships that are harder to characterize with the causal diagrams. These are conscious choices, and we have found it important to help participants understand what is highlighted and what is obscured by different ways of representing the social-ecological system. Similarly, we have pointed out the implications of using particular kinds of categories to organize information, including types of barriers to adaptation. We have also begun to develop a taxonomy of consequences that may be considered by participants, including public health; as well as social, economic, institutional, cultural, and ecological impacts. We also encourage participants to explore different boundary conditions for the system being characterized, including different scales of governance, time horizons, and non-climate stressors.

In addition, technologies of modeling mediate how information is represented and shared. For example, models, as well as information shared by participants, can be displayed and structured in various ways to support individual and group sense-making and transparency (Gaddis et al. 2010). Systems dynamics models help users understand feedbacks in a dynamic system and the sensitivity to system functioning from changes in parameter values, while spatial models and causal pathways, like those used in VCAPS, do not. Visualization influences how people process and share information (Al-Kodmany 2001; Gaddis et al. 2010; Sheppard 2012), but not everyone is a "visual thinker," as a planner involved in one of our cases noted. More generally, the use of different technical means, such as computers, has been shown to influence processes of thinking and group interaction (Stahl et al. 2006).

For example, many participants have commented that they found it useful that we display causal models onto screens in real time so they can both see how scenarios are developing and continue to discuss details by sharing information and perspectives: "They were seeing it put in a diagram and they were following along and they could see the path they were working on and it helped them to come up with new ideas." The generation of new ideas is a basis for learning that goes beyond acquisition of static concepts.

In addition, our approach to projecting causal diagrams on a screen for everyone to see generates particular kinds of interactions that are conducive to learning cultural tools, such as perspective taking, active listening, and collaborative problem solving. For example, a participant in one case stated that:

> "The [facilitator] was constantly asking people to slow down. 'You all are really throwing a lot of things out there.' This gave everyone a feeling of being charged and enthusiastic. He was trying to keep up with the group and the group was surging ahead. There was some great energy in the room. I don't know how to explain that but it seems that the visual aid helped that to happen. People forgot there was a mayor in the room. People forgot that [the facilitator] was a PhD from wherever and that it was city staff and a housewife."

2.5 Conclusion

We introduced a socio-cultural approach to assessment of learning through participatory modeling processes. This approach provides insights about the ways that people learn cultural tools by their participation in joint activities. It also helps to illuminate the ways that contextual features of the participatory modeling process, including institutional, legal, and social features, can impact opportunities for learning. Our focus in this chapter has been on individual learning. Although we did not explore the issue here, the concepts introduced in this chapter can also be applied to groups, as a form of social learning by collectives of individuals (Boreham and Morgan 2004; Lave and Wenger 1991; Rogoff 1991; Wertsch 1998). Learning by groups is defined as occurring when individuals engage in joint, purposive activity in ways that transform a practice. For example, the practice of adaptation planning or participatory modeling can be altered by the introduction of new cultural tools.

A key distinction in our approach is between the internalization or acquisition of facts, values, skills, procedures, and techniques and the mastery and appropriation of *knowing how* to engage in mediated action in specific situations. We have also introduced a way to characterize learning that avoids the pitfall of other approaches that define learning in terms of desired social or political goals (Reed et al. 2010) by proposing a measure of learning associated with cultural tools and their skilled use in socio-culturally situated action.

These distinctions clarify the opportunities and constraints to learning that arise in participatory modeling processes. Their design creates opportunities and constraints for guided learning and deep exchange of knowledge and generative dialogue when scientists, modelers, and local stakeholders interact. Those who design processes can benefit from attention to the different demands and opportunities that shape learning, and be more explicit in their goals for promoting them. Learning cultural tools and using cultural tools to learn can both be accomplished in participatory modeling processes. A socio-cultural approach encourages us to be as reflective about the cultural tools we employ to promote and assess learning as we wish participants to be about their own experiences.

Acknowledgments We developed the VCAPS process with funding from multiple sources, including the National Oceanographic and Atmospheric Administration (NOAA) Sectoral Applications Research Program, the National Oceanographic and Atmospheric Administration (NOAA) Coastal and Ocean Climate Applications Program, and Massachusetts Institute of Technology Sea Grant. Additional cases were funded by separate grants from the NOAA Sectoral Applications Research Program, NOAA National Sea Grant Program Coastal Communities Climate Adaptation Initiative, NOAA National Sea Grant Program Community Climate Adaptation Initiative, Mississippi and Alabama Sea Grant, and the Mobile Bay National Estuary Program. We would also like to thank Nate Kettle and Karly Miller for their role in the early stages of the development and implementation of VCAPS.

References

Al-Kodmany A (2001) Bridging the gap between technical and local knowledge; tools for promoting community-based planning and design. J Archit Plan Res 22(1):110–130

Antunes P, Santos R, Videira N (2006) Participatory decision making for sustainable development—the use of mediated modelling techniques. Land Use Policy 23(1):44–52

Argyris C (1995) Action science and organizational learning. J Manag Psychol 10(6):20–26. doi:10.1108/02683949510093849
Argyris C, Schon DA (1978) Organizational learning: a theory of action perspective. Addison-Wesley, Reading
Bandura A (1971) Social learning theory. General Learning Press, Morristown
Berkhout F, Hertin J, Jordan A (2002) Socio-economic futures in climate change impact assessment: using scenarios as "learning machines". Glob Environ Chang 12(2):83–95
Boreham N, Morgan C (2004) A sociocultural analysis of organisational learning. Oxf Rev Educ 30(3):307–325
Clark G, Moser S, Ratick S, Dow K, Meyer W, Emani S et al (1998) Assessing the vulnerability of coastal communities to extreme storms: the case of Revere, MA, USA. Mitig Adapt Strateg Glob Chang 3:59–82
Cundill G (2010) Monitoring social learning processes in adaptive comanagement: three case studies from South Africa. Ecol Soc 15(3). Available online at: http://www.ecologyandsociety.org/vol15/iss3/art28/
Dow K, Carbone G (2007) Climate science and decision making. Geogr Compass 1(3):302–324
Engestrom Y (2001) Expansive learning at work: toward an activity theoretical reconceptualization. J Educ Work 14(1):133–156
Engestrom Y, Engestrom R, Karkkainen M (1995) Polycontextuality and boundary crossing in expert cognition: learning and problem solving in complex work activities. Learn Instr 5:319–336
Gaddis EJB, Falk HH, Ginger C, Voinov A (2010) Effectiveness of a participatory modeling effort to identify and advance community water resource goals in St. Albans, Vermont. Environ Model Softw 25(11):1428–1438
Ghose R, Elwood S, Ghose UJ (2003) Public participation GIS and local political context: propositions and research directions. URISA J 15:17–24
Jones NA, Perez P, Measham TG, Kelly GJ, D'Aquino P, Daniell K, Dray A, Ferrand N (2009) Evaluating participatory modeling: developing a framework for cross-case analysis. Ecol Soc 44(6):1180–1195
Kaplan B, Crockett WH (1968) Developmental analysis in the problem of cognitive consistency. In: Abelson RP, Aronson EA, McGuire W, Newcomb TM, Rosenberg MJ, Tannenbaum P (eds) Theories of cognitive consistency. Rand McNally, Chicago, pp 661–669
Kasperson JX, Kasperson RE, Turner BLI, Hsieh W, Shiller A (2005) Vulnerability to global environmental change. In: Kasperson JX, Kasperson RE (eds) The social contours of risk, vol II. Earthscan, London
Kates RW, Hohenemser C, Kasperson JX (eds) (1985) Perilous progress: managing the hazards of technology. Westview Press, Boulder
Kettle N, Dow K, Tuler S, Webler T, Whitehead J, Miller K (2014) Integrating scientific and local knowledge to inform risk-based management approaches for climate adaptation. Clim Risk Manag 4-5:17–31
Knapp CN, Fernandez-Gimenez M, Kachergis E, Rudeen A (2011) Using participatory workshops to integrate state-and-transition models created with local knowledge and ecological data. Rangel Ecol Manag 64(2):158–170
Lave J, Wenger E (1991) Situated learning: legitimate peripheral participation. Cambridge University Press, New York
Lotman Y (1990) Universe of the mind: a semiotic theory of culture. Indiana University Press, Bloomington
Mendoza GA, Prabhu R (2005) Combining participatory modeling and multi-criteria analysis for community-based forest management. For Ecol Manag 207(1-2):145–156
Muro M, Jeffrey P (2008) A critical review of the theory and application of social learning in participatory natural resource management processes. J Environ Plan Manag 51(3):325–344
Muro M, Jeffrey P (2012) Time to talk? How the structure of dialog processes shapes stakeholder. Ecol Soc 17(1):3
National Research Council (1996) Understanding risk. National Academies Press, Washington

National Research Council (2008) Public participation in environmental assessment and decision making. National Academies Press, Washington
Pahl-Wostl C, Hare M (2004) Processes of social learning in integrated resources management. J Community Appl Soc Psychol 14:193–206
Reed MS, Evely AC, Cundill G, Fazey I, Glass J, Laing A, Newig J, Parrish B, Prell C, Raymond C, Stringer LC (2010) What is social learning? Ecol Soc 15(4). Available online at: http://www.ecologyandsociety.org/vol15/iss4/resp1/
Rogoff B (1991) Apprenticeship in thinking: cognitive development in social context. Oxford University Press, New York
Rogoff B (1994) Developing understanding of the idea of communities of learners. Mind Cult Act 1(4):209–229
Rogoff B (1995) Observing sociocultural activity on three planes: participatory appropriation, guided participation, and apprenticeship. In: Wertsch J, Del Rio P, Alvarez A (eds) Sociocultural studies of mind. Cambridge University Press, New York, pp 139–164
Rouwette EA, Vennix JA, Mullekom TV (2002) Group model building effectiveness: a review of assessment studies. Syst Dyn Rev 18(1):5–45
Schon D (1983) The reflective practitioner: how professionals think in action. Basic Books, New York
Schusler TM, Decker DJ, Pfeffer MJ (2003) Social learning for collaborative natural resource management. Soc Nat Resour 16(4):15–26
Sheppard S (2012) Visualizing climate change: a guide to visual communication of climate change and developing local solutions. Routledge, New York
Sieber R (2006) Public participation geographic information systems: a literature review and framework. Ann Assoc Am Geogr 96(3):491–507
Smit B, Wandel J (2006) Adaptation, adaptive capacity and vulnerability. Glob Environ Chang 16:282–292
Stahl G, Koschmann T, Suthers D (2006) Computer-supported collaborative learning: an historical perspective. In: Sawyer RK (ed) Cambridge handbook of the learning sciences. Cambridge University Press, Cambridge, pp 409–426
Tidwell VC, van den Brink C (2008) Cooperative modeling: linking science, communication, and ground water planning. Ground Water 46(2):174–182
Van den Belt M (2004) Mediated modeling-a system dynamics approach to environmental consensus building. Island Press, Washington
Voinov A, Gaddis EJB (2008) Lessons for successful participatory watershed modeling: a perspective from modeling practitioners. Ecol Model 216(2):197–207
Vygotsky L (1986) In: Kouzlin A (ed) Thought and language. MIT Press, Cambridge
Webler T, Tuler SP (2008) Organizing a deliberative planning process: what does the science say? In: Odugbemi S, Jacobson T (eds) Governance reform under real-world conditions: citizens, stakeholders, and voice. World Bank, Washington, pp 125–160
Webler T, Kastenholz H, Renn O (1995) Public participation in impact assessment: a social learning perspective. Environ Impact Assess Rev 15:443–463
Webler T, Tuler SP, Dow K, Whitehead J, Kettle N (2014) Design and evaluation of a local analytic-deliberative process for climate adaptation planning. Local Environ. Published online 17 Jul 2014. doi:10.1080/13549839.2014.930425
Wenger E (1998) Communities of practice: learning, meaning, and identity. Cambridge University Press, New York
Werner H (1957) Comparative psychology of mental development. International Universities, New York
Werner H, Kaplan B (1963) Symbol formation. Lawrence Erlbaum Associates, Inc., Hillsdale
Wertsch J (1991) Voices of the mind: a sociocultural approach to mediated action. Harvard University Press, Cambridge
Wertsch J (1998) Mind as action. Oxford University Press, New York

Chapter 3
Values in Participatory Modeling: Theory and Practice

Alexey Voinov and Erica Brown Gaddis

3.1 Introduction

The popularity of participatory modeling has grown in recent years (Voinov and Bousquet 2010) since it is particularly compatible with new environmental management paradigms that focus on ecosystem-based management, integrated water resources management, and adaptive management. All of these incorporate systems theory and aim to protect and improve ecological resources while considering economic and social concerns in the community. New inclusive modeling approaches have emerged that have been adopted by, among others, the Water Framework Directive of the European Commission, the Malawi Principles in the Convention on Biological Diversity (UNEP), and the National Center for Environmental Decision-Making Research (NCEDR) in the United States. The latter recommends that the processes of analysis and deliberation be integrated in such a way that systematic analysis is combined with community values critical to decision-making. This is because participatory modeling provides a platform for integrating scientific knowledge with local knowledge, and when executed well, provides an objective, transparent and flexible workplace for a diverse group of stakeholders to contribute information regarding an ecosystem of interest. Recognition that effective ecological management requires input from both scientific and social processes is key to developing effective partnerships between scientists who know the theory and research methods and stakeholders who live and work within an ecosystem.

A. Voinov (✉)
University of Twente, Faculty of Geo-information Science and Earth Observation, Enschede, The Netherlands
e-mail: aavoinov@gmail.com

E.B. Gaddis
Division of Water Quality, Utah Department of Environmental Quality, Salt Lake City, UT, USA

S. Gray et al. (eds.), *Environmental Modeling with Stakeholders*,
DOI 10.1007/978-3-319-25053-3_3

Participatory modeling (also known as "mediated modeling," "shared vision planning," "group model building," etc.) draws on the theory of post-normal science, which dictates that in problems characteristic of highly complex systems—when facts are uncertain, values are in dispute, stakes are high and decisions are urgent—there is no one, correct, value-neutral solution (Funtowicz and Ravetz 1993). Many ecological and environmental problems are characterized by these challenges. Under such circumstances, standard scientific activities are inadequate and must be reinforced with local knowledge and iterative participatory interactions to derive solutions that are well understood, politically feasible, and scientifically sound.

Stakeholder participation in these types of situations—such as those common to ecological research and management—has, therefore, been justified for multiple reasons. This is because stakeholder participation in the decision-making process supports democratic principles, promotes learning, integrates information about social and natural processes, adds legitimacy to the process, and can lead participants in moving forward toward an agreed agenda. The extent to which the public or representative stakeholder group can effectively participate in ecological research and management is determined by the methods employed in engaging stakeholders, inclusion of diverse groups, group size, incorporation of local knowledge and expertise, and the time available for the process to develop. The development of unique, practical, and affordable solutions to ecological problems is often best accomplished by engaging stakeholders and decision makers in the research process (Seidl et al. 2013; Tàbara and Chabay 2013).

However, we still see little progress is solving some of today's most urgent environmental problems. Even with an increase in the popularity of participation in environmental decision-making in general and the use of participatory modeling specifically, there are still questions about how to best structure the model-building process with stakeholders. In this chapter, we reflect on some of our experiences as modelers engaged in participatory modeling by outlining some of the lessons we learned through our experiences and (1) reflect on some problems of the science-policy interface that we see as preventing the solution to some crucial problems humanity faces, (2) outline best practices for modelers seeking to engage in the process and (3) conclude by presenting an example of a project that uses some of the more innovative techniques of participation.

3.2 Philosophy of Participatory Modeling: Integrating Values, Not Just Knowledge

One of the main promises of participatory modeling has always been the idea that by bringing modeling to the hands of stakeholders and by making sure that they understand and appreciate the modeling tool developed, we can actually expect better decisions and management practices to be implemented and better policies to be adopted. However, in many cases we do not see this outcome.

On a large scale, Rockstrom et al. (2009) have clearly shown that several of the planetary boundaries, or critical indicators such as biodiversity loss, or climate change, or the nitrogen biogeochemical flow, have been already exceeded, while several others are about to be passed. There is evidence that new conflicts are emerging because of limited resources such as food, water, energy, and land (Daily and Ehrlich 1996; Homer-Dixon 1999). It has been shown that climate change, and loss of biodiversity and ecosystem function can be detrimental to our life-support systems (Balvanera et al. 2006). There has been some success fixing smaller issues on local scales (Ettiene 2014; Bousquet and Voinov 2010), but the results are hardly encouraging despite broad stakeholder involvement. While certain watersheds get improved riparian zones and point pollution is contained, the Chesapeake Bay does not show much improvement (Paolisso et al. 2013); while large reductions of phosphorus inputs are attained in the St. Albans Bay, Lake Champlain still has increasing levels of eutrophication (Gaddis et al. 2010a, b).

There are probably different reasons each time the management practices or policies do not work out as intended. On large-scale projects, participatory efforts may be quite prohibitive; it remains unproven that large-scale participation can work. We have not yet developed appropriate tools to allow participation of numerous stakeholders rather than a dozen people meeting for a workshop. There are many technical issues that need to be resolved to provide adequate facilitation and information sharing in big groups. In fact, it is yet to be shown that such large-scale participation can even work. Certainly in all scales, projects are dependent upon funding and funding is rarely available to monitor project outcomes and to follow-up on projects after they are finished. In many instances we find that the participatory process that goes well *during* the study is mostly forgotten *afterwards* when the funding has ended, reports are written, papers are published and researchers are back home.

We argue that the problem is not only how the participatory process is organized and conducted, but concerns the larger issue of how science and policy interact and what role is left for science in this interaction. Many still believe that science is, and should be, value neutral. For example, Robert Lackey, former chief of the US Environmental Protection Agency (EPA) Lab in Corvallis, Oregon, states: "science, although an important part of policy debates, remains but one element, and often a minor one, in the decision-making process," and that "scientists can assess the ecological consequences of various policy options, but in the end it is up to society to prioritize those options and make their choices accordingly" (Lackey 2013). The idea is that society—directly or through its elected or appointed representatives—formulates a task for science. Science—which is expected to act as though removed from society—then takes action by solving the problem and presenting the answer, which society, through its representatives, will consider and either implement or ignore.

This sequence does not seem to work for today's complex problems that are controversial, have no single and simple solutions, and, most importantly, infringe on and depend upon the values and priorities of the parties involved. Such is the case with climate change, with shale oil and fracking, or with alternative renewable energy (wind, solar), etc. Today's environmental problems go beyond technological solutions and mostly depend upon the behavioral choices that the society assumes, the priorities and values that drive those choices, and the way those values are communicated and understood.

Direct engagement in the value-setting process is necessary to instigate action and change. We argue that including value-setting in an iterative cycle of co-design of knowledge with users and stakeholders is crucial for the success of any exercise in participatory modeling. If we want models to be useful, we must acknowledge that their users exist within a socio-political system, and, therefore, including users' values both in models and in the modeling process and providing results based on stakeholder requirements becomes essential. In doing so we must admit that modelers are also stakeholders in the modeling process and have their own values (Voinov et al. 2014). In fact, people are more likely to acquire their scientific knowledge by consulting those who share their values and whom they, therefore, trust and understand (Kahan 2012). How will we expect them to associate with scientifically laden values if science is expected to be value neutral? Trying to convince people only with scientific arguments can be an uphill battle against their values and intentions that were set by the media and advertising and is prone to societal inertia. There is no reason to hide our values when engaging in a participatory modeling exercise, and even less reason to pretend that we, as scientists, hold no values as human beings. We do, and the fact that our human values are usually influenced by the many facts at our disposal as scientists, only makes our scientifically grounded values that much more important to share.

In participatory modeling, we make some important steps in the direction of transparent, science-based tools for decision-making. The definition of project goals becomes one of the stages of the modeling process, which is revisited as many times as needed with active interaction between scientists and stakeholders. Modeling helps define these goals and clarify values, intentions, and actions; potentially, changing them at the same time. Modeling engages stakeholders in a process of social learning (Tàbara and Chabay 2013) and co-design of knowledge (Glaser 2012) that includes a critical self-control feedback. Similarly, in the analysis of the model results, stakeholders are engaged to ensure that their expectations are met and the results can be used in a transdisciplinary framework (Seidl et al. 2013). This helps to bridge different disciplines and appropriately account for human values in modeling (Valkering et al. 2009). Yet, in most cases of participatory modeling, the scientists and modelers still are assumed to be "objective" and "value-neutral" (Voinov and Gaddis 2008). Moreover, they are expected to remain so and the value-neutrality of scientists in the modeling process is recommended as a prerequisite of "good" science.

Science in general, and modeling in particular, still rarely lead to action and is not expected to do so: policy makers are now supposed to make things happen (Lackey 2013). Stakeholders, when left on their own, quickly realize that running models—not to mention improving and adjusting them—is much more difficult than when scientists did the work. Policy makers become interested in their next fund-raising cycle very quickly, and forget to take action recommended by scientists.

Despite the realization of the importance of visualizations and the progress in developing persuasive and powerful visualization tools, presenting scientific evidence from model-based future scenarios and reflecting upon the need for changing societal values, intentions and actions remains elusive. Hence, one of the reasons that scientific understanding and knowledge does not readily translate into actions may be the target audience. Communication of model results should not be limited to the final stage of decision-making and the small group of policy decision makers.

We argue that by divorcing the modeling process from the problem formulation stage and by ending our modeling exercises with a delivery of a solution then disengaging from the actual implementation of this solution, we are not helping the overall decision-making process. Modeling is not an end, it is an evolutionary process of learning to better adapt to the continuing change that societies and ecological systems face (Tàbara and Pahl-Wostl 2007). If we expect actual decisions to be made outside of the modeling process, we are ignoring the power that models have: first, in framing the problems, asking the questions, comparing alternatives, identifying the contexts and boundaries; and second, in determining the actual value sets that lead to action through successful management or governance.

In fact, problem framing and definition are already results of modeling and the problem is most likely to be modified as a result of further modeling. Values and intentions are not static; instead they are constantly changing, and can and should be influenced by the results of models that we build. It is the responsibility of modelers to communicate the results in such a way that they can be understood by the public and are best framed to influence the values in an appropriate way.

To make sure that it is not only knowledge that is integrated in the participatory modeling process but also the values of stakeholders, including scientists, that should be incorporated and should inform the process, we suggest an amended version of the participatory workflow that ensures that scientists play a role in defining the problems to be solved and stay involved until actual action is taken to implement the solutions (Fig. 3.1).

In this regard, the participatory modeling process offers excellent opportunities for such engagement of scientists. However it is still important that scientists are ready to accept this role of setting the values and communicating the results of the modeling process in such a way that they can be understood by the public and are best framed to influence values in an appropriate way.

3.3 Revisiting Best Practices of Participatory Modeling

Participatory modeling is a practice that continues to evolve as it is applied to new, complex problems. Previously, Voinov and Gaddis (2008) presented a series of lessons based on experience working with stakeholder groups to develop watershed and water quality models to address water resource issues in a variety of locations. These lessons in participatory modeling, discussed from our perspective as scientists and modelers engaged in applied watershed issues, are informative to others working to achieve successful participatory modeling efforts elsewhere. Here, we review these lessons as they relate to a wider, more general audience that describes considerations for those seeking to engage in the modeling process with stakeholders, and explore how they may be amended to recognize the importance of values in modeling and participatory science.

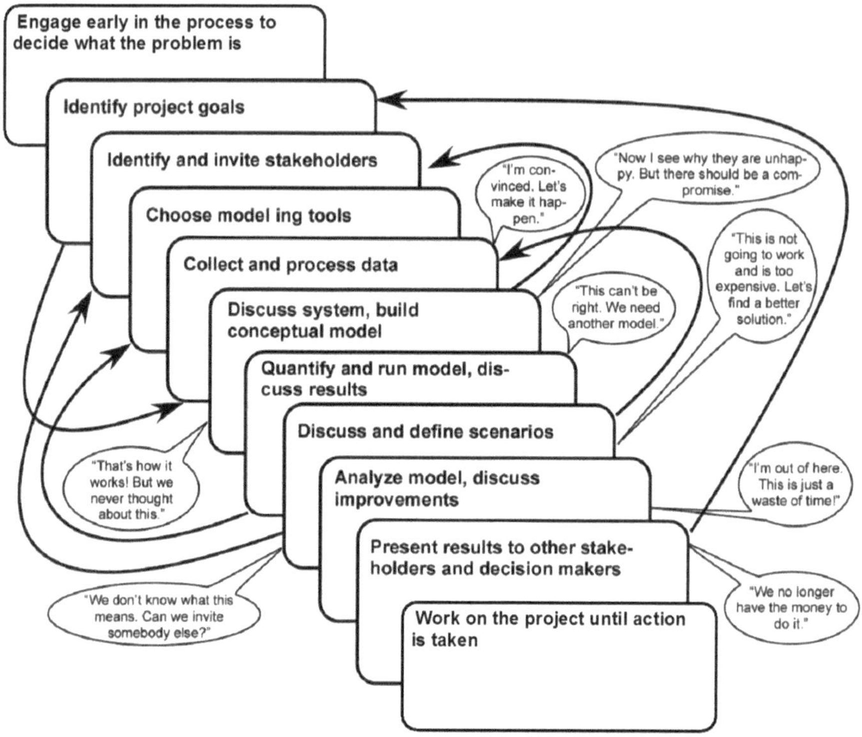

Fig. 3.1 A revised workflow in participatory modeling. Based on Voinov and Bousquet (2010). Scientists and modelers are expected to take a more proactive role in defining the problems and tasks for scientific inquiry rather than only serve the policy makers in providing answers to questions asked. At the end, also more participation in the actual action-taking is essential

- **Identify a Clear Problem and Lead Stakeholders**

  Although most natural resource management decisions benefit from stakeholder input and involvement, some issues might not have raised the interest of a wide group of stakeholders. If the problem is not understood or considered to be important by stakeholders, then it will be very difficult to solicit involvement in a participatory exercise. In many cases, the problem identification stage should go beyond just understanding what stakeholders want. Realization of a problem comes with education; with learning about facts and data. This is the role that science should play early in the process, perhaps even before the problem is defined.

- **Engage Stakeholders as Early and Often as Possible**

  A key to success with any participatory approach, is that the community participating in the research be consulted from the initiation of the project and help to set the goals for the project and specific issues to be studied (Beirele and Cayford 2002). Engaging participants in as many phases of the work as possible and as early as possible—beginning with setting the goals for the project—drastically improves the value of the resulting model in terms of its usefulness to decision makers, its educational potential for the public, and its credibility within the community (Korfmacher 2001).

- **Create an Appropriately Representative Working Group**

  Participatory modeling may be initiated by local decision makers, governmental bodies, citizen activists, or scientific researchers. In some projects, stakeholders are sought for their known "stake" in a problem or decision and invited to join a working group. In other cases, involvement in the working group may be open to any member of the public. Regardless of the method used to solicit stakeholder involvement, every attempt should be made to involve a diverse group of stakeholders who represent a variety of interests regarding the question at hand. This adds to the public acceptance and respect of the results of the analysis.

- **Gain Trust and Establish Neutrality as a Scientist**

  Although participatory modeling incorporates values, the scientific components of the model must adhere to standard scientific practice and objectivity. This criterion is essential for the model to maintain credibility among decision makers, scientists, stakeholders, and the public. Thus, while participants may determine the questions that the model should answer and may supply key model parameters and processes, the structure of the model must be scientifically sound and defensible. This does not necessarily mean that scientists will and should come into the process as value-neutral and totally "objective" players. Scientists are also human, which means that they are always driven by a certain set of values and preferences. Concealing them and pretending to be value-less would be dishonest and can result in loss of trust in the process. On the contrary, admitting adherence to certain values, while demonstrating willingness to discuss them and being open to criticism and disagreement, can only help in the process of co-learning and co-education.

- **Know Your Stakeholders and Acknowledge Conflict**

  In some cases, stakeholders may have historical disagreements with one another. One purpose of engaging in participatory modeling is to provide a neutral platform upon which disputing parties can contribute information and see the perspectives of other stakeholders engaged in the decision-making process. However, it is important to watch for historic conflicts and external issues that may overshadow the process.

- **Select Appropriate Modeling Tools to Answer Questions That Are Clearly Identified**

  A critical step, early in the participatory modeling process, is the development of research questions and goals for the process. The questions identified should be answerable given the time and funding available to the process. In addition, it is important that all stakeholders agree on the goals of the process such that a clear research direction is embraced by the entire group before detailed modeling begins. Selecting the correct modeling tool is one of the most important phases of any modeling exercise (Kelly et al. 2013). Model selection should be determined based on the goals of the participants, the availability of data, project deadlines, and funding limitations rather than determined by scientists' preferred modeling platform and methodology (which, unfortunately is often the case). Modelers should have a robust set of tools available for the process and be clear with stakeholders about the trade-offs of using tools with varied spatial and temporal resolution and complexity.

- **Incorporate All Forms of Stakeholder Knowledge**

  The knowledge, data, and priorities of stakeholders should have a real—not just cursory—impact on model development both in terms of selecting a modeling platform and in setting model assumptions and parameters. Stakeholders contribute existing data to a research process or actively participate in the collection of new data. Some stakeholders, particularly from governmental agencies, may have access to data that is otherwise unavailable to the public because of privacy restrictions or confidentiality agreements. These data can often be provided to researchers if it is aggregated to protect privacy concerns or if permission is granted from private citizens. Stakeholders may be aware of data sources that are more specific to the study area such as locally collected climatic data. Stakeholders can also be very helpful in identifying whether there are important processes or factors that have been neglected in the model structure or verify basic assumptions about the dynamics, history, and patterns of both the natural and socio-economic system. The stakeholders themselves may be important elements of the model, representing the behavior choices and patterns that are important to include in the model. The modeling process should be flexible and adjustable to accommodate new knowledge and understanding that comes from the stakeholder workshops.

- **Gain Acceptance of Modeling Methodology Before Presenting Model Results**

  Giving stakeholders the opportunity to contribute and challenge model assumptions before results are reported also creates a sense of ownership of the process and gives them more confidence in model-based results. This can only occur, however, if the models developed are transparent and well understood by the public or stakeholder group (Korfmacher 2001). Transparency is not only critical to gaining trust among stakeholders and establishing model credibility with decision makers, it is also key to the educational goals often associated with participatory modeling.

- **Engage Stakeholders in Discussions Regarding Uncertainty**

  Many scientific questions, especially those that incorporate socio-economic processes, require analysis of complex systems. As problem complexity increases, model results become less certain. Understanding scientific uncertainty is critically linked to the expectations of real world results associated with decisions made as a result of the modeling process. This issue is best communicated through direct participation in the modeling process itself. Stakeholders who participated in all stages of the model-building activities develop trust in the model and generally have more confidence in model results. Primarily that is because they know all the model assumptions, know the extent of model reliability, know that the model incorporated the best available knowledge and data, and acknowledge that there is always uncertainty associated with scientific model results.

- **Interpret Results with Stakeholders and Develop Scenarios That Are Politically Feasible**

  A primary goal of a participatory modeling exercise is to resolve the difference between perceived and more objective understanding of issues associated with environmental problems (Korfmacher 2001). Given that stakeholders may propose scenarios based on their perceptions of the problem, they may be adept at proposing new policy alternatives following initial model results from a scenario modeling exercise (Carr and Halvorsen 2001). The participatory modeling process can further facilitate development of new policies through development of a collaborative network of stakeholders throughout the research process (Beirele and Cayford 2002). Stakeholders are important communication agents to deliver the findings and decision alternatives to decision makers in the federal, state, or local governments. Stakeholders are best placed to pose solutions to a problem. Many of them have decision-making power and influence in the community. They understand the relative feasibility and cost-effectiveness of proposed solutions. In addition, engaging local decision makers in the scenario modeling stage of the research process can lead to development of more innovative solutions that may not have been considered using scientific knowledge alone (Carr and Halvorsen 2001).

- **Involve Stakeholders When Presenting Results to Decision Makers and the Public**

  An important final step in the participatory modeling method is dissemination of results and conclusions to the wider community. Presentations to larger stakeholder groups, decision makers, and the press should be made by a member of the stakeholder working group. This solidifies acceptance of the model results and cooperation between stakeholders that was established during the participatory modeling exercise.

3.4 An Example: Can Optimization Help with Value-Setting?

Consider the following example of employing a participatory modeling approach in the St. Albans Bay watershed, Vermont to identify new solutions to water resource problems that have historically been locally controversial and divisive (Gaddis et al. 2010a, b). Lake Champlain has received excess nutrient runoff for the past 50 years (VTANR and NYDEC 2002) due to modern agricultural practices and rapid development of open space for residential uses (Hyde et al. 1994). The dramatic effect of excess nutrients has been especially prominent in St. Albans Bay, which exhibits eutrophic algal blooms every August (Hyde et al. 1994). The Lake Champlain Total Maximum Daily Load (TMDL), established by the Vermont Agency of Natural Resources and the New York Department of Environmental Conservation, allocated a phosphorus load to the St. Albans Bay watershed that would require a 33 % reduction of total phosphorus input.

The watershed feeding St. Albans Bay is dominated by agriculture at the same time that the urban area is growing. In the 1980s, urban point sources of pollution were reduced by upgrading the St. Alban's sewage treatment plant. During this period, agricultural non-point sources were also addressed through implementation of "Best Management Practices" (BMPs) on 60 % of the farms in the watershed at a cost of $2.2 million (USDA 1991). Despite the considerable amount of money and attention paid to phosphorus loading into St. Albans Bay, it remains a problem today. The historic focus of those working on this problem has been primarily on agricultural practices in the watershed. This has caused considerable tension between farmers, city dwellers, and landowners with lake-front property.

In this case, participatory modeling was considered not only as a means for integrating scientific knowledge with local knowledge but also as a place for a diverse group of stakeholders to share varied forms of knowledge and as a platform for stakeholder interaction and dispute resolution. An objective of this study was to determine if participatory modeling facilitated more cooperation and reduced conflict between stakeholders in the St. Albans Bay watershed.

There are several places where stakeholder values and perceptions played an important role. All stakeholders came to the process with their perceived knowledge about the system, vested interests, and priorities. These made the stakeholders biased and subjective. For example, the committee was dominated by citizen volunteers and agency representatives; this led to solutions that would be implemented either through volunteer efforts or funded through existing agency programs. The transparency of the modeling process revealed these biases and helped to find common ground. Giving stakeholders the opportunity to contribute and challenge model assumptions before results are reported created a sense of engagement in, and ownership of, the process that made results more credible in the future. This can only occur, however, if the models developed are transparent and well understood by the stakeholder group and, later, the public. Some stakeholders complained that the modeling tools where too complex for them to grasp.

We came to the project believing that facilitators of a participatory modeling exercise must be trusted by the stakeholder community as being objective and impartial, and therefore should not themselves be direct stakeholders. In this regard, facilitation by university researchers or outside consultants, if established as neutral parties, was meant to reduce the incorporation of stakeholder biases into the scientific components of the model. It was also assumed to be essential that stakeholders trust the science used in the project. A track record in the local area and perhaps even recognition of researchers by the local stakeholders based on past research or involvement was helpful in building relationships between the stakeholders and the facilitators. However, it was apparent quickly that scientists could not be totally devoid of certain values and priorities. Even when starting the monitoring part of the project, which was conducted with local school students and their teacher, it quickly became obvious that scientists were deeply concerned about the state of Lake Champlain and held certain values. On the positive side, we had no preferences regarding the major conflict in the project: the standoff between the farmers and the urban residents.

We made every effort to make the model development process transparent to the stakeholders. The stakeholder working group discussed and agreed on model assumptions for some parameters and validated other model assumptions. Stakeholders were asked to verify assumptions about the dynamics, history, and patterns of the watershed system. This approach is based on the assumption that those who live and work in a system or watershed may be better informed about its processes and may have observed phenomena that would not be captured by scientists who live elsewhere. Farmers and homeowners possessed important local knowledge about the biophysical and socio-economic system.

Stakeholders identified processes or pollutant sources that had been neglected in past research for the watershed. For example, farmers identified field drainage of lowland fields as a potentially important process for understanding the flow of water and nutrients through the agricultural landscape. In addition, community stakeholders provided information about typical human behavior in the watershed. Many were important inputs to the simulation model (i.e., frequency of lawn fertilizer application) and have helped us formulate various scenarios for the model. Scenarios in this case were combinations of control factors (BMPs) administered at various spatial and temporal allocations. These scenarios could be then compared in terms of their efficiency by running them through the model. Stakeholders were especially instrumental in formulating these scenarios, since they had a very good feel for what was and was not possible in the watershed.

Again, we as modelers also had values at stake, which we tried not to involve in the discussions at first. We had an overall understanding from previous studies that the phosphorus budget of the watershed was vastly skewed and that more had to be done by all parties to improve the situation. Fortunately, these feelings were not contrary to any particular group among the stakeholders, which allowed us to maintain some "middle ground." Also, it helped that stakeholder-derived scenarios were supplemented by an optimization routine applied to a spatially explicit dynamic model of phosphorus transport.

Optimization, if considered from the point of view of the values involved, has the advantage of internalizing some of the values that may be driving the choice of the scenarios. On the one hand, optimization makes certain values implicit when the objective function and the conditions and constraints are set. For example, we can optimize for the lowest cost, while deciding that certain environmental conditions are to be met. Alternatively, we can optimize for the best possible environmental conditions to be achieved while the maximum allowed expenditures are fixed. On the other hand, once selected, the rest is composed of entirely internal computer computations where values are no longer involved.

This is in contrast to the more widely used scenario-based approach, where management scenarios are chosen as a result of stakeholder deliberations and can be heavily value-laden and contain vested interests that are not necessarily clearly exposed. Whereas stakeholder-derived scenarios represented the most obvious or socially accepted solutions to the problem, model results suggested that they were less cost-effective than solutions derived using an optimization algorithm. In fact, although the stakeholder-developed watershed solution showed similar phosphorus reduction, the cost of their preferred management plan was almost 3.5 times the cost of the solution generated by the optimization algorithm. The optimal solutions ranged in total cost for the watershed from $418,400 to 976,417 ($138 to 321 USD/ha) and represented a range in diffuse phosphorus load reduction from 0.89 to 1.13 mtP/year (0.29 to 0.38 kg/ha). The maximum diffuse phosphorus load reduction was found to be 1.25 mtP/year using the most cost-effective technologies for each diffuse source at a cost of $3,464,260. However, 1.13 mtP/year could be reduced at a much lower cost of $976,417 using the interventions selected by the optimization routine. This solution represented the practical upper limit of achievable diffuse phosphorus reduction for the Stevens Brook watershed. That is, there is a clear threshold of cost-effectiveness around $1 million, after which additional spending would not result in substantially more phosphorus reduction. Selecting solutions from the steep side of the Pareto curve provides the most cost-effective approach to reduce phosphorus at the watershed scale. On the steep slope, the marginal costs for additional phosphorus reduction are the lowest (Fig 3.2).

Of course, the results of the optimization runs are by no means binding. In fact there are numerous assumptions and uncertainties in the model, which mean that the modeling results should be always treated with some skepticism, and the optimization results are good only as an estimate of what is possible under certain ideal conditions. The next step is to reconcile stakeholder preferences and model results—a kind of critical assessment of what has been produced so far.

Watershed managers could use the results of the optimization runs to select the best combinations of watershed interventions along a Pareto optimal curve based on

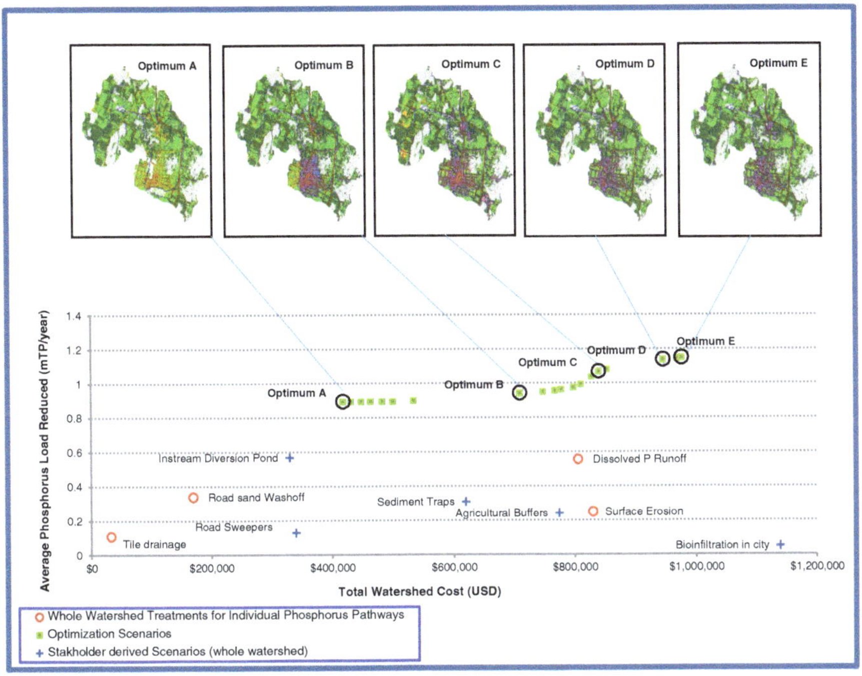

Fig. 3.2 The cost efficiency of various strategies of watershed management. The figure clearly indicates the differences between the stakeholder-selected solutions formulated as scenarios and the optimal solutions derived from the optimization procedure with the objective of minimizing the phosphorus load to the estuary. Connecting the optimum solutions creates the so-called Pareto optimal curve that shows what could be achieved under some ideal optimal conditions

a water quality goal or available funds. Each solution could also be used to inform where in the landscape implementation will be most cost-effective through detailed analysis of the BMP map output with each optimum. In our case, the results demonstrated the power of using spatial optimization methods to arrive at a cost-effective distribution of BMPs across a landscape. However the stakeholders should always be—and were in our case—informed that the "optimal" solutions we produced are good only as idealized targets that can inform the process of decision making, but by no means are actually guaranteed to produce exactly the kind of outcomes that the model showed.

While there is a big difference between solving applied problems using scenario modeling vs. optimization, and while this is something yet to be appreciated by stakeholders who are rarely involved in modeling exercises that include an optimization component, there is not much difference in terms of the associated uncertainties. In both cases we base our decisions on model runs, and models are always built on approximations, simplifications, assumptions, and always contain imprecise data.

In fact, a model that has been used within an optimization framework is likely to be more robust than a model that has been used only to run scenarios. That is because optimization requires that the model performs well over a much more densely populated parameter space—instead of only a few points described by a few scenarios, we now run hundreds or thousands of parameter combinations to choose the optimal one.

Although many of the stakeholders involved in the St. Albans Bay watershed participatory modeling process were decision makers who influence policy and implementation of watershed interventions at multiple scales, there was no direct mechanism by which model results would be used in any decision-making process. Through qualitative discussions, however, several stakeholders indicated that they intended to use the information gleaned from the project to direct existing funding sources and adapt policies to the extent possible to address the most significant phosphorus transport processes and sources in the watershed. Clearly, stakeholders are often limited in appropriating money and influence towards new projects, since other projects may have support for other reasons or are mandated by policies developed at higher bureaucratic levels, especially in the case of federally funded projects. Changing programs and policies of governmental agencies, especially to adapt to local conditions and problems takes time.

The issue of future use of the model was a focus of concern during several interviews. Initially the model was to be put on the Internet so community members could continue to use it after the modeling process concluded. Due to a lack of resources, this did not occur. Although the future use of the model by the community will be extremely limited due to its complexity and lack of continued support by the university, many of the stakeholders were under the impression that they would be able to use it. Unfortunately, since the end of the participatory modeling process, the stakeholder group has not had the capacity to work with the model. However, they have continued to draw on results from the modeling exercise conducted over the course of the project. Several stakeholders participated in the presentation of model results to the local press and general public in May 2006.

There are several specific examples of watershed management changes that have emerged from this project. In addition, several partnerships have been created or strengthened and trust developed between previously opposing groups as a result of the participatory-modeling exercises. In addition to management changes, stakeholders offered other recommendations and observations. A new focus on local decision making was suggested by a state employee as well as a town official. A member of the watershed alliance suggested a move away from adversarial relationships with the farming community. Another focus, echoing others' sentiments, is that information should be expressed in terms that people can understand. Several stakeholders suggested that education of the public was necessary in order to make important community-wide changes to deal with diverse water pollution issues.

3.5 Conclusions

Recent focus on ecological management that is adaptive, participatory, and collaborative has given rise to new approaches to scientific research and the incorporation of stakeholder knowledge and values into scientific models used for decision making. Participatory modeling incorporates input from stakeholders and decision makers into scientific models that support decisions involving complex ecological questions. The process supports democratic principles, is educational, integrates social and natural processes, can legitimize a local decision-making process, and can lead participants to be instrumental in implementing an agreed upon agenda. Modeling tools employed include indices, statistical models, spatial models, temporal models, and spatially explicit dynamic models. Stakeholder participants engage in the modeling research process in the form of model selection and development, data collection and integration, scenario development, interpretation of results, and development of policy alternatives. Variations of participatory modeling are distinguished by who initiates the process, how stakeholders are enlisted and engaged in the process, the breadth of research questions addressed, and the mechanism by which modeling results are incorporated into decision making—all of which can significantly influence model-based and social outcomes. Criteria of successful participatory modeling include scientific credibility, objectivity, transparency, understanding uncertainty, model adaptability, representative involvement, incorporation of stakeholder knowledge, and usefulness in decision making.

Both policy makers and academic researchers frequently engage the public and stakeholders in an outreach process that aims to inform or educate about a new policy or application of a scientific finding. Public comments may be solicited on agency-developed documents that bridge the science-policy interface; but responses to such comments are too often dismissive therefore not resulting in meaningful changes to policy. Such outreach efforts are substantively different than genuine participation in a modeling process. The best practices outlined above, if adhered to, should result in a process by which stakeholders feel that they have been heard, their knowledge objectively considered, and that the final results reflect a deliberative process that has been inclusive of multiple perspectives and all available data. The goal should be a bidirectional process resulting in true collaboration rather than an effort to "teach" the public and stakeholders. The learning should be mutual and not only address knowledge sharing, but also value sharing which has been an area of participatory modeling that is vastly understudied.

Most importantly, we expect true participation to play an important value-setting role, which becomes quite crucial in the state of the world today.

In addition to general recommendations related to practices associated with participatory modeling, we have experienced fine-scale issues that, to date, have not been considered adequately by the literature. For example: What kind of models should be built in the participatory process? How detailed, or how simple they should be? Should stakeholders be able to understand all aspects of the model or just key inputs and outputs? What should stakeholders be exposed to and what can stay "behind the scenes" (Voinov and Bousquet 2010)? Answers to these questions depend upon the resource management problem and the stakeholder group involved. We found, however, from our own experience that even some very complicated modeling tools that include optimization can still be successfully employed and provide important information for the stakeholder process, while also benefiting from the collaboration that takes place (Gaddis et al. 2010a, b, 2014).

In conclusion, it appears that science in general, and modeling in particular, are assigned a certain niche in society and are tolerated as long as they stay within that niche. In fact, many scientists are quite comfortable with this role because it may safeguard them from direct responsibility alternatives, identifying the contexts and boundaries, and determining the actual value sets that lead to action through successful management or governance.

Participatory modeling has the potential to integrate meaningful input from stakeholders and decision makers into the modeling process. When executed well it provides an objective, value-neutral place for a diverse group of stakeholders to contribute information regarding an ecosystem of interest. Even more important is the flow of information from science towards stakeholders, from theory to practice, and to action. One of the main problems facing society today is our lack of action on some of the crucial issues that have been identified by scientific research, but science fails to communicate the urgency and need for action to the rest of society. This disconnect remains serious and threatening in several contexts that endanger our future (e.g., climate change, biodiversity, etc.).

We argue that nowhere else can science and practice come as close together as in the process of participatory modeling. When stakeholders are already involved in the scientific process, as in the participatory modeling process, and when scientists are already directly and actively communicating and collaborating with stakeholders, it takes only a few more steps to directly engage in the political and decision-making process. Scientists should not shy away from taking a more proactive role in identifying the most urgent problems, and then making sure that action is taken to implement the solutions they have identified in real life.

References

Balvanera P et al (2006) Quantifying the evidence for biodiversity effects on ecosystem functioning and services. Ecol Lett 9(10):1146–1156

Beirele TC, Cayford J (2002) Democracy in practice: public participation in environmental decisions. Resources for the Future, Washington

Bousquet F, Voinov A (2010) Modeling with stakeholders. Thematic issue. Environ Model Softw 25:1268

Carr DS, Halvorsen K (2001) An evaluation of three democratic, community-based approaches to citizen participation: surveys, conversations with community groups, and community dinners. Soc Nat Resour 14:107–126

Daily GC, Ehrlich PR (1996) Global change and human susceptibility to disease. Annu Rev Energy 21:125–144

Étienne M (2014) Companion modelling. Springer, Dordrecht, p 403

Funtowicz SO, Ravetz JR (1993) Science for the post-normal age. Futures 25:739–755
Gaddis E, Adams C, Voinov A (2010) Effective engagement of stakeholders in total maximum daily load development and implementation. In: Proceedings of the international congress on environmental modelling and software, Ottawa, Ontario, Canada, 5–8 July 2010
Gaddis EJB, Falk HH, Ginger C, Voinov A (2010b) Effectiveness of a participatory modelling effort to identify and advance community water resource goals in St. Albans, Vermont. Environ Model Softw 25(11):1428–1438
Gaddis EJB, Voinov A, Seppelt R, Rizzo DM (2014) Spatial optimization of best management practices to attain water quality targets. Water Resour Manag 28(6):1485–1499
Glaser G (2012) Policy: base sustainable development goals on science. Nature 491(7422):35 http://dx.doi.org/10.1038/491035a.
Homer-Dixon TF (1999) Environment, scarcity, violence. Princeton University Press, Princeton/Oxford
Hyde K, Kamman N, Smeltzer E (1994) History of phosphorus loading to St. Albans Bay, 1850–1990. Lake Champlain Basin Program. Technical Report No. 7B
Kahan D (2012) Why we are poles apart on climate change. Nature 488(7411):255. http://www.ncbi.nlm.nih.gov/pubmed/22895298.
Kelly (Letcher) R, Jakeman AJ, Barreteau O, Borsuk ME, ElSawah S, Hamilton SH, Henriksen HJ et al (2013) Selecting among Five common modelling approaches for integrated environmental assessment and management. Environ Model Softw 47:159–181. doi:10.1016/j.envsoft.2013.05.005. http://linkinghub.elsevier.com/retrieve/pii/S1364815213001151
Korfmacher KS (2001) The politics of participation in watershed modelling. Environ Manag 27(2):161–176
Lackey R (2013) Normative science: it is easy – and wrong – for scientists to become stealth policy advocates. Terra Mag 1–15. http://oregonstate.edu/terra/2013/01/normative-science/
Paolisso M, Trombley J, Hood RR, Sellner KG (2013) Environmental models and public stakeholders in the Chesapeake Bay watershed. Estuar Coasts. doi:10.1007/s12237-013-9650-z
Rockstrom J et al (2009) A safe operating space for humanity. Nature 461(7263):472–475
Seidl R et al (2013) Science with society in the anthropocene. Ambio 42(1):5–12
Tàbara JD, Chabay I (2013) Coupling human information and knowledge systems with socialecological systems change: reframing research, education, and policy for sustainability. Environ Sci Pol 28:71–81
Tàbara JD, Pahl-wostl C (2007) Sustainability learning in natural resource use and management. Ecol Soc 12(2):3
USDA (1991) St. Albans Bay Rural CleanWater Program. United States Department of Agriculture, Vermont Water Resources Research Center, Burlington
Valkering P et al (2009) Modelling cultural and behavioural change in water management: an integrated, agent based, gaming approach. Integr Assess J 9(1):19–46
Voinov A, Bousquet F (2010) Modelling with stakeholders. Environ Model Softw 25:1268–1281
Voinov A, Gaddis EJB (2008) Lessons for successful participatory watershed modelling: a perspective from modelling practitioners. Ecol Model 216:197–207
Voinov A, Seppelt R, Reis S, Nabel JEMS, Shokravi S (2014) Values in socio-environmental modelling: persuasion for action or excuse for inaction. Environ Model Softw 53:207–212
VTANR, NYDEC (2002) Lake Champlain Phosphorus Total Maximum Daily Load (TMDL). Vermont Agency of Natural Resources, and New York State Department of Environmental Conservation, Waterbury and Albany. http://www.anr.state.vt.us/dec/waterq/lakes/docs/lp_lctmdl-report.pdf

Chapter 4
Eliciting Judgments, Priorities, and Values Using Structured Survey Methods

Marc A. Nelitz and Ben Beardmore

4.1 Introduction

Increasingly there is a trend in civil society to include the perspectives of scientists, managers, and stakeholders in environmental management through participatory engagement. Despite some skepticism about benefits, there are a variety of reasons for doing so (Bulkeley and Mol 2003; Reed 2008). Such approaches can allow for inclusion of a diversity of knowledge and perspectives that lead to more robust information. In situations where data or knowledge is limited, expert input can allow for the use of experience and judgments while waiting for field data collection. Inclusive processes can increase the level of buy-in and trust from stakeholders leading to more acceptable outcomes. Participatory approaches can also allow for more defensible decision making and a cleaner separation of evidence (scientific facts) from preferences (policy choices), which can commonly be intertwined and create challenges in environmental management (Lackey 2009).

The judgments, preferences and values of different audiences are foundational to addressing today's environmental challenges. Scientists use their expert judgments when deciding on research questions, developing conceptual models, designing monitoring programs, or estimating the likelihood of different outcomes based in part on their mental models about how social-ecological systems are organized. Managers decide on priorities by demonstrating their preferences about policy interventions, allocations of resources, and limits on human uses that reflect trade-offs among competing social, ecological, and economic objectives. Other stakeholders

M.A. Nelitz (✉)
ESSA Technologies Ltd, Vancouver, BC, Canada
e-mail: mnelitz@essa.com

B. Beardmore
Bureau of Science Services, Wisconsin Department of Natural Resources, Madison, WI, USA

S. Gray et al. (eds.), *Environmental Modeling with Stakeholders*,
DOI 10.1007/978-3-319-25053-3_4

and the public behave in ways that reflect their opinions or values related to their desired protection or consumption of natural resources.

Examples of participatory engagement on environmental issues can be found throughout civil society, ranging from a few (e.g., scientific advisory panels or environmental review boards to provide technical oversight) to hundreds (e.g., crowd-sourced science to gather data) to thousands of individuals (e.g., public opinion surveys on pipelines or climate change). Underlying these examples is the notion of a "*Wisdom of Crowds*" (Surowiecki 2004) in which the collective intelligence of a group can be superior to that of a few individuals. An early example of this phenomenon can be found in a study published in *Nature* more than 100 years ago (Galton 1907). The article describes a researcher's interest in examining the trustworthiness of judgments from a crowd. To study this issue, he examined results from a weight-judging competition at an agricultural exhibition where 800 competitors were asked to guess the weight of an ox. Competitors included people with a wide range of expertise from butchers and farmers to those with no specialized knowledge. Though the range of individual estimates was from 10 % lower to 8 % higher, the median of the group was <1 % of the true value, illustrating that the estimate of the group was more accurate than that of most individuals.

Structured survey methods include a set of tools that can facilitate participatory engagement of diverse audiences in a wide range of settings. Most commonly, these methods enable the consistent and repeatable collection of information by asking many respondents the same closed-ended questions to elicit their underlying preferences. A variety of established approaches are available for asking questions and analyzing the corresponding answers (Carson and Louviere 2011; Huang et al. 2011). In simpler situations, participants may be asked to respond using a single measurement scale, for instance when assigning a probability, ranking, or degree of importance. In more complex situations, respondents may be asked to make a choice which requires making trade-offs among multiple attributes having different measurement scales (e.g., cost, time, effectiveness).

Entire articles and books have been dedicated to the description of specific aspects and methods required to apply these techniques. Moreover, because every case involves a unique set of actors and issues, any participatory engagement, including structured survey methods, cannot be overly prescriptive. Therefore, the purpose of this chapter is to provide a primer on the key considerations when designing, deploying, and developing models from survey data that allow researchers and managers to prioritize alternatives or predict human behavior. The intended audience includes social and natural scientists who are interested in applying these techniques and learning more about their value-added potential. To support this intent, we point to resources to find additional information and provide a set of real-world examples that illustrate the way in which we have applied these methods when eliciting knowledge from different types of respondents.

4.2 Survey Respondents

At the heart of applying structured survey methods is a need to understand the knowledge to be elicited. Our experience shows that the knowledge of scientists, managers, and the public differs in important ways. These differences affect the means by which knowledge should be elicited and used in addressing environmental problems. Core differences include their varying roles in environmental management, levels of technical understanding, and underlying values. Given the focus of natural and social scientists in conducting evidence-based and value-neutral research, surveys should be focused on understanding technical judgments about the structure of and interactions within social-ecological systems and the related level of agreement among experts (e.g., Hagerman and Satterfield 2013). Managers tend to serve a bridging role between scientists and stakeholders by integrating technical evidence with a diversity of societal values to make choices often requiring complex trade-offs among competing objectives. Thus, their knowledge relates to understanding the set of available choices, decision rules (influencing attributes), and priorities (trade-offs and weightings) that underlie their choice among alternative courses of action (e.g., Tutsch et al. 2010). Lastly, the public and stakeholders behave in ways that reflect the diversity of values related to the protection or consumption of resources and the environment. As a result, their knowledge relates to understanding what values are important to society, how these values align among different stakeholder groups, and what level of support exists for interventions that affect the values stakeholders care about (e.g., Peterson et al. 2012).

When using structured survey methods, there are a variety of cross-cutting considerations that will affect the number and quality of responses from these audiences. Barriers to participation can lead to a refusal to engage and have impacts on response rates. A lack of engagement can be because of previously formed aversions to participation, a lack of incentive, the level of burden/time required, consultation fatigue, and perceptions that there is little or no value in the information being provided (Rogelberg and Stanton 2007; Reed 2008). Just as important is a need to ensure that surveys are robust, because poorly designed surveys can introduce bias into results.

Challenges in survey design can be attributed, in part, to the frailties in human judgment and decision making which are well documented in cognitive psychology (Tversky and Kahneman 1974). These frailties can include a range of cognitive/emotional traps (Hammond et al. 1999), such as the "anchoring trap" in which answers are biased by some initial value (e.g., perception of life satisfaction can be influenced by the point of comparison, Schkade and Kahneman 1998). Additional examples include the "framing trap" in which answers are influenced by a question's frame of reference (e.g., people perceive equivalent gains and losses unequally, Kahneman et al. 1990) and the "recallability trap" in which answers are influenced by a respondent's ability to recall an experience (e.g., the perceived likelihood of an event can be influenced by the event's pattern of occurrence, Estes 1976).

Another consideration is that respondents' perceptions can be biased depending on their level of expertise. For instance, expert opinions can be poorly calibrated or self-serving (Tversky and Kahneman 1974), while non-experts can be unaware of their poor performance (Ehrlinger et al. 2008). Moreover, people tend to have a difficult time accurately understanding uncertainties as illustrated by people's nonlinear interpretations about the likelihood of actual events (i.e., overestimating infrequent events and underestimating frequent events, Patt and Schrag 2003). As well, there needs to be practical considerations given to the complexity of a question type, which can affect the repeatability and credibility of a response, and the total number of questions posed. The latter of which can lead to survey fatigue (de Vaus 2002). Given the potential implication of these biases on results, it is important to address these challenges to the extent possible without crippling a survey's design (Kynn 2008).

4.3 Survey Design and Deployment

Because each stakeholder group brings their own set of perspectives, knowledge, and understandings to the table, the success of any survey approach depends on meeting the needs of the target group. While several references are available to guide survey efforts (de Vaus 2002; Dillman et al. 2008), elicitation of preferences that will provide more defensible weights to decision-making processes or more accurate predictions in behavioral models require that respondents be engaged and fully understand the context and terminology. To this end, framing the question correctly is particularly important, and may require a substantial portion of text in the survey. The goal of clear framing is to remove variation in responses due to misunderstanding the issue or the question. In applications where respondents are asked to consider novel scenarios, they may need guidance to suspend their disbelief to evaluate the scenarios seriously. When the goal of such an evaluation is to quantify the contribution of various individual components, each component must be well-defined and salient, and variations with each component must be meaningful. Scenario descriptions are, by necessity, simplifications of a potential reality. How simple such descriptions must be depends on the respondent group, and their prior understanding of the issue under investigation. Consequently, researchers must consider their target audience and tailor the complexity of scenarios to reflect the most important trade-offs.

Other considerations relate to the sampling strategy and sample size. The quality of results depends on successfully achieving a representative sample of the target population. When done well, the resulting models have the potential for including preference information from the "silent" majority who would otherwise not be part of a participatory process (Hunt et al. 2010). Budgets often dictate the number of participants, and researchers are therefore required to trade off the complexity of the desired outputs from a survey against the desire to minimize error around estimates. For a given sample size, the diversity of values, preferences and mental models within the sample, and the number of attributes and their levels affect the reliability of the final model.

Another factor that can influence the quality of the resulting model is the mode of survey deployment. While several textbooks on mail, internet and mixed mode survey methods exist (Dillman et al. 2008), researchers must understand their target population well enough to ensure that the selected mode meets the needs of the participants. For example, surveys of older populations may achieve better response rates if delivered on paper rather than over the internet. Computer-based surveys, however, facilitate rapid data collection and analysis, which can make them a suitable and powerful tool to deploy in a workshop environment.

While it is critical to try to address everyone's needs, the aggregative nature of modeling makes it often impossible to reflect the personal preferences on survey design for each individual participant. In applications with a larger sample size, statistical approaches can be used to account for heterogeneity. These approaches include developing separate models for different segments of participants (Dorow et al. 2010; Oh and Ditton 2006), mixed models that describe the variance around each estimate (Hensher and Greene 2003; Hunt 2008) or latent class (also known as finite mixture) models that probabilistically assign individuals to different groups based on their preferences may be valuable (Beardmore et al. 2013; Boxall and Adamowicz 2002). When sample sizes are smaller, accounting for diverse preferences within the sample becomes more difficult, and attaining group consensus may need to become an important component of the modeling process.

4.4 Elicitation Approaches

Traditional structured survey methods have often relied on evaluations of individual items using a rating or ranking exercise (Vaske 2008). For example, a group of scientists could be asked to provide opinions on a quantitative model by rating their agreement with a statement regarding a predicted outcome or to indicate their preferences by ranking possible outcomes. These approaches may be useful for assigning weights to different elements in a model (e.g., ratings of individual items do not require respondents to assign values relative to one another). Similarly, frequency formats can be used in a survey question for eliciting the probability distribution of a parameter of interest (e.g., productivity or harvest rate) from a group of scientists. In cases where participants can evaluate discrete alternatives, the frequency distribution of chosen outcomes from a simple "pick one" task may provide adequate information, such as in democratic elections.

As decision-making exercises become more complex, however, approaches that specifically address multivariate trade-offs become more attractive as they often better reflect the nature of real-world decisions. Several books are available that describe these methods in detail (Louviere et al. 2000; Hensher et al. 2005), so we constrain our discussion to a brief overview of some of the more common multivariate approaches used to elicit stakeholder preferences that have emerged from the field of discrete choice modeling. This field originated in transportation research to predict demand for traffic routes and modes of transportation based on actual behavior (i.e., revealed preferences), and has since been applied extensively in the fields of applied decision making and market research based on stated preferences (i.e., individuals' stated intentions or evaluations of hypothetical scenarios provided in a survey, Adamowicz et al. 1994).

These stated preference methods, often referred to as conjoint analysis, ask respondents to evaluate hypothetical scenarios that vary across multiple components. Consider for example, the scenario presented in Fig. 4.1, in which the ecological effect of an unspecified event is described by five distinct attributes labeled A through E. Such a scenario may be varied by changing the level, or specification of each of these five attributes. A defining feature of conjoint analysis is to use this variation to statistically model the influence of each individual component on respondents' evaluations.

Conjoint methods are additionally appealing because they can be developed using experimental design considerations that enable researchers to elicit the information from a given sample efficiently. When the number of possible combinations of attributes comprising a scenario is few, one can rely on a full factorial design in which every possible combination is evaluated by respondents. However, as the number of attributes present in a scenario increase, it is quickly apparent that a more efficient approach is required. Therefore conjoint methods typically rely on experimental design theory to ensure that reliable estimates of each desirable parameter may be derived (Raktoe et al. 1981). In the absence of prior information about respondents' preferences, one typically relies on fractional factorial orthogonal designs to ensure that all parameter estimates remain uncorrelated (Hensher et al. 2005). More efficient designs rely on prior knowledge to minimize variances and covariances of parameter estimates for a given number of evaluated scenarios (Kuhfeld et al. 1994). When the design calls for more sets than is reasonable for a single respondent, the sets are randomly allocated among several survey versions. Analysis using some variant of a multinomial logistic regression can then provide estimates of the relative contribution of each attribute to respondents' evaluations. Furthermore, these so-called part-worth

Situation leading to incidental take		Vegetation Removal
COMBINED ECOLOGICAL EFFECT OF 100 SITUATIONS WITHIN A SECTOR		
Chance that 100 situations will cause incidental take	A	**75 out of 100 situations** (75%)
Immediate effects of 100 situations on current populations	B	**BCR effects** (Affects populations at BCR level)
Long-term effects of 100 situations on populations	C	**Regional, delays reaching population objective** (Small but detectable regional effect, delays recovery towards BCR population objective).
Confidence in our understanding of the effects described above	D	**Moderate** (Certain about immediate effects, comfortable with estimate of long term effects)
Effects apply to one or more species at risk	E	**No**

Part (1) In the space provided, please indicate the value of the factor to the right (A - E) that...

A)... MOST concerns you about the effects across 100 situations. ***Letter (Most):*** ____

B)... LEAST concerns you about the effects across 100 situations. ***Letter (Least):*** ____

Part (2) On a scale from 1 (least concern) to 10 (most concern) please rate your level of concern about the impact on migratory birds given the effects (A-E) presented to the left.

Number: ____

1-2	3-4	5-6	7-8	9-10
Minor	Mild	Moderate	Large	Very Large

Fig. 4.1 Example of a conjoint profile to as conjoint analysis illustrating both a best-worst scaling task (Part (1) task) and a rating of the entire scenario (Part (2) task)

utilities are dimensionless and therefore allow the relative importance of each component to be placed on an interval scale.

Several ways of presenting choices to respondents have been used in conjoint analyses, ranging from ratings of single profiles, to choices of preferred scenarios. Ultimately, the appropriate approach depends on the type of outputs that are required. Respondents may be asked to rate the appeal of the provided scenario (Fig. 4.1), an approach that is particularly relevant when the objective of the model is to classify scenarios based on the collective wisdom of a group.

In a discrete choice experiment (DCE), the respondent is presented with two or more alternative scenarios (one of which often involves maintaining the status quo), and is required to indicate his/her preference for one of the alternatives, assuming these are the only alternatives available to her/him. A primary advantage of such formats over rating tasks is its grounding in a robust theory of human behavior—Random Utility Theory. This Nobel Prize-winning contribution to economics assumes that people will choose the single option that maximizes their benefit, and that the probability of choosing one alternative over another is proportional to the difference in their benefits (McFadden 1974). When the scenarios of a DCE present situations where respondents would typically make repeated decisions, such as choosing a nearby lake for a recreational fishing trip, it can make sense to ask respondents to allocate more than one choice among each alternative. By allocating ten fishing days among the alternatives presented in each choice set (Fig. 4.2), additional information becomes available to refine the preference model. In this case, each alternative (i.e., fishing location chosen by an angler) is treated as an observation, whose replication weight is equal to the choice frequency (Vermunt and Magidson 2005).

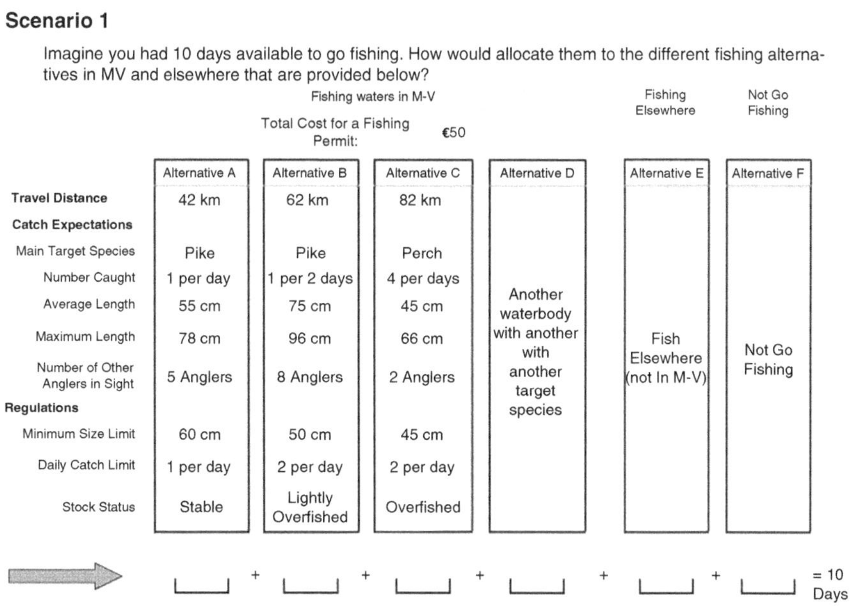

Fig. 4.2 Example of a discrete choice experiment using an allocation task to elicit preferences

Best-Worst scaling (BWS) is a type of partial ranking exercise with discrete choice methods used in the analysis. Also known as maximum difference conjoint (MDC; Finn and Louviere 1992), respondents are presented with a set of four or more items from a larger list. The respondents' task is to choose the two issues from each set that is (a) the best and (b) the worst (Fig. 4.1). A primary advantage of this approach is that it allows the investigator to establish the relative importance of each attribute separate from the relative importance of levels within each attribute. BWS offers several benefits over rating or ranking approaches. For example, identifying the most distinct pair of most and least concern from a subset of items places considerably less of a burden on respondents than ranking each item in an exhaustive list (Marley and Louviere 2005). In contrast to rating individual items presented in a list, BWS prevents scale biases that might arise when respondents rate all items similarly (Haider and Hunt 1997), by forcing trade-offs among the items. Finally, choosing both a best and worst item captures more information than the "pick one" task common among discrete choice experiments (Flynn et al. 2007). To accommodate the multiple choices, the dependent variable is treated as a single choice in a sequential process (Kamakura et al. 1994). Thus, the first choice is eliminated from the set when assessing the second choice. Another important consideration is that unlike a ranking of the two most important issues, the choice probability of the least important issue is assumed to be negatively related to its utility (Cohen 2003).

4.5 Applications in Environmental Science and Decision Making

Below we describe three examples to which we have applied the above structured survey methods for different audiences. They are informative for illustrating how these methods can improve our understanding of technical judgments for assigning significance of adverse environmental effects (scientists in example #1), preferred courses of action in a regulatory setting (managers in example #2), and diversity of opinions on management priorities within a single resource user group (stakeholders in example #3).

4.5.1 Characterizing the Significance of Adverse Events Across a Large-Scale Hydropower System in British Columbia, Canada

The operation of large hydro-electric facilities inevitably leads to impacts on the environment; for instance, dewatering of streams through normal dam operations, accidental releases of contaminants from equipment, or mortality of wildlife through electrical contact. Recognizing this reality, hydro operators can use risk management systems to proactively avoid and reactively respond to adverse events beyond their regulatory requirements for managing environmental impacts. Such systems can require characterizing the significance of adverse events to guide field operators on appropriate responses and inform corporate decision makers on aspects of their operational and business risks.

A challenge for staff within BC Hydro (a large-scale hydro-electric operator in British Columbia, Canada), is that consequence scales tend to involve a subjective appraisal of risk which can lead to variable interpretations of the same event that depend, among other considerations, on a person's technical expertise and the amount of information available. To help resolve this dilemma, we identified a defensible set of decision criteria for characterizing the full range of risks to which BC Hydro could hypothetically be exposed. These criteria included: (1) *sensitivity* of the receiving environment; (2) *magnitude* of impact; (3) *spatial extent* of exposure; and (4) *temporal duration* of recovery. Working with an internal group of scientists, we then defined four levels to further discriminate variations within each criterion (e.g., discriminate between short-term vs. irreversible events). This effort resulted in a framework that included four criteria, each with four levels. To distinguish between impacts on different receptors (aquatic vs. terrestrial environments and habitats vs. species), different definitions were used to represent similar levels of sensitivity and magnitude of impact. This approach ensured an acceptable level of equivalency in the way impacts were described for different receptors (e.g., a contaminated site and highly altered watercourse were categorized as having the same "class" of sensitivity). This framework was then used for assigning events into one of six categories of significance with S1 events being minor and S6 events being catastrophic.

Internal scientists were issued a structured exercise as a homework assignment (Martin et al. 2011) to explore the combinations of decision criteria and levels that would lead to these different categories of significance. The exercise presented a set of hypothetical incidents representing a different combination of sensitivity, magnitude, spatial extent, and temporal duration. They were then asked to use their technical judgments to assign an appropriate level of significance to the scenario. An orthogonal fractional factorial experimental design (Louviere et al. 2000) was used to identify 60 of 108 plausible scenarios that best represented the contrasting combinations of criteria and levels without the need to present all scenarios (Raktoe et al. 1981). These 60 scenarios were used as the basis for randomly assigning 30 to each participant. To account for possible biases in responses related to learning or fatigue, we randomized the order in which scenarios were presented. Data were analyzed using a Classification and Regression Tree Analysis (Brieman et al. 1984).

A robust decision tree resulted which represented the decision rules that internal scientists use to assign levels of significance to adverse events (see Fig. 4.3). For instance, when considering an incident in which reservoir drawdown for maintenance increases turbidity across many kilometers in a downstream reach, the event would be described as having a Class B sensitivity, Category 1 magnitude of impact, short-term recovery, and regional spatial extent. Following the corresponding pathways in the decision tree leads to an S2 level of significance. Hence, by characterizing new incidents and following decision pathways using a structured process, users were able to achieve a higher degree of consistency and were better able to describe the supporting rationale for assigning significance.

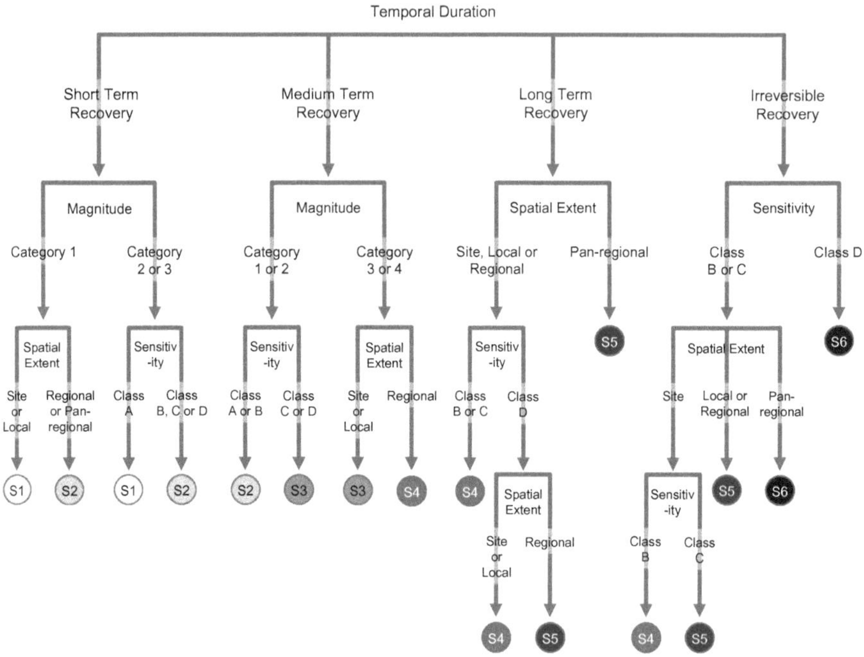

Fig. 4.3 Decision tree reflecting the criteria (sensitivity, magnitude, duration, and spatial extent) and levels (short-, medium-, long-term, or irreversible) that a group of scientists used for assigning different categories of significance to events with impacts on the environment (S1 through S6). Note that this decision tree was developed using specific definitions of criteria applied in a specific context. As such, the decision rules are not intended to be applied to other settings

4.5.2 Selecting Regulatory Options for Managing Incidental Take of Migratory Birds from Human Development Across Canada

Canada and the United States jointly support the long-term conservation of migratory birds through a treaty signed in 1916. In Canada, the *Migratory Birds Convention Act*, 1994 (MBCA) provides the supporting federal legislation prohibiting the "*incidental take*" of migratory birds, defined as the:

> inadvertent harming, killing, disturbance or destruction of migratory birds, nests and eggs.[1]

Current regulations state that no person shall hunt a migratory bird without a permit—where "hunt" is very broadly described—and no one shall disturb, destroy, or take a nest or egg. These prohibitions are broad reaching, implicating many common development and public activities as potential sources of incidental take. This

[1] Extracted from: http://www.ec.gc.ca/paom-itmb/.

situation is difficult for anyone expected to not violate the law because there are no means for permitting incidental take to occur.

To address this challenge the Canadian Wildlife Service (CWS) of Environment Canada had proposed developing a new system to help industry and the public assess risks and allow managers to use regulatory tools, including permits, that could help achieve conservation goals. While development of this framework was halted in 2010, at the time it was being designed to semi-quantitatively identify and minimize risks in a way that aligned with conservation objectives of the MBCA, allowed for a feasible level of oversight by managers to ensure effective implementation and enforcement, and imposed an acceptable burden on proponents to encourage compliance. To promote regulatory efficiency, a variety of options were proposed which provided contrasting levels of oversight and compliance based on the level of risk to migratory birds. These regulatory options included: (1) using published *advice* to help proponents avoid incidental take (no permit would be issued), (2) providing *class authorizations*, with conditions, for classes of similar projects, (3) allowing for a *simplified review* in situations where risks and conditions are understood, and (4) allowing for a *thorough review* in situations where risks and conditions are unclear. These options vary in how unique project configurations and environmental settings are considered, the flexibility to tailor advice or conditions to distinct circumstances, the level of administrative burden, and public acceptability. For instance, a thorough review might be preferable to managers since it allows for a consideration of a project's unique context, but it requires a higher administrative burden and would not be feasible to implement across all sectors and project types. Hence, explicit trade-offs among their features was needed when aligning these four options with different project and impact scenarios.

A stated preference survey was electronically deployed to managers within CWS to help identify these alignments (see sample in Fig. 4.1). Respondents were asked to consider a unique situation with different spatial and temporal impacts on birds, similarity to other projects, mitigation effectiveness, and acceptability of permit conditions by proponents. Managers had to choose their preference among four regulatory options. A fifth choice was also provided representing a situation in which the scenario could be rejected (no permit issued) following a review of the project. Of the >32,000 scenarios available (based on seven attributes with each having 2–8 levels), a manageable subset of 32 was created using statistical design principles to ensure orthogonality (Raktoe et al. 1981). Once data were compiled, a multinomial logit model was specified which allowed for an estimate of regression parameters, along with standard errors, for each attribute and level influencing a manager's choice around regulatory options (see Hensher et al. 2005). The resulting model was then used to reflect the decision rules of managers and estimate the proportion that would choose a regulatory option in a given scenario.

Sample results from this model are provided for two hypothetical scenarios in Fig. 4.4. Scenario A represents a situation in which it is unlikely that incidental take would occur, though if it does it would have a regional effect on a species at risk. Further, there would be a high level of variation across comparable projects with permit conditions having an unknown level of effectiveness, although these condi-

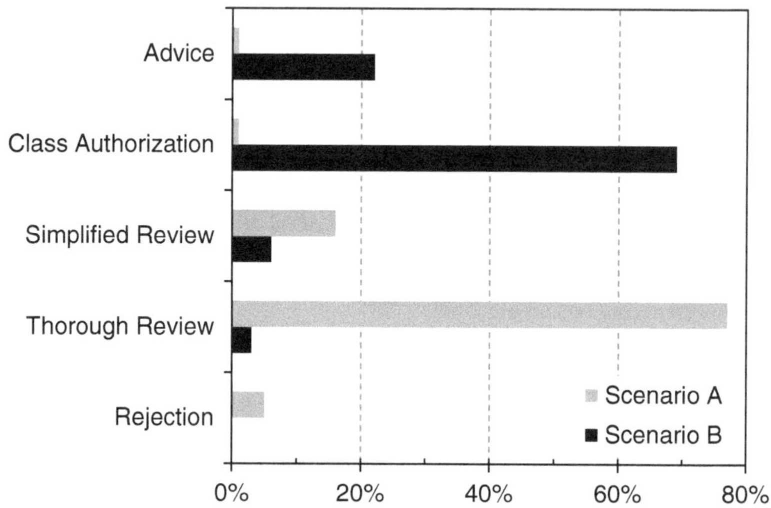

Fig. 4.4 Bar chart representing the proportion of managers who would choose different regulatory options when responding to two contrasting hypothetical scenarios that could result in incidental take on migratory birds. Note that the final option (rejection) represents an outcome of a review, not a regulatory option itself

tions would be acceptable to proponents. For this situation the model estimates that a thorough review would be preferred by 77 % of managers. Scenario B represents a situation in which incidental take is about as likely as not, would result in a negligible project scale effect, have a high degree of similarity across comparable projects, and result in effective and acceptable permit conditions. In this instance, the model estimates that 69 % of managers would prefer the use of class authorizations with conditions.

Though the results do not provide definitive answers on how real project activities and sectors across Canada should be aligned with different regulatory options, the model was useful for clarifying the influencing factors when making final determinations on these alignments. This process also allowed for the broader views of geographically dispersed managers as inputs into developing the permitting system, and was the first time the organization's risk profile for development was quantified.

4.5.3 *Understanding Boater Perceptions of Environmental Issues Affecting Lakes in Northern Wisconsin, USA*

The Northern Highlands Lake District of Wisconsin (NHLD) is a largely rural and forested region, and is one of the densest lake regions in the world, with over 7000 lakes covering 13 % of the landscape (Buffram et al. 2011). The amenities provided by these lakes have attracted summer residents over the last century, and in that time the region has seen a 4.6-fold increase in population (Carpenter et al. 2007). Boat-based recreational activities (e.g., recreational fishing and waterskiing) have had a conspicuous impact on lakes in the NHLD (Carpenter et al. 2007) including decreased water quality because of increased runoff (Carpenter et al. 1998) as well as loss of habitat (e.g., coarse woody debris in littoral zones), the introduction of aquatic invasive species, and diminished fish populations (Carpenter et al. 2007). While these four issues have been identified by researchers as pressing concerns, public perceptions of issues that affect freshwater systems are much less known. Not only are members of the public holders of local knowledge, but they also bear both the costs and benefits of environmental policy, the success of which is often dependent on their compliance with imposed regulations. Identifying stakeholder concerns is critical if one is to address any issues, by clarifying common concerns and providing opportunities to discuss apparent differences.

Recognizing that shared goals and perceptions among managers, scientists, and the public are important for successful collaborative ecosystem-based management (Gray and Jordan 2010), and that public values should provide the framework for outreach efforts, we engaged in a participatory process to understand public perceptions of the above issues. Participatory processes, however, often favor a vocal minority of stakeholders, and managers are challenged to ensure that resulting policies and actions reflect public concerns more broadly, while acknowledging diversity within and among stakeholders (Hunt et al. 2010). Many boaters, for example, engage in very different pastimes on these lakes, from the consumptive (fishing) to the appreciative (bird watching), from the tranquil (canoeing) to the exhilarating (water skiing). Therefore, the aim of this study was to identify issues of concern perceived to be most important by the boating public in a way that allowed for the discrimination of different perceptions among them. To do so, best-worst scaling (Finn and Louviere 1992; Louviere and Woodworth 1983) was used to estimate the relative importance of 16 potential issues of concern to freshwater aquatic systems (Beardmore 2015). An experimental design provided 16 groups of four issues that were blocked into four survey versions. Each respondent therefore evaluated four groups of issues, indicating the most and least important issues from each group. The responses were analyzed using a latent class conditional logit (Swait 1994) to identify sub-groups of boaters whose concerns differed most.

This analysis clearly discriminated among the issues of concern (Fig. 4.5), highlighting the overall importance of controlling pollution and the spread of invasive species in contrast to issues related to over- or under-regulation, crowding, and water levels that were of lesser concern (Fig. 4.5f). That said, the latent class analysis revealed that these overall preferences masked systematic differences in preferences among distinct groups of boaters (Fig. 4.5a–e). While primary concerns—namely, point source pollution, declining fishing quality, or over-development—dominated within one or more boater groups, these same concerns were much less important to other groups. These results underscore the diversity among boaters related to the ways in which these users interact with the lake environment and highlight a challenge for resource managers to balance issues of greatest overall concern against priorities of special interests within stakeholder groups who share divergent

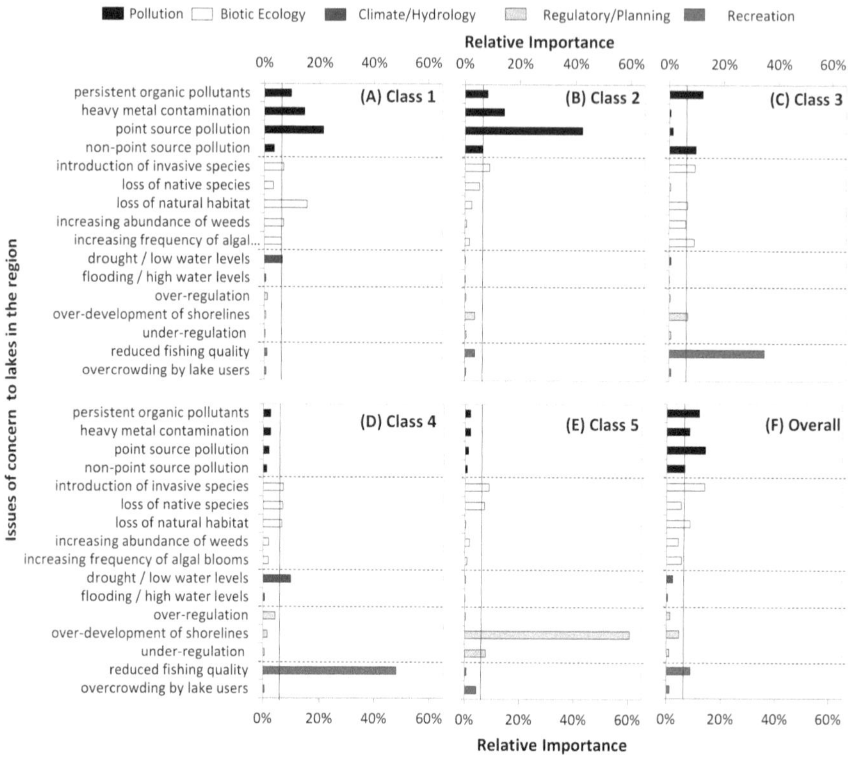

Fig. 4.5 Interval scale ranking of 16 issues concerning waterways of the Northern Highlands Lake District for five classes of boater and the overall sample. The relative importance of each issue is shown on a scale that sums to 100 % within each panel. The *vertical line* at 6.5 % indicates the expected value if all issues were considered equally important, while the *horizontal dashed lines* distinguish among broad categories of issues (modified from Beardmore 2015)

perceptions and preferences. In this case, structured surveys consolidated information on the preferences from a large sample who might not have participated in a more hands-on process. Through statistical modeling, key differences among groups of users were identified that would likely not have been so apparent using *a priori* clusters.

4.6 Final Thoughts

As demonstrated by these examples, structured survey methods can be useful for providing a clear and consistent structure to elicit knowledge and developing insights about the collective intelligence of a group; in these examples the heuristics of scientists to characterize environmental significance, preferred permitting options for managers, and perceptions of priority issues affecting stakeholders. More generally, our experience shows that these approaches can be very informative for providing insights into measures of central tendency (e.g., majority opinion) or dispersion (e.g., level of consensus or diversity of opinions), both of which can be informative for different purposes. As demonstrated, analyses of data like these can then allow for the development of statistically robust models. We acknowledge, however, that survey methods may not be ideal and may even be unacceptable in some situations where group facilitation (Kaner et al. 2007) or Delphi methods (Linstone and Turoff 2002), for instance, may be more appropriate. Nonetheless we have found great value in applying these techniques and developing models that serve as powerful tools for providing rigorous inputs into technical discussions and decision making that may not be credibly provided by other means.

References

Adamowicz W, Louviere JJ, Williams M (1994) Combining revealed and stated preference methods for valuing environmental amenities. J Environ Econ Manag 26:271–292

Beardmore B (2015) Boater perceptions of environmental issues affecting lakes in northern Wisconsin. J Am Water Resour Assoc 51(2):537–549

Beardmore B, Haider W, Hunt LM, Arlinghaus R (2013) Evaluating the ability of specialization indicators to explain fishing preferences. Leis Sci 35(3):273–292. doi:10.1080/01490400.2013.780539

Boxall PC, Adamowicz WL (2002) Understanding heterogeneous preferences in random utility models: a latent class approach. Environ Resour Econ 23(4):421–446

Brieman L, Friedman JH, Olshen RA, Stone CJ (1984) Classification and regression trees. Wadsworth, Belmont

Buffram I, Turner MG, Desai AR, Hanson PC, Rusak JA, Lottig NR, Stanley EH, Carpenter SR (2011) Integrating aquatic and terrestrial components to construct a complete carbon budget for a north temperate lake district. Glob Chang Biol 17:1193–1211

Bulkeley H, Mol APJ (2003) Participation and environmental governance: consensus, ambivalence and debate. Environ Values 12:143–154

Carpenter SR, Caraco NF, Correll DL, Howarth RW, Sharpley AN, Smith VH (1998) Nonpoint pollution of surface waters with phosphorus and nitrogen. Ecol Appl 8:559–568

Carpenter SR, Benson BJ, Biggs R, Chipman JW, Foley JA, Golding SA, Hammer RB, Hanson PC, Johnson PTJ, Kamarainen AM, Kratz TK, Lathrop RC, McMahon KD, Provencher B, Rusak JA, Soloman CT, Stanley EH, Turner MG, Vander Zanden MJ, Wu C-H, Yuan H (2007) Understanding regional change: a comparison of two lake districts. Bioscience 57:323–335

Carson RT, Louviere JJ (2011) A common nomenclature for stated preference elicitation approaches. Environ Resour Econ 49(4):539–559

Cohen S (2003) Maximum difference scaling: improved measures of importance and preference for segmentation. Research Paper Series. Sawtooth Software Inc, Sequim

de Vaus D (2002) Surveys in social research, 5th edn. Allen & Unwin, Crows Nest

Dillman DA, Smyth JD, Christian LM (2008) Internet, mail, and mixed-mode surveys: the tailored design method. Wiley, Hoboken

Dorow M, Beardmore B, Haider W, Arlinghaus R (2010) Winners and losers of conservation policies for European eel, *Anguilla anguilla*: an economic welfare analysis for differently specialised eel anglers. Fish Manag Ecol 17(2):106–125

Ehrlinger J, Johnson K, Banner M, Dunning D, Kruger J (2008) Why the unskilled are unaware: further explorations of (absent) self-insight among the incompetent. Organ Behav Hum Decis Process 105(1):98–121. doi:10.1016/j.obhdp.2007.05.002

Estes WK (1976) The cognitive side of probability learning. Psychol Rev 83(5):37–64
Finn A, Louviere JJ (1992) Determining the appropriate response to evidence of public concern: the case of food safety. J Public Pol Market 11:12–25
Flynn TN, Louviere JJ, Peters TJ, Coast J (2007) Best-worst scaling: what it can do for health care research and how to do it. J Health Econ 26:171–189
Galton F (1907) Vox populi. Nature 75:450–451
Gray SA, Jordan R (2010) Ecosystem-based angling: incorporating recreational anglers into ecosystem-based management. Hum Dimens Wildl 15:233–246
Hagerman SM, Satterfield T (2013) Entangled judgments: expert preferences for adapting biodiversity conservation to climate change. J Environ Manag 129:555–563
Haider W, Hunt LM (1997) Remote tourism in northern Ontario: patterns of supply and a motivational segmentation of clients. J Appl Recreat Res 22:49–78
Hammond JS, Keeney RL, Raiffa H (1999) Smart choices: a practical guide to making better life decisions. Harvard Business School Press, Boston
Hensher DA, Greene WH (2003) The mixed logit model: the state of practice. Transportation 30(2):133–176
Hensher D, Rose J, Green W (2005) Applied choice: a primer. Cambridge University Press, Cambridge
Huang IB, Keisler J, Linkov I (2011) Multi-criteria decision analysis in environmental sciences: ten years of applications and trends. Sci Total Environ 409:3578–3594
Hunt LM (2008) Examining state dependence and place attachment within a recreational fishing site choice model. J Leis Res 40(1):110–127
Hunt LM, Gonder D, Haider W (2010) Hearing voices from the silent majority: a comparison of preferred fish stocking outcomes for Lake Huron by anglers from representative and convenience samples. Hum Dimens Wildl 15:27–44
Kahneman D, Knetsch J, Thaler R (1990) Experimental test of the endowment effect and the Coase Theorem. J Polit Econ 98(6):1325–1348
Kamakura WA, Wedel M, Agrawal J (1994) Concomitant variable latent class models for conjoint analysis. Int J Res Mark 11:451–464
Kaner S, Lind L, Toldi C, Fisk S, Berger D (2007) Facilitator's guide to participatory decision making, 2nd edn. John Wiley & Sons, San Francisco
Kuhfeld WF, Tobias RD, Garratt M (1994) Efficient experimental design with marketing research applications. J Mark Res 31(4):545–557
Kynn M (2008) The 'heuristics and biases' bias in expert elicitation. J R Stat Soc Ser A 171:239–264
Lackey RT (2009) Is science biased toward natural? Northwest Sci 83(3):291–293
Linstone HA, Turoff M (eds) (2002) The Delphi method: techniques and applications. Addison-Wesley Publishing Company, Reading
Louviere JJ, Woodworth G (1983) Design and analysis of simulated consumer choice or allocation experiments: an approach based on aggregate data. J Mark Res 20:350–367
Louviere JJ, Hensher DA, Swait JD (2000) Stated choice methods: analysis and applications. Cambridge University Press, Cambridge
Marley AAJ, Louviere JJ (2005) Some probabilistic models of best, worst, and best–worst choices. J Math Psychol 49:464–480
Martin TG, Burgman MA, Fidler F, Kuhnert PM, Low-Choy S, McBride M, Mengersen K (2011) Eliciting expert knowledge in conservation science. Conserv Biol 26(1):29–38
McFadden D (1974) Conditional logit analysis of qualitative choice behaviour. In: Zarembka P (ed) Frontiers in econometrics. Academic, New York, pp 105–142
Oh C, Ditton RB (2006) Using recreation specialization to understand multiattribute management preferences. Leis Sci 28:369–384
Patt A, Schrag DP (2003) Using specific language to describe risk and probability. Clim Chang 61:17–30
Peterson MN, Hartis B, Rodriguez S, Green M, Lepczyk CA (2012) Opinions from the front lines of cat colony management conflict. PLoS ONE 7(9), e44616. doi:10.1371/journal.pone.0044616

Raktoe BL, Hedayat A, Federer WT (1981) Factorial designs. Wiley, New York
Reed MS (2008) Stakeholder participation for environmental management: a literature review. Biol Conserv 141:2417–2431
Rogelberg SG, Stanton JM (2007) Introduction: understanding and dealing with organizational survey nonresponse. Organ Res Methods 10(2):195–209
Schkade DA, Kahneman D (1998) Does living in California make people happy? A focusing illusion in judgments of life satisfaction. Psychol Sci 9:340–346
Surowiecki J (2004) The wisdom of crowds. Doubleday Publishing, New York
Swait J (1994) A structural equation model of latent segmentation and product choice for cross-sectional revealed preference choice data. J Retail Consum Serv 1(2):77–89
Tutsch M, Haider W, Beardmore B, Lertzman K, Cooper AB, Walker RC (2010) Estimating the consequences of wildfire for wildfire risk assessment: a case study in the southern Gulf Islands, British Columbia, Canada. Can J For Res 40(11):2104–2114
Tversky A, Kahneman D (1974) Judgment under uncertainty: heuristics and biases. Science 185(4157):1124–1131
Vaske JJ (2008) Survey research and analysis: applications in parks, recreation and human dimensions. Venture Publications, State College
Vermunt JK, Magidson J (2005) Technical guide for Latent GOLD 4.0: basic and advanced. Statistical Innovations Inc, Belmont

Chapter 5
Participatory Modeling and Structured Decision Making

K.F. Robinson and A.K. Fuller

5.1 Introduction

Structured decision making (SDM) is a process that breaks complex problems into their component parts, allowing evaluation of each piece separately, and leads to a more defensible, rigorous, and transparent decision. Since the focus of this chapter is about decision making, it is important to be explicit about what we consider a decision. In this context, we define a decision as an "irrevocable allocation of resources" (Howard 1966). The distinction is that a decision requires an investment of resources (e.g., time, money, etc.), rather than just a "mental commitment." The decision-making approach that we describe is used as a form of decision-aiding to guide collection of appropriate information to make robust decisions, and is not meant to be prescriptive.

The process of SDM provides a framework for evaluating decision problems that contain both technical complexity (e.g., complex and uncertain population dynamics) and difficult group dynamics (e.g., stakeholder groups with competing values systems; Gregory et al. 2012). Structured decision making combines applied ecology and decision theory to evaluate the multiple aspects of a decision problem, drawing from the subjects of cognitive theory, negotiation theory and practice, and group dynamics (Gregory et al. 2012). Through SDM, stakeholder values are identified prior to the other components of the decision process, thus ensuring that the science-based information included is directly relevant to the decision. This separation of values and science makes the process more transparent. Structured decision

K.F. Robinson
Quantitative Fisheries Center, Department of Fisheries and Wildlife,
Michigan State University, East Lansing, MI, USA

A.K. Fuller (✉)
Department of Natural Resources, US Geological Survey, New York Cooperative Fish and Wildlife Research Unit, Cornell University, Ithaca, NY, USA
e-mail: akf3y@cornell.edu

S. Gray et al. (eds.), *Environmental Modeling with Stakeholders*,
DOI 10.1007/978-3-319-25053-3_5

making as a process provides the decision maker with relevant information about uncertainty, multiple objectives, and difficult tradeoffs, with the end goal of helping the decision maker to find an optimal decision alternative that best meets the stated objectives (Clemen 1996; Gregory et al. 2012). The resulting decision framework developed using the SDM process is often built through a collaborative effort involving the decision maker (e.g., an agency commissioner or board of commissioners), managers, and stakeholders, and depending on the complexity of the problem often includes ecologists, decision analysts, facilitators, and policymakers.

The steps of SDM provide multiple opportunities for the inclusion of stakeholder values. These steps encourage decision makers and stakeholders to describe the context of the decision, establish objectives—including attributes of each objective that can be measured effectively—and then create a set of alternative management actions that have the ability to achieve outcomes in line with their stated values. The consequences of these alternatives, measured as the outcome of implementing each alternative on each objective, as well as key uncertainties that affect the decision, are evaluated, and tradeoffs are made among the objectives (Hammond et al. 1999; Gregory et al. 2012). The components of SDM are based on more than half a century of research in the decision sciences, having been applied in many fields, including natural resources management. Examples include decisions pertaining to recreational fisheries (Peterson and Evans 2003; Irwin et al. 2008), threatened and endangered species protection (Conroy et al. 2008; Gregory and Long 2009; Tyre et al. 2011), invasive species management (Runge et al. 2011), migratory bird harvest (Williams and Johnson 1995), estuarine habitat management (Robinson and Jennings 2012), and competing water uses (Gregory and Failing 2002).

5.1.1 Collaborative Decision Making: A Participatory Process

Inclusion of stakeholders is quite useful and very often essential for the decision-making process. In this chapter, we define stakeholders as those who are affected by or can affect the outcome of a decision process (Susskind and Cruikshank 1987; Decker et al. 1996; Reed 2008). There are both pragmatic and normative reasons to include stakeholders in the decision process (Reed 2008). In terms of pragmatism, the inclusion of stakeholders can create a higher quality decision. For example, participatory processes like SDM create perceptions of transparency and fairness and increase accountability in decision making, which can lead to greater trust and satisfaction in management of natural resources (Lauber and Knuth 1997; Decker et al. 2012). The normative reasons for stakeholder participation focus more on the process, especially when making decisions about trust resources (i.e., resources owned by the public). Participation in environmental decision making is a democratic right (Conroy and Peterson 2013), and the public has a right to evaluate the decision process for management of trust resources (Leong et al. 2012). By including stakeholders in the decision process, marginalization of different stakeholder groups is reduced (Reed 2008).

Different levels of engagement of stakeholders might be appropriate depending on the context and objectives of the problem (Reed 2008). In some decision processes, the scope may require only a subset of all stakeholder groups (Reed 2008), or there may be governmental or agency mandates of specific responsibility for decision making, such as in endangered species listing decisions in the United States (Cochrane et al. 2012). Additionally, a country's laws, such as the United States Federal Advisory Committee Act, could inhibit non-governmental stakeholders from fully participating in the decision process (Cochrane et al. 2012). However, even in decision contexts in which full stakeholder participation is not possible or practical, the decision process often will require participatory responsibilities from multiple entities, and decision makers can use participatory practices, as described in this chapter, to elicit stakeholder concerns. Stakeholders can be engaged at multiple steps in the process (Ascough et al. 2008), but are most commonly involved during the problem-framing stage.

Structured decision making provides a formal framework for a collaborative process to create the decision model, as well as for providing input in the consequences stage, where more quantitative approaches to predictive modeling often are implemented. The participation possible through the steps of SDM meets the needs of a public participatory process as laid out by Reed (2008): emphasizing equity and empowerment of all participants, enhancing learning, and creating trust in natural resources management decisions. Through SDM, facilitators can use the insights provided by the elicited values and objectives to create a decision framework that can lead to a more optimal management decision (Gregory 2000).

5.2 The Structured Decision Making Framework

The SDM framework provides a roadmap for thoughtfully considering each step (Hammond et al. 1999). This framework is flexible, allowing the decision maker to revisit previous steps as new ideas or information emerge (Fig. 5.1). In addition to these steps in the framework, uncertainty, risk tolerance, and linked decisions can be evaluated through SDM (Hammond et al. 1999).

5.2.1 Problem

The first step in decision making is to clearly define the problem. The problem statement, or description of the decision context, drives the entire decision process. This statement provides the background for the objectives and alternatives that will be considered in the decision framework and ensures that all subsequent analyses are relevant to the decision at hand. Laying out a clear and complete problem statement takes time and effort, but is essential and will pay off in the end. Often, natural resource management conflicts can be traced to a poorly framed problem definition (Riley and Gregory 2012). Different stakeholders with varying interests likely will frame a problem in different ways, and the collaborative nature of SDM ensures that

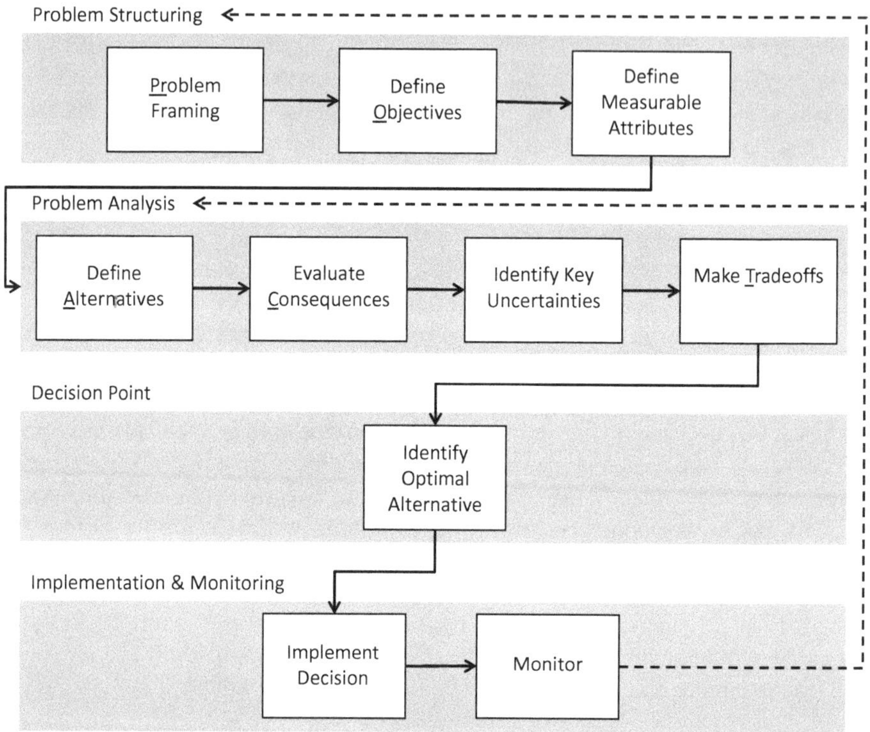

Fig. 5.1 The framework of structured decision making. Adapted from Marcot et al. (2012)

all perspectives are taken into account. Participation of various stakeholder groups in this step also provides more ideas for defining objectives and helps uncover unnecessary constraints (Hammond et al. 1999). Additionally, involvement of stakeholders at this initial step establishes an investment in the decision that will ultimately be made. The backbone of a successful decision model in SDM requires a well-framed problem statement to guide the rest of the process.

Key Elements of Formulating a Problem Statement
Sources: Hammond et al. 1999; Cochrane et al. 2012

Determine the triggers for the problem
Define and question constraints
Establish the essential elements
Identify any linked decisions
Determine a workable scope for the problem
Establish who the decision maker(s) and stakeholders are
Determine the spatial and temporal scale of the decision context
Establish the frequency and timing of the decision
Identify the key uncertainties that might affect the decision

5.2.2 *Objectives*

The next step is to determine the objectives that should be met for the optimal decision to be made. Objectives are succinct statements that describe the values and preferences of those involved in the decision process (Cochrane et al. 2012). Eliciting objectives from multiple stakeholders often provides more depth and breadth to the problem and allows participants to see how their concerns regarding natural resources management are being taken into account (Keeney 1992). These concerns can range from ecological and environmental issues, to economic, human health, and cultural topics. The requirement in this step is that the set of objectives is complete, concise, sensitive to management actions, understandable, independent, and appropriately scaled to the decision context (Keeney 1992; Cochrane et al. 2012; Gregory et al. 2012).

In SDM, objectives can fall into one of four categories: fundamental, means, process, and strategic objectives (Table 5.1). Eliciting the set of fundamental, or ends, objectives from the group is essential. Fundamental objectives describe what the decision maker fundamentally cares about—what must be achieved to solve the problem. The optimal management alternative best achieves each of the fundamental objectives.

Means objectives answer the question “How?” describing ways to achieve the fundamental objectives. For example, a fundamental objective of a natural resource management problem might be to minimize costs associated with the management action to be taken. The means objectives for minimizing costs might be to minimize staff hours spent on this issue and minimize dollars spent on implementing the new management action. These means objectives must be met in order to achieve the fundamental objective of minimizing costs.

Process objectives refer to the process of making the decision, or how the decision will be made, such as objectives to promote inter-agency collaboration, use data only of high quality, or involve stakeholders. Process objectives are meant to provide guidance for how best to achieve the other types of objectives, and therefore can directly influence the other three types of objectives (Keeney 2007).

Table 5.1 The four types of objectives

Category	Definition
Fundamental	Objectives that describe what the decision maker fundamentally cares about and wants to achieve through the decision process
Means	Objectives that must be met for the fundamental objectives to be achieved. They represent the means to an end
Process	Objectives that refer to the process of making the decision, and are not concerned with the decision that is made
Strategic	Objectives that guide all of the decisions made by an individual or organization over time

Adapted from Keeney (2007)

Strategic objectives are met by achieving the other three levels of objectives, as they guide all of the decisions made by an organization over time, such as maximizing public consent and trust or carrying out agency mandates (Keeney 2007; Cochrane et al. 2012). Strategic objectives are used to ensure that the body of decisions made by an organization adheres to the values and mandates of the organization. By working with groups to elicit these objectives, an objectives hierarchy (Maguire 2004) can be created that will guide the decision process.

In the objectives step, measurable attributes are created to evaluate the performance of each alternative at achieving each fundamental objective. The participatory approach to SDM provides a means of collaborating to determine how best to measure values that are hard to quantify, such as the spiritual quality of a river to First Nations (Gregory et al. 2012) or being respectful of the relationships between human and "non-human" beings (i.e., non-native fish), as expressed by Native American tribes (Runge et al. 2011). Participatory modeling of the objectives and the related attributes for a natural resources management decision provides a way to determine shared values and build common ground among stakeholders early in the creation of the decision framework (Gregory et al. 2012).

5.2.3 *Alternatives*

The alternatives describe the set of management actions that could be taken to achieve the fundamental objectives. The creation of this set of alternatives is unique to SDM. In an evaluation of published decision processes, Nutt (2004) found that just 4 % of these processes used objectives to search for multiple decision options or alternatives. Through SDM, an entire new set of creative management actions focused on the achievement of the fundamental objectives may be produced (Cochrane et al. 2012). Alternatives provide a way for participants in the decision process to see how their values are expressed in terms of a management action (Gregory 2000). Moreover, as a collaborative process, the creation of management alternatives can help to alter participants' perspectives on tough choices, or tradeoffs, that must be made. Through elicitation of these alternatives, participants can see how difficult tradeoffs can or cannot be avoided (Gregory et al. 2012).

Good alternatives are critical to SDM, as the optimal decision will only be as good as the best alternative in the set. A good alternative has at least five hallmark qualities: it is complete and comparable, value-focused, fully specified, internally coherent, and distinct (Table 5.2; Gregory et al. 2012). Creating a set of good alternatives requires a great deal of thought, good elicitation skills, the suspension of judgment, and a willingness to probe the unknown (Nutt 2004). Additionally, participants must be aware that cognitive biases can influence the development of alternatives (Chap. 6, this book). Groups tend to anchor on the first alternative that is suggested, using this as the benchmark for any future alternatives (Keeney 1992). Availability, or focusing on recent or memorable events, can minimize the potential set of actions (Keeney 1992). Finally, there can be a tendency to rely on sunk costs,

Table 5.2 Qualities of a good alternative

Qualities	Definition
Complete and comparable	All alternatives address the same aspects of the problem, at the same temporal and spatial scale
Value-focused	Alternative addresses the fundamental objective(s)
Fully specified	Alternative is unambiguous, logically consistent, and sufficiently detailed
Internally coherent	The pieces that make up the alternative are feasible and logical
Distinct	Alternatives are different, providing different methods of achieving the objectives

Adapted from Gregory et al. (2012)

such that alternatives are based on past expenditures (Cochrane et al. 2012). Overall, alternatives play a critical role in the decision-making process, as they are the means to giving decision makers meaningful choices to achieve their stated objectives (Gregory et al. 2012). As such, participatory methods for eliciting a creative set of management actions from the group are essential at this step.

5.2.4 Consequences

The next step in the SDM framework is to evaluate the consequences of each alternative on each of the objectives. In other words, how well does each alternative work towards achieving the fundamental objectives? Participatory modeling is widely used in this step, as objectives often are diverse and expertise from many different stakeholders is necessary to evaluate each alternative's outcome on multiple objectives. Evaluation of the consequences of management actions can be both a qualitative and a quantitative modeling process, depending on the problem at hand, or even on the objective being considered. The main goal of these models is to predict the outcomes of each alternative in terms of the measurable attributes. The data needed for these predictions can come from a variety of sources, including literature-based studies, empirical field data, experiments, experts in the field, and indigenous knowledge, making participatory modeling important in this step (Failing et al. 2004, 2007; Irwin et al. 2008; Conroy and Peterson 2013).

These quantitative predictions can be made through a process of expert elicitation (Martin et al. 2012). Whether the necessary predictive models are qualitative or quantitative, incorporating group model-building can provide information about data availability and give participants a better understanding of the definition of the problem, objectives, and alternatives in the model (Irwin et al. 2011). This section provides an overview of some of these methods and of the value of stakeholder participation for data gathering. Other authors cover model parameterization methods, especially with regards to data and empirical modeling, in more detail (e.g., Kendall 2001; Williams et al. 2001; Dorazio and Johnson 2003; Conroy and Carroll 2009; Conroy and Peterson 2013).

5.2.4.1 Influence Diagrams

Qualitative modeling in the SDM process either can provide a starting point for creating a fully quantitative predictive model, or provide the necessary framework to make a distinction among competing alternatives. Influence diagrams are one common type of qualitative model used in the consequences step. Influence diagrams graphically represent the decision (Clemen 1996; Conroy and Peterson 2013). Through the creation of an influence diagram, groups are able to visually describe the causal relationships among alternative actions, key uncertainties that might affect the decision, events not under the control of the decision maker, fundamental and means objectives, and utility values (Fig. 5.2). Utility values represent the relative value of different combinations of decisions and outcomes, and can be represented as a monetary unit, a score (such as 0–1 or 0–100 %), or some other scale that has been defined by the group (Conroy and Peterson 2013). For example, when deciding on whether to close an area to fishing, stakeholders would assign a greater utility value to a high stock size than a low stock size under the same management alternative (Fig. 5.2). Influence diagrams provide a conceptual roadmap for evaluating the consequences of each alternative on the objective(s), explicitly laying out the uncertainties embedded in how the decision will affect the values of the stakeholders. Creating influence diagrams with stakeholder groups provides an opportunity for stakeholders to put their mental model of the system down on paper. Therefore, including various stakeholder groups in the creation of influence diagrams ensures that all aspects of the consequences of a management action on a

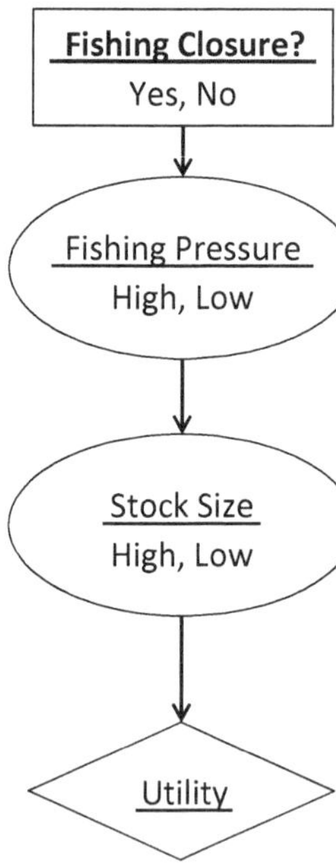

Fig. 5.2 Simple influence diagram describing the hypothetical relationships between a decision of whether to close an area to fishing, the amount of expected fishing pressure on a fish stock, the predicted stock size, and the utility associated with the resulting stock size

fundamental objective are considered. A well thought out influence diagram can be used in more quantitative approaches to evaluating the consequences of the management alternatives at hand, such as decision trees and Bayesian belief networks.

5.2.4.2 Decision Trees

Decision trees display the causal relationships of the decision framework as a series of tree branches, similar to the way that an influence diagram displays information (Fig. 5.2; Clemen 1996; Failing et al. 2004; Blomquist et al. 2010; Conroy and Peterson 2013). In a decision tree, the first set of branches represents the decision node, with each branch corresponding to a different alternative action (Fig. 5.3). The following sets of branches each represent a key uncertainty related to the decision (uncertainty nodes), much like the intermediate nodes of an influence diagram. Each of the alternative states of nature in an uncertainty node has an assigned probability of happening. The end points in the branches on the decision tree correspond to utility values. To identify the optimal decision, the probabilities and corresponding utility value for each complete branch of the decision tree first are multiplied together. Then, the scores for each branch that corresponds to a particular management action are added, providing the expected (i.e., probability weighted) utility for each management action. For the example in Fig. 5.3, the expected utility value of closing the area to fishing ("Yes") is (0.1*0.2*0.9)+(0.1*0.8*0.1)+(0.9*0.75*0.9)+(0.9*0.25*0.9), or 0.656. The action with the greatest expected utility value is the optimal alternative. Because of its structure, a decision tree can quickly become unwieldy as external drivers and uncertainties are added to the decision framework (Clemen 1996). For simple decisions, though, decision trees can be very helpful for graphically depicting the choices and key uncertainties in a decision to stakeholders.

5.2.4.3 Bayesian Belief Networks

A Bayesian belief network (BBN) is a graphical and quantitative representation of an influence diagram, describing correlation and causation among variables (Aalders 2008). Similar to an influence diagram, the nodes of a BBN include (1) a decision node: the set of alternative management actions, (2) uncertainty nodes: all key variables that are directly or indirectly influenced by the decision, and (3) a utility node: the utility values for the outcome of each management action on the fundamental objective (Fig. 5.4). BBNs can include many more nodes of uncertainty than can decision trees, making them useful for more complicated decision problems. Like decision trees, BBNs are most useful for single-objective problems, although a multi-objective utility score can be created in some instances. Unlike decision trees, in which joint probability distributions are calculated, BBNs use Bayesian statistics to estimate probabilities associated with the alternative states of nature in each uncertainty node (McCann et al. 2006).

BBNs have many qualities that make them flexible and useful for decision analysis. They allow for both categorical and continuous data, the various nodes can be populated with both empirical data and expert opinion, and the outcomes of the

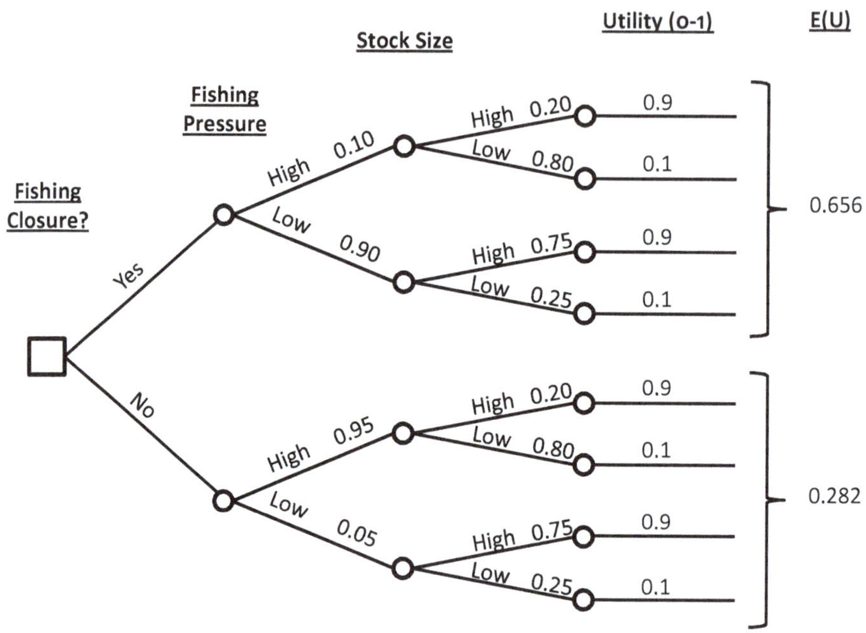

Fig. 5.3 Example of a decision tree for the influence diagram from Fig. 5.2. The branches at the *far left* represent the set of alternative actions (decision node). The branches at the *far right* represent the utility value associated with each potential outcome of the management actions. The branches in the middle represent the key uncertainties that are influenced by the choice of action and that influence the potential outcome (uncertainty nodes). The states of each uncertainty node have associated probabilities that are dependent on the influencing branches. The expected utility values [E(U)] are calculated by multiplying through each branch of the tree and adding the scores for all outcomes related to each alternative. $E(U)_{yes} = 0.656$, $E(U)_{no} = 0.282$

Fig. 5.4 Simple Bayesian belief network depicting the decision problem of whether to close fishing areas. Probabilities for each node are the same as in the decision tree in Fig. 5.3. The decision node at the *top* contains the expected utility value for each management decision, suggesting the optimal decision is to close fishing areas

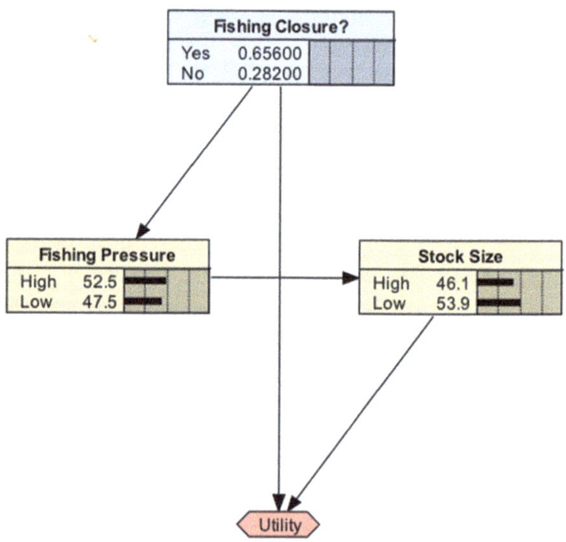

BBN are expressed as likelihoods (Marcot et al. 2001). Additionally, they can take into account prior knowledge and missing data, and they use Bayesian updating to incorporate new data into the model (McCann et al. 2006). With the flexibility afforded by BBNs, participatory modeling with these networks allows for the simultaneous building of the decision model and parameterization of the model with multiple types and sources of data. All participants can visualize how each decision alternative will affect each of the key uncertainties in the decision model and how their interests are affected by management decisions.

Although BBNs can by very useful for participatory modeling in SDM, they have their limitations. For example, each uncertainty node has a corresponding conditional probability table. In this table, the probability of the node being in each state must be assigned for each combination of all states of the influencing (or parent) nodes. When these nodes have many states and many influencing factors, the conditional probability table can become quite large (McCann et al. 2006). Additionally, describing temporal dynamics in the BBN framework is difficult, and requires that the BBN essentially be replicated for each temporal step in the framework (McCann et al. 2006; see Robinson and Jennings 2012 for example). Although this is possible for short timespans, the complete replication of the network for multiple time steps can quickly become computationally intensive and difficult to manage. Finally, BBNs can appear complicated to those who are unfamiliar with them. Careful explanation of the network is required, and BBNs are most useful when a facilitator familiar with the software can elicit the required information from the group of stakeholders. BBNs are an excellent participatory modeling tool, but often become just one tool in the decision process for a complicated decision problem.

5.2.4.4 Empirical Models

Rarely are data available that specifically predict the outcomes of each of the proposed alternative actions on each of the fundamental objectives. Most often, different types of models will be necessary to predict the outcomes of various parts of the decision model, based on some combination of data collected from previous studies, the scientific literature, or expert elicitation. These predictive models can range from simple linear regressions relating influencing factors on a parameter of interest (Conroy and Peterson 2013), to stochastic simulation of populations of the organism of interest (Ralls and Starfield 1995), to optimization of dynamic decisions through stochastic dynamic programming (Conroy and Moore 2001) and Markov decision processes (Williams 2009). We focus here on methods for parameterizing decision models for static decisions, in which the decision will be applied once, or the system can be assumed to be in equilibrium (Conroy and Peterson 2013). Adaptive management, which employs the same steps as SDM and often uses methods for optimization of dynamic decisions, is used for recurrent decisions (Conroy and Peterson 2013).

Often, data might not be available to exactly predict the outcomes or consequences for a decision problem, but data likely are available to inform the alternative states of nature for the key uncertainties represented in a BBN or influence diagram. In these cases, statistical modeling can be quite useful. Depending on the data available and the parameter being estimated, regression models can be used to determine how variables in observed data are related (e.g., how water temperature affects dissolved oxygen content in estuarine waters, Robinson and Jennings 2012) or to predict the changes in the response variable based on observed data (e.g., how fish mortality in one estuary might be influenced by water temperature based on observations from another estuary; Conroy and Peterson 2013). In addition to these regression models, Bayesian approaches, which are especially helpful for incorporating uncertainty, often are used for developing parameter distributions based on previously observed data (Conroy and Peterson 2013). When using statistical models, group participation might seem daunting, as some group members might feel ill equipped to work through these models. However, the benefits of group model building include probing all participants for data sources to parameterize the model and revisiting models with the full group once parameterized to ensure that the models accurately represent the system.

In some cases, more complex dynamics might be needed to predict the consequences of management actions. For instance, BBNs and decision trees cannot adequately incorporate temporal dynamics. When consequences must be predicted over time, such as in plant or animal population dynamics, stochastic simulation models can be created. These models have been used in many SDM processes, including fish population dynamics (Peterson and Evans 2003; Irwin et al. 2008; Robinson and Jennings 2012), Hawaiian monk seal populations (Starfield et al. 1995), and horseshoe crab and red knot population relationships (McGowan et al. 2011). These models easily represent uncertainties in the parameters through the use of statistical distributions. The output from these simulation models can be incorporated into the nodes of a BBN, which helps to represent the uncertainty of this output (Conroy and Peterson 2013), or into the cells of a consequence table (Gregory et al. 2012; described below).

Building a stochastic simulation model with the group can be beneficial for the model itself, as well as for providing group members with a better understanding of how plant or animal populations specifically will be affected by proposed management actions. Collaborative model building can ensure that specific model assumptions are identified, the data available for model parameterization are acquired, and unknowns related to the model are determined (Irwin et al. 2008). Use of a collaborative model-building approach can provide checks and balances for the modeling team, as the larger group can evaluate the model throughout the process (Irwin et al. 2008). Finally, collaboratively building the simulation model can provide insights into the objectives and measurable attributes that can lead to refinement of these elements of the decision model.

5.2.4.5 Expert Elicitation

In some cases, data relevant to a management decision might be lacking. In the case of data-poor parameters, the elicitation of expert judgment can help fill in the gaps. The process of expert elicitation, which includes among other techniques, the

Delphi method, was developed by the RAND Corporation in the 1950s to evaluate how technology would impact warfare (Brown 1968). One of the first implementations of expert elicitation in environmental science was by the United States Environmental Protection Agency (USEPA) for the National Ambient Air Quality Standards (Hetes et al. 2011). Eliciting data from experts for predicting outcomes is a formal, structured process that follows explicit protocols (Cochrane et al. 2012). These protocols are in place to ensure that the elicitation process is transparent, defensible, repeatable, and not subject to cognitive and motivational biases (Cochrane et al. 2012). Expert elicitation is an exceedingly important part of the participatory modeling approach to SDM. Many different methods of elicitation exist (Goodwin and Wright 2009; Ayyub 2001), and often the data need dictates the elicitation method. For elicitation of single parameter values, such as a harvest rate or nest abandonment rate, the four-step elicitation method is a robust method that can reduce the overconfidence of experts (Speirs-Bridge et al. 2010). The four-step elicitation method asks experts to provide a lower limit, upper limit, and best guess of the parameter in question, as well as their confidence that the true value falls within the interval they provided. This last piece of information was found to be crucial in reducing the over-confidence that experts typically exhibited when asked for the other three values (Speirs-Bridge et al. 2010). Additionally, the likelihood point method is useful for eliciting probabilities necessary for BBNs (Cochrane et al. 2012). Most decision problems will benefit from expert knowledge, and structured methods for eliciting this knowledge from diverse groups of experts are necessary for a defensible decision model.

5.2.4.6 Consequence Tables

The consequence table, in which the consequences of each alternative on each objective are laid out in terms of the values of the measurable attributes, is a tool that can consolidate the information gathered from the approaches listed above. Unlike the other frameworks discussed, consequence tables do not represent the pathways of how the alternatives influence the consequences (i.e., the uncertainty nodes). Instead, the objectives are laid out in the rows and the alternatives in the columns, and the predicted values for each measurable attribute under each alternative action populate the rest of the table (Table 5.3). The consequence table provides a simple layout for all participants to see the outcome of taking any particular management action and how that helps in achievement of the fundamental objectives, making it an excellent tool for decision problems with multiple objectives.

The simplified nature of consequence tables is very useful in a group setting, but also means that characterizing and evaluating uncertainty is more difficult than in a framework such as a BBN. Uncertainty can be incorporated by reporting the expected value of each measurable attribute. The use of the expected value can obscure important information about risk, though, if the distribution about the expected value (i.e., probability weighted average) is quite large or skewed, making it less helpful for decision makers to consider their risk tolerance (Gregory et al. 2012). Another way to

Table 5.3 Example of a consequence table for a multi-objective decision of whether to close areas to fishing

Fundamental objective	Measurable attribute	Alternatives			
		Close area 1	Close area 2	Close areas 1 and 2	No closures
Maximize fish stock size	Spawning stock biomass (kg/km^2) after 5 years	30	15	50	5
Maximize commercial angler satisfaction	Constructed scale: 1 (no satisfaction)–10 (complete satisfaction)	1	7	3	4
Maximize recreational angler satisfaction	Constructed scale: 1 (no satisfaction)–10 (complete satisfaction)	5	4	6	5

represent uncertainty in a consequence table is to report confidence intervals, though the relative probabilities of the upper and lower bounds might differ, and participants often assume that all values within the range of confidence intervals are equally probable. Additionally, assigning weights in the tradeoffs step becomes more difficult with a range of values (Gregory et al. 2012). Inclusion of uncertainty throughout the entire consequence table also can overwhelm participants. Sensitivity analyses can be performed to perturb each of the key sources of uncertainty, one source at a time, to determine how the optimal decision is affected by that uncertainty. These analyses can ensure that only uncertainties that directly affect the optimal decision are included. Participatory modeling throughout the SDM process provides all participants with a better understanding of each element within the consequence table, which can lead to group members more confidently expressing their risk tolerance for those elements of uncertainty that are most important for the decision (Gregory et al. 2012). By using this approach to understanding the relative importance of different sources of uncertainty, participants can better assess their risk tolerance for the overall decision.

5.2.5 *Tradeoffs*

The final step in the SDM framework is to evaluate the tradeoffs among the objectives. In some instances, a decision problem might have one objective, but in natural resources management problems, multiple objectives are the norm. One goal of the tradeoffs step is to determine how much value the decision maker or group places on each objective, based on the expected outcomes of each of the alternatives on each objective. This differential achievement of the objectives is an integral part of the tradeoffs process, as decision makers must take into account the predicted consequences such that these values judgments are not global, but rather specific to the decision (Monat 2009). For example, in Table 5.3, recreational angler satisfaction ranges from 4 to 6 among the alternatives, but commercial angler satisfaction ranges from 1 to 7. Globally, the decision maker might place equal value on commercial and recreational angler satisfaction. However, within the bounds of the decision, the decision maker likely would place more weight on the objective to maximize commercial angler satisfaction, as the difference among the outcomes of the alternatives on this objective is much greater. Including multiple stakeholders in the tradeoffs step allows each participant to see the quantification of their values over each of the objectives (Keeney 1992). By evaluating the tradeoffs as a group, each participant can provide their value judgments in the SDM process, and these clearly articulated differences of opinion provide avenues for communication about these points (Keeney 1992).

Quantitative weights to be placed on each objective for the calculation of a linear utility function are elicited in the tradeoffs step. In the linear utility function, the expected utility value for a given alternative is calculated as the sum of the weighted normalized measurable attribute outcomes for each of the objectives,

$$U_i = \sum_{j=1}^{j} W_j X_{i,j}, \tag{5.1}$$

such that U_i is the utility score for alternative i, $X_{i,j}$ is the normalized score of the measurable attribute predicted under alternative i for objective j, and W_j is the weight assigned to objective j (Keeney 1992). The outcome with the greatest expected utility value is the optimal decision. The measurable attribute scores ($X_{i,j}$) are normalized to a common scale (e.g., 0–1) to ensure that each attribute is on the same scale. In the consequence table framework, the addition of a column with objective weights (W_j) provides all the necessary information to calculate the linear utility function, with the normalized measurable attribute values populating the interior of the table ($X_{i,j}$; Table 5.4). There are many tools available to elicit objective weights (W_j) from

Table 5.4 Example of a consequence table in which the objective weights (W_j) are located on the far right column and the expected utility value of each management alternative (U_i) is located in the bottom row

		Alternatives				
Fundamental objective	Measurable attribute	Close area 1	Close area 2	Close areas 1 and 2	No closures	W_j
Maximize fish stock size	Spawning stock biomass (kg/km^2) after 5 years	0.56	0.22	1	0	0.5
Maximize commercial angler satisfaction	Constructed scale: 1–10	0	1	0.33	0.50	0.35
Maximize recreational angler satisfaction	Constructed scale: 1–10	0.50	0	1	0.50	0.15
U_i		0.35	0.46	0.77	0.25	

The outcomes have been normalized to a 0–1 scale. The optimal alternative is to close areas 1 and 2, with an expected utility value of 0.77

participants for use in multi-criteria decision analysis (Belton and Stewart 2002), including swing weighting (Edwards and Barron 1994) and direct ranking on the global scale (Monat 2009). The overall goal of the tradeoffs step is to help decision makers to make what are potentially very difficult tradeoffs for complex natural resources management problems. Employing multiple elicitation methods likely will help decision makers to see how they value these objectives relative to one another in a more complete way (Gregory et al. 2012).

5.2.6 *Implementing the Decision*

There are some considerations, as well as some suggestions from other SDM practitioners, regarding ways to ensure successful implementation of the optimal alternative. The considerations should be taken into account throughout the SDM process, but they will become more important in the implementation phase. First, the group should consider tractability of the project, as a very large scale or a complex array of issues can lead to decisions that are difficult to implement (Failing et al. 2004). Groups should consider time, cost, benefits, risks, and degree of impact associated with implementation (Marcot et al. 2012). If there are multiple management entities who are responsible for the implementation step, the goals and scope of the project should be compared with those of the diverse decision-maker and stakeholder pool (Marcot et al. 2012), or ideally, these stakeholders should be included throughout the participatory SDM process. Being aware of these potential pitfalls can help the group to create implementation strategies that are specific to their particular decision framework.

There are also hallmarks of successful implementation of the SDM process itself. The participatory modeling aspect, especially engaging stakeholders and managers for the quantitative modeling of the consequences of the management actions, promotes trust in these models and ease of implementation by these managers (Moore and Runge 2012). When regulatory agencies call for the use of new methods to make decisions that would avoid litigation and build stakeholder support, the product of the SDM process can be implemented more easily (Wilson and McDaniels 2007). Additionally, starting with a small problem, such as hydropower generation on one river in British Columbia, and then working up to the larger scale (e.g., watershed) provides groups with a better sense of the process (Wilson and McDaniels 2007). Finally, making the case for the benefits of SDM, such as identifying new, potentially less costly, management alternatives, and the structuring and transparency of the management decisions, can help when implementing the results of an SDM project for the first time (Wilson and McDaniels 2007).

5.3 Conclusions

Structured decision making is a framework with a specific process that enables decision makers to carefully consider all aspects of a decision individually and build a complete decision framework to make a defensible, transparent, and rigorous decision based on values and preferences. In natural resources management, complex decisions that have high degrees of uncertainty, as well as multiple stakeholder groups representing many different points of view, have benefitted greatly from this framework. As a values-based process, SDM is particularly amenable to participatory modeling, both at the level of creating the overall decision model and within the consequences step, where more quantitative predictive models often are necessary. Through a collaborative process, participants' values are taken into account in the objectives-setting phase; these objectives are turned into actionable management alternatives, and the effectiveness of the management alternatives at achieving these objectives is predicted. Each participant can use the information gained in the preceding steps to make important, and often difficult, tradeoffs among competing objectives. Collaborative efforts to create decision models through SDM can lead to greater trust and satisfaction in the process of making natural resources decisions (Lauber and Knuth 1997; Decker et al. 2012), which in turn can make management decisions more acceptable to a wider variety of stakeholders.

Acknowledgments We thank M. Eaton, W. Siemer, and two anonymous reviewers for insightful comments on this chapter.

References

Aalders I (2008) Modeling land-use decision behavior with Bayesian belief networks. Ecol Soc 13:16

Ascough JC II, Maier HR, Ravalico JK, Strudley MW (2008) Future research challenges for incorporation of uncertainty in environmental and ecological decision making. Ecol Model 219:383–399

Ayyub BM (2001) Elicitation of expert opinions for uncertainty and risks. CRC Press, Boca Raton

Belton V, Stewart TJ (2002) Multiple criteria decision analysis: an integrated approach. Kluwer Academic Publishers, Norwell

Blomquist SM, Johnson TD, Smith DR et al (2010) Structured decision-making and rapid prototyping to plan a management response to an invasive species. J Fish Wildl Manage 1:19–32

Brown BB (1968) Delphi process: a methodology used for the elicitation of opinions of experts. The RAND Corporation, Santa Monica

Clemen R (1996) Making hard decisions: an introduction to decision analysis. Duxbury Press, Pacific Grove

Cochrane JF, Haynes MA, Holcombe T et al (2012) Decision analysis: elicitation and facilitation. US Fish and Wildlife Service. National Conservation Training Center, Shepherdstown

Conroy MJ, Carroll JP (2009) Quantitative conservation of vertebrates. Wiley-Blackwell, Chichester

Conroy MJ, Moore CT (2001) Simulation models and optimal decision making in natural resource management. In: Shenk T, Franklin A (eds) Modeling in natural resource management: development, interpretation, and application. Island Press, Washington, pp 91–104

Conroy MJ, Peterson JT (2013) Decision making in natural research management: a structured, adaptive approach. Wiley, Chichester

Conroy MJ, Barker R, Dillingham P et al (2008) Application of decision theory to conservation management: recovery of Hector's dolphin. Wildl Res 35:93–102

Decker DJ, Krueger CC, Baer RA Jr, Knuth BA, Richmond ME (1996) From clients to stakeholders: a philosophical shift for fish and wildlife management. Hum Dimens Wildl 1:70–82

Decker DJ, Riley SJ, Siemer WF (2012) Human dimensions of wildlife management. In: Decker DJ, Riley SJ, Siemer WF (eds) Human dimensions of wildlife management. The Johns Hopkins University Press, Baltimore, pp 1–14

Dorazio RM, Johnson FA (2003) Bayesian inference and decision theory- a framework for decision making in natural resource management. Ecol Appl 13:556–563

Edwards W, Barron F (1994) SMARTS and SMARTER: improved simple methods for multiattribute utility measurement. Organ Behav Hum Decis Process 60:306–325

Failing L, Horn G, Higgins P (2004) Using expert judgment and stakeholder values to evaluate adaptive management options. Ecol Soc 9:13

Failing L, Gregory R, Harstone M (2007) Integrating science and local knowledge in environmental risk management: a decision-focused approach. Ecol Econ 64:47–60

Goodwin P, Wright G (2009) Decision analysis for management judgment. Wiley, Chichester

Gregory R (2000) Using stakeholder values to make smarter decisions. Environ Sci Pol Sustain Dev 42:34–44

Gregory R, Failing L (2002) Using decision analysis to encourage sound deliberation: water use planning in British Columbia, Canada. J Pol Anal Manag 21:492–499

Gregory R, Long G (2009) Using structured decision making to help implement a precautionary approach to endangered species management. Risk Anal 29:518–532

Gregory RS, Failing L, Harstone M et al (2012) Structured decision making: a practical guide to environmental management choices. Wiley-Blackwell, Chichester

Hammond JS, Keeney RL, Raiffa H (1999) Smart choices: a practical guide to making better life decisions. Broadway Books, New York

Hetes B, Richmond H, Pekar Z (2011) U.S. Environmental Protection Agency Expert Elicitation Task Force White Paper. Science and Technology Policy Council, U.S. Environmental Protection Agency, Washington

Howard RA (1966) Decision analysis: applied decision theory. In: Herty DB, Melese J (eds) Proceedings of the fourth international conference on operational methods. Wiley-Interscience, New York

Irwin BJ, Wilberg MJ, Bence JR, Jones ML (2008) Evaluating alternative harvest policies for yellow perch in southern Lake Michigan. Fish Res 94:267–281

Irwin BJ, Wilberg MJ, Jones ML, Bence JR (2011) Applying structured decision making to recreational fisheries management. Fisheries 36:113–122

Keeney RL (1992) Value-focused thinking: a path to creative decision making. Harvard University Press, Cambridge

Keeney R (2007) Developing objectives and attributes. In: Edwards W, Miles R Jr, von Winterfeldt D (eds) Advances in decision analysis: from foundations to applications. Cambridge University Press, New York, pp 104–128

Kendall WL (2001) Using models to facilitate complex decisions. In: Shenk T, Franklin A (eds) Modeling in natural resource management: development, interpretation, and application. Island Press, Washington, pp 147–170

Lauber T, Knuth B (1997) Fairness in moose management decision-making: the citizens' perspective. Wildl Soc Bull 25:776–787

Leong K, Decker DJ, Lauber TB (2012) Stakeholders as beneficiaries of wildlife management. In: Decker D, Riley SJ, Siemer WF (eds) Human dimensions of wildlife management. The Johns Hopkins University Press, Baltimore, pp 26–40

Maguire LA (2004) What can decision analysis do for invasive species management? Risk Anal 24:859–868

Marcot B, Holthausen R, Raphael MG et al (2001) Using Bayesian belief networks to evaluate fish and wildlife population viability under land management alternatives from an environmental impact statement. Forest Ecol Manag 153:29–42

Marcot BG, Thompson MP, Runge MC et al (2012) Recent advances in applying decision science to managing national forests. Forest Ecol Manag 285:123–132
Martin TG, Burgman MA, Fidler F et al (2012) Eliciting expert knowledge in conservation science. Conserv Biol 26:29–38
McCann RK, Marcot BG, Ellis R (2006) Bayesian belief networks: applications in ecology and natural resource management. Can J Forest Res 36:3053–3062
McGowan CP, Smith D, Sweka J et al (2011) Multispecies modeling for adaptive management of horseshoe crabs and red knots in the Delaware bay. Nat Resour Model 24:117–156
Monat JP (2009) The benefits of global scaling in multi-criteria decision analysis. Judgm Decis Mak 4:492–508
Moore JL, Runge MC (2012) Combining structured decision making and value-of-information analyses to identify robust management strategies. Conserv Biol 26:810–820
Nutt P (2004) Expanding the search for alternatives during strategic decision-making. Acad Manage Exec 18:13–28
Peterson JT, Evans JW (2003) Quantitative decision analysis for sport fisheries management. Fisheries 28(1):10–21
Ralls K, Starfield AM (1995) Choosing a management strategy: two structured decision-making methods for evaluating the predictions of stochastic simulation models. Conserv Biol 9:175–181
Reed MS (2008) Stakeholder participation for environmental management: a literature review. Biol Conserv 141:2417–2431
Riley SJ, Gregory RS (2012) Decision making in wildlife management. In: Decker DJ, Riley SJ, Siemer WF (eds) Human dimensions of wildlife management. The Johns Hopkins University Press, Baltimore, pp 101–112
Robinson KF, Jennings CA (2012) Maximizing age-0 spot export from a South Carolina estuary: an evaluation of coastal impoundment management alternatives via structured decision making. Mar Coast Fish 4:156–172
Runge MC, Bean E, Smith DR, Kokos S (2011) Non-native fish control below Glen Canyon Dam - Report from a structured decision-making project. United States Geological Survey, Reston
Speirs-Bridge A, Fidler F, McBride M et al (2010) Reducing overconfidence in the interval judgments of experts. Risk Anal 30:512–523
Starfield AM, Roth JD, Ralls K (1995) "Mobbing" in Hawaiian monk seals (*Monachus schauinslani*): the value of simulation modeling in the absence of apparently crucial data. Conserv Biol 9:166–174
Susskind L, Cruikshank JL (1987) Breaking the impasse: consensual approaches to resolving public disputes. Basic Books, New York
Tyre AJ, Peterson JT, Converse SJ et al (2011) Adaptive management of bull trout populations in the Lemhi Basin. J Fish Wildl Manage 2:262–281
Williams BK (2009) Markov decision processes in natural resources management: observability and uncertainty. Ecol Model 220:830–840
Williams BK, Johnson FA (1995) Adaptive management and the regulation of waterfowl harvests. Wildl Soc B 23:430–436
Williams BK, Nichols JD, Conroy MJ (2001) Analysis and management of animal populations. Academic, San Diego
Wilson C, McDaniels T (2007) Structured decision-making to link climate change and sustainable development. Clim Pol 7:353–370

Chapter 6
Ensuring that Ecological Science Contributes to Natural Resource Management Using a Delphi-Derived Approach

Amy K. Wolfe, Virginia H. Dale, Taryn Arthur, and Latha Baskaran

6.1 Introduction

The management of natural resources builds from site-specific knowledge about the spatial and temporal distribution and abundance of water, air, soil, and organisms, and the services they provide. Ecosystem services fall into four categories: provisioning, such as the production of food and fuel; regulating, such as the control of climate and disease; supporting, such as nutrient cycles and crop pollination; and cultural, such as spiritual and recreational benefits (Millennium Assessment 2005). Social services include jobs, and economic services relate to the monetary benefits provided. Resource management aims to enhance long-term benefits from the environment by employing means by which humans can utilize natural assets in a way that does not deplete or degrade them over time.

Many research studies are undertaken with an explicit purpose of improving resource management practices. However, the question of how emerging scientific findings are integrated into existing management practices is rarely addressed. Stated another way, it is unclear how to evaluate if management practices are appropriate or based on the latest scientific understandings. These issues are at the crux of this chapter. Our challenge was to integrate a set of ongoing scientific studies in a way that would prove useful for resource managers, specifically those at the United States (US) Army military installation at Fort Benning, Georgia, in the southeastern

A.K. Wolfe (✉) • V.H. Dale • L. Baskaran
Environmental Sciences Division, Oak Ridge National Laboratory,
P.O. Box 2008, Bethel Valley Road, Oak Ridge, TN 37831-6036, USA
e-mail: wolfeak@ornl.gov

T. Arthur
Environmental Sciences Division, Oak Ridge National Laboratory, P.O. Box 2008, Bethel Valley Road, Oak Ridge, TN 37831-6036, USA

Fox Run Dr., Plainsboro, NJ 08536, USA

S. Gray et al. (eds.), *Environmental Modeling with Stakeholders*,
DOI 10.1007/978-3-319-25053-3_6

US. Here, we focus on the participatory methods we used—intentionally and opportunistically—to achieve this integration, describing how they helped disparate parties achieve consensus over time.

Our challenge reflects broader issues that emerge from the persistent gap between the production of science and the use of science in practice,[1] which plays out differently in the two arenas. The "science" arena is one in which there are continuing calls for "science-based" decision making and greater science literacy, as well as expressions of frustration about the gulf between science and policy or practice (e.g., Aber et al. 2000; Carnegie Commission 1992; Sigma XI 1993; National Academies 2005). In contrast, data, studies, and models that do not prove useful to practitioners simply are not used (see, as examples, Jones et al. 1999; National Research Council 2009; Rayner et al. 2001; Steel et al. 2000–2001).

There are many explanations for why the disjunction between science and practice endures. Explanations range from the questionable view that science is objective and divorced from social influences, to the solicitation-plus-peer-review process that defines and constrains what science is funded, to the incentives or markers of success for scientists versus practitioners, to stereotypic motivations for scientists (seek knowledge) and practitioners (resolve problems). Other explanations emphasize cultural and sociological factors that influence scientists and practitioners working within their organizational settings to adopt particular goals, objectives, and constraints. While this chapter does not explore these explanations, they all factor into the challenges of reconciling science with practice. The need to develop a consistent means whereby science provides information that is useful for management was the impetus for this analysis. Our practitioner-oriented goal led us to use direct queries of key players as a mechanism for going beyond implicit assumptions about what constitutes "usefulness."

6.2 Resource Management and Environmental Research at Fort Benning

Our research focused on Fort Benning, Georgia, home to a 75,533 hectare (181,626 acre) military facility. The installation includes a 5759 hectare (14,231 acre) cantonment area, which houses residential, office, and other similar infrastructure that must be managed and maintained. Fort Benning's prime mission is military training and testing. Portions of the installation are used for—and managed to allow—such

[1] This gap is evident in many contexts. As one example, Sedlacko et al. (2013) assess the use of science by policy makers in the realm of sustainability in their discussion of CORPUS (Enhancing the *Co*nnectivity between *R*esearch and *P*olicy-Making in *S*ustainable Consumption) in Europe. As another example, in the medical arena, the science-practice gap falls under the umbrella of "translational research." In fact, the US National Institutes of Health National created the Center for Advancing Translational Sciences (NCATS) in 2011 to help assure that the results of medical research aid patients more quickly than has occurred in the past (http://www.ncats.nih.gov/, accessed in December 2014). And a 2005 National Research Council (National Research Council 2005) report makes recommendations to close the gaps between social science research and its use in environmental decision making.

activities as tank maneuvering, firing ranges, drop zones, and bivouac areas. In addition, Fort Benning is subject to a variety of state and federal natural resource guidelines and regulations and is managed accordingly. As examples, resource managers thin upland pine forests; use fire to control understory growth; restore ecological conditions in the understory; and protect rare species like the red-cockaded woodpecker (*Picoides borealis*) and gopher tortoise (*Gopherus polyphemus*). Taking all of these elements together, the installation is faced with sometimes competing and conflicting planning and management objectives.

The US Department of Defense (DoD) is responsible under the Sikes Act (16 U.S.C. 670a-670f, as amended) for implementing management strategies that conserve and protect biological resources on its lands. Because military lands and waters often are protected from human access and impact, they contain some of our nation's most significant remaining large tracts of land with valuable natural resources. Integrated Natural Resource Management Plans (INRMPs) are developed for each military installation to define management of natural resources, while allowing multipurpose uses of resources including appropriate public access and uses, without any net loss in the capability of an installation to support its military mission. INRMPs are living documents that provide direction for management and evolve as more is learned about the particular system. The US National Environmental Policy Act (NEPA) requires federal agencies to integrate environmental values into their decision-making processes by considering the environmental impacts of proposed actions and reasonable alternatives to those actions. The INRMP integrates conservation actions with military operations, serves as a principal information source for NEPA documentation, and aids planners and facility managers. We label as "practitioners" the Fort Benning resource managers, a group that includes both military personnel and staff of The Nature Conservancy, who develop and update the INRMP for Fort Benning.

The environmental science used in this project consisted of several years of indicator and threshold studies at Fort Benning, designed with the broad intent of assisting installation resource managers. Ecological indicators and thresholds are intended to ascertain and help forecast ecological conditions and thus can be important tools in resource management. Indicators provide information about potential or realized effects on phenomena of concern and can be used to assess environmental conditions of a system, to monitor trends in conditions over time, or to provide an early warning signal of change (Cairns et al. 1993). Thresholds, where they exist, in effect are tipping points beyond which ecosystems cannot recover from disturbances naturally or through interventions. Beginning in the late 1990s, the DoD Strategic Environmental Research and Development Program (SERDP) Ecosystem Monitoring Project (SEMP) funded five projects at Fort Benning intended to identify indicators (three projects) or thresholds (two projects) that signal ecological change. These projects explicitly aimed to be useful for planning, implementing, and monitoring the impacts of military land-management practices at military installations. Once identified, the concept was to determine how indicators and thresholds can be incorporated effectively into Fort Benning's monitoring and management programs (Dale and Beyeler 2001). These findings, then, should be applicable to other military installations with similar ecological conditions.

Aspects of the five Fort Benning projects center on plot-scale investigations, but each project had different goals and field-investigation sites (see the SEMP website: https://www.serdp-estcp.org/Program-Areas/Resource-Conservation-and-Climate-Change/Natural-Resources/Species-Ecology-and-Management/RC-1114[2]).

One threshold project compared military training compartments that are open or closed to tracked vehicles (e.g., tanks), where the underlying sandy or clay soils experimentally are subjected to different forest management practices (different burn cycles, thinning regimes, etc.) (Dilustro et al. 2002; Duncan et al. 2004). A second threshold project emphasized soil integrity by focusing on soil organic matter and soil nitrogen dynamics (Garten et al. 2003). The indicator projects identified indicators that mark ecological change in intensely versus lightly used ecological systems by identifying the suite of variables needed to measure changes at several scales (Dale et al. 2004); by investigating forest understory, stream chemistry and aquatic biology, and soil microorganisms (Peacock et al. 2001; Dale et al. 2002, 2008; Maloney et al. 2005); by taking a multi-indicator approach to evaluate a set of soil, understory vegetation, and surface hydrology parameters (Reddy et al. 2003); or by using classifications of ecological indicators to assess and monitor ecological changes and thresholds (Krzysik et al. 2005).

6.3 Participatory Methods for Addressing Integration Goals

Our goal was to integrate these five indicator and threshold projects in a manner that facilitated their contribution to existing Fort Benning resource management documents, tools, and practices. In its entirety, this integration activity involved multivariate statistical analyses of SEMP project-derived indicators and geographic information system (GIS) mapping of analytical results. Although SERDP funded this integration project for the purpose of assuring that the results of the ecological research projects it funded at Fort Benning would be translated into use by resource managers at the installation, the integration effort was initiated well after the five multi-year projects were underway. In fact, some projects were nearly completed. The push for integration came more from SERDP, the organization funding the environmental research projects, than from the land managers at Fort Benning. Nevertheless, integrating projects that collected different kinds of data, using different units of measurement, sampled with varying frequencies from disparate field locations and conditions poses obvious challenges. Conducting this integration in a way that simultaneously proves useful for resource management amplified those challenges.

A first step in the larger integration was to create a common framework, depicted by a matrix, within which to operate. We initially planned to delineate a suite of defined, discrete Fort Benning land-use categories acceptable to all SEMP researchers,

[2] Website accessed in November 2014.

thinking that "land use" would be an effective backdrop for integration because it influences ecological conditions and shapes land management goals and interventions. Agreed-upon land-use categories then would provide a framework that focuses and guides the integration of disparate indicators across the Fort Benning reservation. In this context, "integration" refers to an evaluation of the several proposed indicators to ensure that, collectively, they provide comprehensive and useful metrics that can serve as a basis for improved environmental management. The final result was intended to be a set of land-use categories for Fort Benning, effective for its land management activities and likely transferable to other installations in the region to which similar land-use and management practices are applied. However, as will be detailed below, because "land-use categories" proved inadequate as an integrator, we shifted to what we label "land-management categories." As will be described, this shift is far more than a semantic adjustment; it represents a considerably different basis for integration, one that reflects the perspectives and needs of both ecological scientists and resource managers.

We originally selected a Delphi approach to achieve consensus on the integration framework with the expectation that it would be an efficient and effective method that, pragmatically, suited our project needs. The Delphi approach is a well-established, structured communication technique for eliciting information from experts (Dalkey and Helmer 1962; Linstone and Turoff 1975; Taylor and Ryder 2003). Experts answer questions in two or more rounds, and a facilitator provides a summary of each round. As part of the process, experts are encouraged to revise their earlier answers in light of the collective replies, which fosters a decrease in the range of the answers and convergence of the group opinion. The Delphi approach adheres to the principle that decisions from a structured group of individuals are more accurate than those from unstructured groups (Rowe and Wright 2001).

It has multiple pragmatic advantages, too. For example, the ability to elicit information at a distance instead of face-to-face saves money and time for everyone involved (particularly before the ability to conduct meetings online became commonplace) and does not require logistical coordination among participants. The Delphi approach seeks everyone's thoughtful participation individually and not in a group setting, so attributes such as dominating or shy personalities do not affect levels of participation or the kinds of information obtained. Iteration is built into the Delphi approach, which allows participants to reflect on what others have said and fine-tune, correct, or expand. In addition, the Delphi emphasis on expert opinion (representatives of the five SEMP research teams) underscored the credibility of its results and may have contributed to SERDP and participants' acceptance of the approach. Overall, the Delphi approach suited our needs, available time, and resource constraints. We mistakenly thought it would be the only formal approach we would use for this activity.

In retrospect, a key reason for the challenges we faced was the fact that we actually were engaging two sets of experts, Fort Benning resource managers and SEMP researchers. At first we framed resource managers' contributions as being more in the realm of end users than Delphi contributors. Thus, because we wanted the land-use categories to be useful and clear to Fort Benning resource managers, we originally

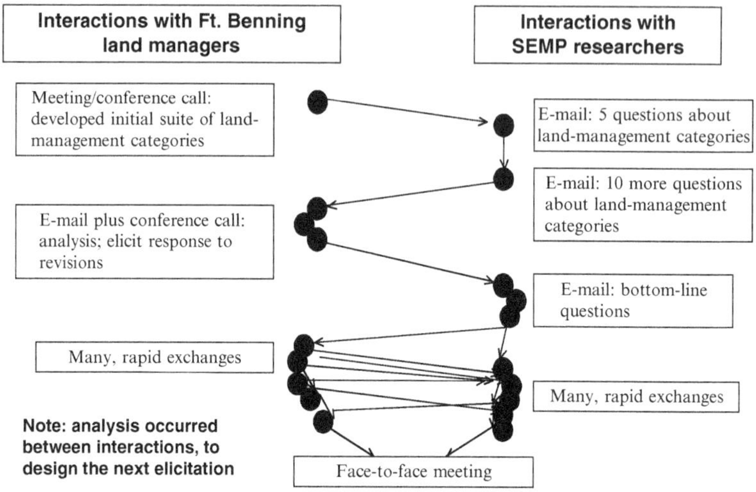

Fig. 6.1 Schematic view of the Delphi method as implemented

planned to engage them at two stages—before initiating the Delphi process with SEMP researchers to help develop our first set of questions and after the Delphi process was completed to check that the resulting integration framework made sense. However, we ended up consulting the land managers much more often than anticipated and injecting their input into the interactions with SEMP researchers. As a consequence, and as we will discuss, the Delphi process itself started to morph. It became a facilitated (by us), iterative information elicitation and negotiation process that occurred primarily by email, occasionally by telephone. That facilitated-negotiation process predominated when we took advantage of a face-to-face meeting that was to be convened for other purposes. That meeting proved to be a critical one that dramatically changed the integration framework that previously was developed through Delphi and modified Delphi participatory approaches. Figure 6.1 schematically shows the evolution of these approaches. Though the figure ends with the face-to-face meeting with researchers and resource managers (one of whom participated by phone), our project team also continued interacting with both groups via email to fine-tune the resulting framework, a land-management category matrix.

6.4 Participatory Approaches and Their Outcomes

The following sections describe the interactions in some detail to provide information and insights about: the nature of the participatory approaches we used; the heterogeneity of small groups of participants who fit within small seemingly homogeneous categories; the difficulty in reconciling viewpoints and achieving consensus, the importance of details and subtleties in that process; and the product that resulted from researchers' and land managers' participation. For us, process and outcome necessarily were intertwined. Our goal was to produce an effective integration framework, where "effective" encompassed the best available ecological science and real-world land management issues and needs. The participatory approaches we used were means to achieve these ends; participation was not the end in and of itself.

6.4.1 Preliminary Consultation with Land Managers: Developing an Initial Land-Use Framework

Our first discussion with Fort Benning land managers made clear that, though we categorized them as a single group, it was not a homogeneous group. Each individual had his or her professional objectives and perspectives. Furthermore, our meeting prompted a rare circumstance in which the group met face-to-face. And we found that "stove piping" of those making management decisions was linked to poor communication across resource areas, which hampered holistic, integrated planning and management.

This first meeting raised many of the issues that we grappled with throughout the course of developing a consensual integration framework. At their core, many of these issues centered on articulating precisely what "entity" to use as the integrator. For instance, while ecological conditions have been defined for *land-cover* groups at Fort Benning, land cover (ecological state as conveyed by physical appearance—closed forests, open forests, grasslands, etc.) may mask land uses and the influence of natural versus human-caused elements. Participants in the initial meeting decided, instead, to focus the integration of *land use* (purpose to which land is put by humans, such as protected areas, forestry for timber products, pastures, etc.). The rationale was that some ecological indicators may be able to distinguish among land uses and signal when a particular area is becoming degraded.

However, the meeting made clear that some land-use issues were important to resolve during the process of developing a limited set of land-use categories (the integration framework). As one example, some land areas are subjected to multiple uses, such as timber management and military training. In a different vein, the resource managers discussed the difficulties in determining when natural disturbance impacts and subsequent management actions should differ according to land use. Further, they highlighted elements that operate simultaneously at Fort Benning and that are necessary to consider in distinguishing and making management decisions about particular parcels of land. These elements are (1) military uses of land, (2) the frequency of those uses, and (3) land-management goals.

Because military uses of land together with their frequency can dramatically influence ecological effects (e.g., tank traffic versus occasional wheeled-vehicle traffic, versus foot traffic), the Fort Benning resource managers underscored the importance of distinguishing kinds of military use and their frequency. There also was considerable discussion of land-management goals and practices. The managers decided that the installation's land-management goals for particular areas are more stable than either the specific management practices undertaken in those areas or land-cover types. Therefore, participants suggested categorizing land areas within Fort Benning according to land-management goals. In addition, practitioners noted that different land goals can involve varying kinds of land-management activity, ranging from light ("extensive," in their language) to heavy ("intensive").

Based on this meeting, military use and land-management dimensions became a cornerstone for land-use category development. Rather than delineate a list of land-use categories, the group juxtaposed the dimensions and created a land-use category matrix (see Table 6.1 for the initial version, which showed all possible combinations rather than those specifically relevant to Fort Benning).

6.4.2 *Round 1 with SEMP Researchers: Raising Challenging Issues*

The matrix developed with Fort Benning resource managers became the focus for the first round of the Delphi process with SEMP researchers at which researchers addressed five questions (Fig. 6.2). The responses of the researchers raised three

Table 6.1 Land-use categories as determined by military training and land management practices, initial version

	Military uses of land							
Land management goals	Tracked vehicles	Wheeled vehicles	Foot traffic	Bivouac areas	Firing ranges	Impacts areas	Drop zones	Not used
Extensively managed areas	0	0	I,F	0	0	I,F	0	+
Intensively managed areas								
Upland pine forests								
– Set-aside areas	0	I	I	0	0	0	0	+
– Modified management areas	0	I	I,F	0	0	0	0	+
– Standard management	I	I,F	I,F	I,F	0	0	0	+
Mowed areas	0	I	I,F	0	I	0	I	0
Wildlife openings	0	I	I	0	0	0	I	+
Erosion control areas	I,F	I,F	I,F	I,F	I,F	I,F	I,F	+

Key: '0' = which *military uses* do NOT occur in areas managed in specified ways

'I' and 'F' = the relative frequency with which *military uses* occur in areas managed in specified ways (I = infrequent and F = frequent)

'+' = *land management* options in areas not used by the military

1. As a set, are the proposed land-use categories

    a. Well-defined?

    b. Comprehensive?

    Please explain your answers, providing as much specific detail as possible.

2. Are each of the land-use categories

    a. Sufficiently discrete?

    b. Focused appropriately (neither too broad nor too narrow)?

    Please explain your answers, providing as much specific detail as possible.

3. Do the proposed land-use categories capture the differences among field research plots about which your research team is concerned? Explain your answer, providing as much specific detail as possible.

4. Give a rough approximation of how your research team's field plots are distributed across the proposed suite of land-use categories (or, across the suite of categories according to your proposed revisions). Take only a few minutes to complete this question.

5. What land-use categories would you revise, add, or subtract? Please provide all of your suggested revisions.

Fig. 6.2 Questions used in first-round elicitation

issues that remained contentious and unresolved throughout much of the modified Delphi process. One issue previously had been raised by Fort Benning resource managers, namely how to categorize areas in which there are multiple military uses. Researchers also suggested possible solutions such as categorizing according to intensity of military use or by majority use. In preparing questions for the second round of elicitations, we suggested using the label "*predominant* military uses of land" instead of "military uses of land" and asked whether "predominant" should be interpreted in terms of frequency of use or extent of ecological impact. This issue remained unresolved, even after the second round of elicitations.

Researchers also raised two "new" issues about how best to categorize those portions of Fort Benning (a) whose current ecological condition is dominated by past—

but not current—land uses, and (b) that are affected by adjacent land uses. It was only at the face-to-face meeting towards the end of our Delphi-derived process that the group decided that "predominant" military use of land referred to the use with the greatest ecological impact, no matter whether that impact was caused by one of multiple, past, or adjacent land uses. Labels used in successive versions of the evolving integration matrix show the evolution of the group's (both researchers and practitioners) thinking. First, the label was "military use(s) of land" (Table 6.1). "Predominant military use of land" was the interim label (Table 6.2). And, the final version (Table 6.3), though wordier, became quite specific—"cause of predominant ecological effect from military use(s) of land."

6.4.3 Round 2: Refining the Integration Matrix

The second round SEMP researcher elicitation consisted of a summary of Round 1 and a new set of questions (Fig. 6.2) based on the specific suggestions and issues raised during Round 1. Table 6.2 depicts the manner in which we incorporated most suggested revisions and identified questions for SEMP researchers to address. Changes from the initial proposed land-use table were denoted in a heavier, bold font. We emphasized to researchers that Table 6.2 offered one way to respond to their suggestions, and that it was essential for the SEMP integration effort that they all agree that the final suite of land-use categories is acceptable and usable. For researchers, "usable" meant that they would be able to assign one land-use category to each of their field plots, a task they were told by SERDP managers that they would be asked to do (Fig. 6.3).

6.4.4 Round 3 and the Face-to-Face Elicitation: The "Final" Integration Matrix Emerges

It was in preparing this third formal elicitation that we deviated from a typical Delphi approach, looking beyond our group of researcher experts for assistance and did so in an increasingly informal, rapid manner. Our reason for making this deviation was that, to create an integration matrix for Round 3, we needed to make several judgments about how to handle issues researchers raised and variations in their responses to Round 2. Rather than make those judgments alone, we consulted with the Fort Benning resource managers to help assure that the integration process would serve their needs. We contacted Fort Benning resource managers initially by email and then through a conference call, with subsequent email and telephone contacts. This set of interactions evolved partly because the modified matrix and the issues raised by researchers generated considerable discussion among the resource managers. Ultimately, the matrix used in the Round 3 elicitation reflected researchers' and resource managers' input (Table 6.3; again, modifications are in bold). We also provided a summary of the preceding round's results and briefly mentioned our interactions with Fort Benning resource managers.

Table 6.2 Land-use categories as determined by military training and land management practices—second version

Land management goals	Predominant military uses of land								
	Tracked vehicles	Wheeled vehicles	Foot traffic	Bivouac areas	Firing ranges	Impact areas	Drop zones	Forestry	Not used
Extensively managed areas									
• **Upland pine forests** • **Bottomlands** • **Other**? [**need to specify**]	0	0	I,F	0	0	I,F	0		+
Intensively managed areas									
• Upland pine forests									
– Set aside areas	0	I	I	0	0	0	0		+
– Modified management area	0	I	I,F	0	0	0	0		+
■ **Unique ecological area**									
■ **RCW mgmt zone**									
■ **Gopher tortoise recovery zone**									
■ **Other**? [**need to specify**]									
– Standard management	I	I,F	I,F	I,F	0	0	0		+
• **Pine plantations**									
• Mowed areas	0	I	I,F	0	I	0	I		0
• Wildlife openings	0	I	I	0	0	0	I		+
• Erosion control areas	I,F	I,F	I,F	I,F	I,F	I,F	I,F		+

Note: If SEMP researchers agree that the military land use category should reflect the predominant military use in areas where there are multiple uses, then researchers must define "predominant." Two possible options are (a) the most frequent of multiple military uses occurring in a single area; and (b) the military use with the most substantial impact on the land (intensity?)

Key: '0' = which *military uses* do NOT occur in areas managed in specified ways

'I' and 'F' = the relative frequency with which *military uses* occur in areas managed in specified ways (I = infrequent and F = frequent)

'+' = *land management* options in areas not used by the military

Table 6.3 Land-use categories as determined by military training and land management practices—proposed revisions are in bold (August 12, 2003)

	Cause of predominant ecological effect from military use(s) of land								
Land management goals	Tracked vehicles	Wheeled vehicles	Foot traffic	Bivouac areas	Firing ranges	Impact areas	Drop zones	Sedimentation	Not affected
Extensively managed areas									
• **Upland pine forests** • **Bottomlands** • **Other**? [**need to specify**]	0	0	I,F	0	0	I,F	0		+
Intensively managed areas									
• Upland pine forests									
– Set aside areas	0	I	I	0	0	0	0		+
– Modified management area	0	I	I,F	0	0	0	0		+
■ **Unique ecological area**									
■ **RCW mgmt zone**									
■ **Gopher tortoise recovery zone**									
■ **Other**? [**need to specify**]									
– Standard management	I	I,F	I,F	I,F	0	0	0		+
• **Pine plantations**									
• Mowed areas	0	I	I,F	0	I	0	I		0
• Wildlife openings	0	I	I	0	0	0	I		+
• Erosion control areas	I,F	I,F	I,F	I,F	I,F	I,F	I,F		+

Key: '0' = which *military uses* do NOT occur in areas managed in specified ways
'I' and 'F' = the relative frequency with which *military uses* occur in areas managed in specified ways (I = infrequent and F = frequent)
'+' = *land management* options in areas not used by the military

1. What is the best way to categorize land areas on which there are multiple military uses?
2. What is the best way to categorize land areas whose current ecological condition is dominated by past, but not current, land uses?
3. What is the best way to categorize "not used" lands that are affected by adjacent land uses?
4. What is the best way to categorize "modified management area" lands within the upland pine forests?
5. What other categories or subcategories should be merged into "modified area management" lands within the upland pine forests?
   You may wish to refer to the land use and management goal descriptions in the Appendix.*
6. What is the best way to categorize vehicle, foot, and bivouac military uses of land?
7. What is the best way to categorize forestry uses?
8. What is the best way to categorize pine plantation areas?
9. Considering previous responses, Table 2, and your answers to these questions, how would you revise Table 2 to reflect Fort Benning land-use categories?
10. Any additional comments?

*The Appendix to the questionnaire consisted of definitions and descriptions of terms and repeated material disseminated to researchers during the first elicitation.

Fig. 6.3 Questions asked in 2nd-round elicitation regarding the proposed framework (minus answer options provided)

We thought—or, perhaps, hoped—that Round 3 would be the final one. Thus, we asked just a single question, "Do you find the current land-use category matrix acceptable? If not, please provide specific suggestions that will make it acceptable to you." The matrix proved unacceptable, which generated a host of additional interactions via email and telephone, both between SEMP researchers and our project team and between Fort Benning land managers and our team. We endeavored—unsuccessfully—to obtain sufficient clarification from the various parties about the sticking points to make modifications that everyone involved would find acceptable. The pace of interactions was too rapid to allow the formal, iterative summary-and-elicitation processes that marked the early portion of the Delphi process.

These interactions occurred shortly before a previously scheduled face-to-face SEMP Integration Project meeting. We were opportunistic and obtained time on the agenda. That session ended up serving as a venue in which to resolve remaining issues and develop a "final" (in actuality, the penultimate) version of the matrix. Because the larger meeting's objectives were not limited to our integration efforts, participants included representatives of SEMP research teams (including some who had not been direct participants in our process; one individual participated by telephone), a Fort Benning resource manager, and SERDP SEMP managers. Clearly, even if a traditional Delphi round included a face-to-face elicitation, participants would not vary from the original group of experts.

Apparently simple changes to the integration matrix may embody sophisticated thinking and considerable complexity. With that knowledge in mind, the changes to the integration matrix after Rounds 1, 2, and 3 appear to be relatively simple refinements. The matrix that emerged from the face-to-face meeting, in contrast, was markedly different from previous versions (Table 6.4—with changes from preceding versions in bold—includes minor revisions made through email exchanges after the face-to-face meeting). Changes were both substantive and organizational. The label for the "land management goals" dimension was amended to include endpoints as well as land management goals. "Endpoint," a key element of ecological risk assessment (Suter 2008), is a term and concept familiar to ecologists engaged in indicator-related research. Land-management labels shifted from indicating the kind (intensity) of management activity toward specifying the purpose of management activities. Other label and categorization revisions were made with the explicit intention of being more (a) compatible with researchers' and practitioners' perspectives; (b) understandable for individuals who may use the matrix in the future, particularly if they were not involved in the process of matrix creation; and (c) amenable to eventual application across all of Fort Benning. As one example, the "extensively managed" terminology was confusing to most researchers. Land managers distinguished "extensively" managed areas, which required few managerial interventions, from "intensively" managed areas, which required far more oversight and management action. Therefore, "extensively managed" was changed to "minimally managed," to be more readily understandable both to researchers and potential future matrix users. Another illustration is the addition of the "built environment" subcategory, thereby including for future use the cantonment area excluded from consideration for the purposes of this integration project.

6.4.5 Mapping the Land Management Goals Based on the Integration Matrix

The analysis phase of integration continued with development of a map of the land management goals (Fig. 6.4). Briefly, all research teams were asked to assign each of their field plots to a particular cell in the integration matrix. These assignments

Table 6.4 Land-management categories as determined by military training and land management practices—final version

Land management goals and endpoints	Cause of predominant ecological effect from military use(s) of land								
	Tracked vehicles	Wheeled vehicles	Foot traffic	Designated bivouac areas	Firing ranges	Impact areas	Drop or landing zones	No military effect	Administrative use
1. Minimally managed areas									
1.1 Wetlands	I,F	I, F	I	0	0	0	0	+	0
1.2 Vegetation on steep slopes	I, F	I, F	I	0	0	0	0	+	0
1.3 Forests in impact zones	0	0	0	0	0	I,F	0	+	0
2. Managed to restore and preserve upland forest									
2.1 Upland forests **2.1.a Long leaf dominance** **2.1.b Mixed pine** **2.1.c Scrub oak pine mix**	I	I,F	I, F	0	0	0	0	+	0
2.2 RCW mgmt clusters	I	I	I,F	0	0	0	0	+	0
2.3 Sensitive area designated by signs	0	0	I,F	0	0	0	0	+	0
3. Managed to maintain an altered ecological state									
3.1 Intensive military use areas	F	F	0	I,F	F	0	0	0	0
3.2 Wildlife openings	0	I	I	0	0	0	I	+	0
3.3 Mowed fields	0	I	I,F	0	I,F	0	I,F	+	0
3.4 Roads (paved and unpaved)	I, F	I, F	I, F	0	0	0	0	+	0
3.5 Built environment	0	0	0	0	0	0	0	0	+

Key: '0' = which *military uses* do NOT occur in areas managed in specified ways
'I' and 'F' = the relative frequency with which *military uses* occur in areas managed in specified ways (I = infrequent and F = frequent)
'+' = *land management* options in areas not used by the military

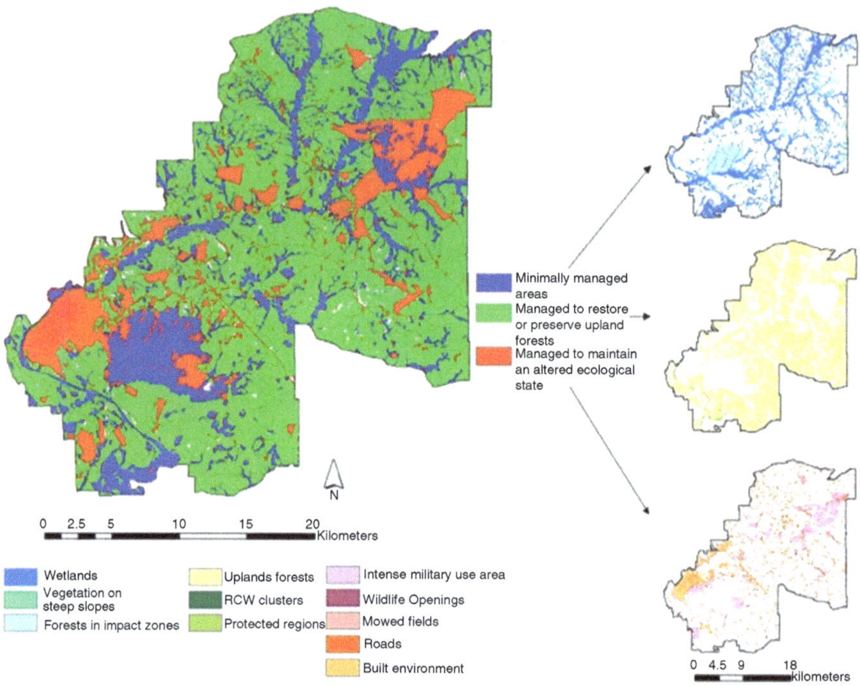

Fig. 6.4 Map of land management goals and endpoints for the Fort Benning military installation

were checked and validated by our integration team and, where questions arose, by a Fort Benning resource manager especially knowledgeable about the installation's ecology. Then, the field data associated with each cell were analyzed through multivariate statistics to determine the suite of indicators best able to describe a set of ecological conditions (Dale et al. 2008).

Results of these sets of analyses were mapped in GIS layers. We created detailed GIS maps of land-management categories (Fig. 6.4) in advance of the integration itself. Maps consisted of two layers derived from the integration matrix: (a) land management goals and endpoints and (b) cause of predominant ecological effects from military use(s) of the land. Existing data were used to create these maps, but it also was necessary to consult with and obtain input from Fort Benning resource managers to assure their accuracy. Likewise, Fort Benning resource managers reviewed both sets of integration results—statistical and GIS—as a form of "ground-truthing." All of these efforts contributed to identification of a set of ecological indicators for Fort Benning that are technically sound (defined by criteria established primarily by ecological researchers) and practically useful (defined by criteria established primarily by Fort Benning resource managers) thereby meeting some of our selection criteria (Dale and Beyeler 2001). The map serves as a means to facilitate interpretation of the land management goals.

6.5 Discussion

This chapter details our efforts to develop a common, consensus-based framework for integrating several research projects, and to do so in a way that would be useful for practitioners. Although participatory in nature, our initial plan (to use a Delphi approach with representatives of the research teams, eliciting input from Fort Benning resource managers before and after to help prepare the first elicitation and as a check on the resulting framework) proved overly simplistic. We anticipated that scientists and practitioners would act in accordance with substantially different perspectives, goals, and objectives. From a pragmatic perspective, we cared more about reconciling these differences than about analyzing underlying explanations for them. Still, we underestimated the diversity of perspectives within both resource manager and researcher groups. And, our decision to introduce an interim check by resource managers had the effect of altering our research approach—and results—substantially. What started as a Delphi approach morphed into a facilitated (by our project team) negotiation within and between groups, producing the desired integration framework.

These experiences made it clear that our overarching approach of consulting both practitioners and researchers in developing a commonly understood and agreed upon integration framework was appropriate. However, part of why reconciling practitioners' and scientists' world views was more challenging than we anticipated was that we were also reconciling varying perspectives and knowledge sets within each group. After our initial meeting with Fort Benning practitioners, some of them commented on how rare it was for that group to get together and talk with one another. Focusing on creating an integration framework revealed differences in participants' roles at the installation and in the kinds of ecological information needed for their jobs. Unlike the practitioners, SEMP researchers met periodically in review or information-sharing meetings to discuss their work. However, the researchers focused on their own work and not on producing a common, synthesized product (documents like annual reports to which researchers contribute usually are more compilations than syntheses). Producing the integration framework had the effect of forcing these researchers to confront how their disparate foci, measures, and findings could be combined to paint an ecological picture of Fort Benning useful as a basis for resource management decision making.

Once deciding that both practitioners and researchers should be involved in the process, the question of what methods to use in accomplishing this integration had to be resolved. This question was not simply one of how to incorporate science into decision making because neither "science" nor "practice" are singular entities. Science is disparate in its goals, measures, and findings; sometimes contradictory; evolving over time; and incomplete. Practice also entails different goals and approaches, even within a single installation. Considering these kinds of complexity together with our experiences made us think about several questions. What methods would we use if we were undertaking a similar project again either after most research was completed, or, better, before research would be undertaken? Would we propose the "Delphi-derived" approach that emerged during our project?

There are multiple factors to consider in answering the previous questions. One factor was how we frame our work. The shift in our methods reflected a shift in how our project team conceptualized our task, although we might not have been able to articulate what that shift was as it was unfolding. When we were in the Delphi mode, we thought of our task as an expert elicitation. The Delphi approach has proven useful for conducting that kind of elicitation, particularly for parties who are geographically dispersed and when time pressures exist. Its iterative aspects were desirable in the context of our project goals because the feedback would allow us to check the accuracy of our interpretations and would prompt new insights and information from participants. However, trying to implement a Delphi or Delphi-like approach simultaneously for two disparate groups of experts was awkward at best, particularly given our time constraints.

Information elicitation was not parallel between researchers and practitioners. We queried researchers as individuals, but because the initial purposes of the two groups were different, we queried practitioners as a group or through a key contact, who then would talk with others at Fort Benning. Thus, practitioners had the opportunity to exchange ideas and discuss matters directly. Two members of our team were parties to the initial meeting with practitioners, benefiting from hearing the interactions and observing the attention paid to the questions posed. We have no way of knowing the amount or kind of attention individual SEMP researchers paid to our inquiries (though they received additional funding for the purpose of assisting our integration effort). Nor do we know whether researchers were in any way upset or put off when we included Fort Benning resource managers and their input during the Delphi process.

It was adding the face-to-face meeting, however, that marked the greatest departure from the traditional Delphi approach. It also was the face-to-face meeting that embodied the shift from iterative knowledge elicitation and consensus building to facilitated negotiation. The meeting evolved from pragmatic project considerations. Though we were opportunistic in taking advantage of a previously planned meeting, we used it as a forcing event that would, in a time-efficient manner, lead the groups to resolve remaining issues. Beyond its venue, several other factors operated to distinguish it from our email elicitations.

First, before initiating discussion, we were asked to give a presentation summarizing our progress and integration matrix to date. This presentation and the question-and-answer session associated with it seemed to generate a deeper understanding of our objectives among some participants than the written background materials we provided with each elicitation. Second, there was a greater number and diversity of meeting participants than Delphi participants. Meeting participants included researcher team members who had, and who had not, participated in the Delphi elicitation; individuals who conducted other related research at Fort Benning; persons involved in Fort Benning resource management and operations; and SERDP managers. This broader group participated actively in developing the penultimate integration matrix. Third, meeting participants talked directly to one another—asking questions of each other (e.g., what do you mean by "x"), of the entire framework (e.g., why exclude the cantonment area), adding different perspectives (e.g., my unit of study is a watershed, not plot), debating points (e.g., should we be looking at management goals or endpoints), and jointly resolving points of contention (e.g., how to categorize impacts to one locale caused by activities in an adjacent locale). Our project team's primary roles were to facilitate the discussion and record results. The extensive modifications of the integration matrix that resulted from this meeting reflect its dynamic and productive interactions.

On the one hand, the results from the face-to-face meeting were dramatically different from the marginal refinements after each Delphi round. The face-to-face meeting also led to consensus, unlike the preceding efforts. Would a face-to-face meeting occurring in the absence of the Delphi build-up have proved so effective? And, could a Delphi approach, alone, have produced the substantial revisions and consensus of the face-to-face meeting? We do not have the luxury of testing these questions systematically through controlled research projects. We would hypothesize, however, that an effective methodological approach would consist of three general stages that combine knowledge elicitation and negotiation:

- an initial and separate, non-confrontational elicitation of information (in our case, a preliminary integration framework or its necessary dimensions and components) from each group;
- documenting and synthesizing each group's position(s), assuring that each group finds its synthesis accurate; and
- sharing syntheses with both groups, and using the syntheses as a basis for negotiating a consensus-based product.

These stages could be operationalized in a variety of ways, perhaps through a combination of methods (see Bañuls and Turoff 2011; Bloor et al. 2013; Landeta et al. 2011 as examples of multiple, or hybrid approaches that involve Delphi methods). For instance, the initial elicitations could be accomplished through a Delphi approach, nominal group process, or other methods. In a practicing (rather than academic) setting, it may be most efficient to "force" within-group consensus by structuring the initial elicitations around the goal of creating a tangible, though interim, product (e.g., a preliminary integration framework). Going through this initial process engages participants and starts them thinking about the issues at hand. Resulting interim products, together with a summary of the thought processes supporting them, give members of each group a glimpse into the other group's world view. The combination of initial consideration plus documentation may help participants articulate the sources of their discomfort or disagreement with the other group's proposition in later, negotiation stages. While it is possible that the facilitated negotiation stage could occur in various venues including online videoconference-style meetings, our success with face-to-face interaction would encourage us to use that process in the future. Working from tangible interim products to create a final, consensus product also may help to focus discussions.

Assuring that science conducted to assist practitioners achieves that goal is deceptively difficult to accomplish. Conducting scientific studies and reporting results is insufficient, even if that science explicitly is aimed at improving practice and especially when the studies produce different bits—and types—of information that do not automatically produce a coherent or comprehensive picture. The framework we sought to develop was intended to serve as an explicit foundation for integrating diverse scientific studies in a way that is useful for practitioners. Thus the framework became a product—a goal—that served to anchor and guide interactions. Our experiences indicate that creating such a framework exposes both implicit and explicit assumptions and subjects them to open discussion and debate. Our experiences also highlight some of the vagaries associated with participatory approaches. No matter the mechanism, getting affected parties together does not magically translate into consensus. The pathway from invitation to engagement to outcomes can be long, winding, and bumpy. Thus, while delineating and conveying one group's perspectives and opinions to the other may be necessary, it is an insufficient step. We propose adding direct, facilitated negotiation to the process of achieving an explicit output or goal. That process helped a set of ecological studies meet the dual objectives of being good science and being useful to the resource managers they are intended to support.

Acknowledgements We appreciate the assistance of many people who helped with this study and who participated in the discussions: Robert Addington, Beverly Collins, John Dilustro, Charles Garten, Thomas A. Greene, Anthony Krzysik, Robert Larimore, Maureen Mulligan, Joseph Prenger, and Peter Swiderek. Jeffrey Fehmi, Bill Goran, Hal Balbach, and Hugh Westbury provided support and helped us focus the effort as it unfolded. The project was funded by the Strategic Environmental Research and Development Program (SERDP) Ecosystem Management Project (SEMP), project CS 1114C, to Oak Ridge National Laboratory (ORNL). Oak Ridge National Laboratory is managed by the UT-Battelle, LLC, for the U.S. Department of Energy under contract DE-AC05-00OR22725.

References

Aber J et al (2000) Applying ecological principles to management of the U.S. national forests. Issues Ecol 6:1–20

Bañuls VA, Turoff M (2011) Scenario construction via Delphi and cross-impact analysis. Technol Forecast Soc Chang 78(9):1579–1602

Bloor M, Sampson H, Baker S, Dahlgren K (2013) Useful but no Oracle: reflections on the use of a Delphi Group in a multi-methods policy research study. Qual Res 0(0):1–14

Cairns J, McCormick PV, Niederlehner BR (1993) A proposed framework for developing indicators of ecosystem health. Hydrobiologia 236:1–44

Carnegie Commission on Science, Technology, and Government (1992) Enabling the future: linking science and technology to societal goals. A report. Available via http://carnegie.org/fileadmin/Media/Publications/PDF/Enabling%20the%20Future%20Linking%20Science%20and%20Technology%20to%20Societal%20Goals.pdf

Dale VH, Beyeler SC (2001) Challenges in the development and use of ecological indicators. Ecol Indic 1(1):3–10

Dale VH, Beyeler SC, Jackson B (2002) Understory vegetation indicators of anthropogenic disturbance in longleaf pine forests at Fort Benning, Georgia, USA. Ecol Indic 1(3):155–170

Dale, VH, Mulholland P, Olsen LM, Feminella J, Maloney K, White DC, Peacock A, Foster T (2004) Selecting a suite of ecological indicators for resource management. In: Kapustka LA, Gilbraith H,

Luxon M, Biddinger GR (eds) Landscape ecology and wildlife habitat evaluation: critical information for ecological risk assessment, land-use management activities and biodiversity enhancement practices, ASTM STP 11813. ASTM International, West Conshohocken, pp 3–17
Dale VH, Peacock AD, Garten CT Jr, Sobek E, Wolfe AK (2008) Selecting indicators of soil, microbial, and plant conditions to understand ecological changes in Georgia pine forests. Ecol Indic 8(6):818–827
Dalkey N, Helmer O (1962) An experimental application of the Delphi method to the use of experts. Manag Sci 9(3):458–467
Dilustro JJ et al (2002) Soil texture, land-use intensity, and vegetation of Fort Benning upland forest sites. J Torrey Bot Soc 129(4):289–297
Duncan LK, Dilustro JJ, Collins BS (2004) Avian response to forest management and military training activities at Ft. Benning, GA. Georgia J Sci 62(2):95–103
Garten CT Jr, Ashwood TL, Dale VH (2003) Effect of military training on indicators of soil quality at Fort Benning, Georgia. Ecol Indic 3:171–179
Jones S, Fischhoff B, Lach D (1999) Evaluating the usefulness of climate-change research for policy decisions. Clim Chang 43:581–599
Krzysik AJ et al (2005) Development of ecological indicator guilds for land management, final report. Available via: https://www.serdp-estcp.org/Program-Areas/Resource-Conservation-and-Climate-Change/Natural-Resources/Species-Ecology-and-Management/RC-1114
Landeta J, Barrutia J, Lertxundi A (2011) Hybrid Delphi: a methodology to facilitate contribution from experts in professional contexts. Technol Forecast Soc Chang 78:1629–1641
Linstone HA, Turoff M (eds) (1975) The Delphi method: techniques and applications. Addison-Wesley, Reading
Maloney KO, Mulholland PJ, Feminella JW (2005) Influence of catchment-scale military land use on stream physical and organic matter variables in small southeastern plains catchments (USA). Environ Manag 35(5):677–691
Millennium Ecosystem Assessment (MA) (2005) Ecosystems and human well-being: synthesis. Island Press, Washington, p 155
National Academies (2005) Sustainability at the National Academies: strengthening science-based decision making in developing countries. Available via: http://sites.nationalacademies.org/cs/groups/pgasite/documents/webpage/pga_057633.pdf
National Research Council (2005) Decision making for the environment: social and behavioral science research priorities. In: Brewer GD, Stern PC (eds) Panel on Social and Behavioral Science Research Priorities for Environmental Decision Making. Committee on the Human Dimensions of Global Change, Division of Behavioral and Social Sciences and Education. The National Academies Press, Washington
National Research Council (2009) Informing decisions in a changing climate. Panel on Strategies and Methods for Climate-Related Decision Support, Committee on the Human Dimensions of Global Change. Division of Behavioral and Social Sciences and Education. The National Academies Press, Washington
Peacock AD et al (2001) Soil microbial biomass and community composition along an anthropogenic disturbance gradient within a longleaf pine habitat. Ecol Indic 1(2):113–121
Rayner S et al (2001) Weather forecasts are for wimps: why water resource managers don't use climate forecasts, final report. Available via: http://www.isse.ucar.edu/water_conference/CD_files/Additional_Materials/Weather%20Forecasts%20are%20for%20Wimps.pdf
Reddy R et al (2003) Determination of indicators of ecological change. Final report. September 2004. Available via: https://www.serdp-estcp.org/Program-Areas/Resource-Conservation-and-Climate-Change/Natural-Resources/Species-Ecology-and-Management/RC-1114
Rowe G, Wright G (2001) Expert opinions in forecasting the role of the Delphi technique. In: Armstrong JS (ed) Principles of forecasting: a handbook of researchers and practitioners. Kluwer Academic Publishers, Boston
Sedlacko M, Pisano U, Berger G, Lepuschitz K (2013) Bridging the science-policy gap: development and reception of a joint research agenda on sustainable food consumption. Sustain Sci Pract

Policy 9(2):105–123, Available at http://sspp.proquest.com/static_content/vol9iss2/1204-019.sedlacko.pdf
Sigma XI, The Scientific Research Society (1993) Science and public policy: linking users and producers. The Carnegie Commission on Science, Technology and Government, Washington
Steel B et al (2000–2001) The role of scientists in the natural resource and environmental policy process: a comparison of Canadian and American Publics. J Environ Syst 28(2):133–155
Suter GW (2008) Ecological risk assessment in the United States Environmental Protection Agency: an historical overview. Integr Environ Assess Manag 4(3):285–289
Taylor JG, Ryder SD (2003) Use of the Delphi method in resolving complex water resources issues. J Am Water Resour Assoc 39(1):183–189

Part II
The *Application and Products* of Environmental Modeling with Stakeholders

Chapter 7
Fuzzy-Logic Cognitive Mapping: Introduction and Overview of the Method

Žiga Malek

Chapter Highlights

Approach: This chapter presents fuzzy-logic cognitive mapping as a participatory method for modeling social-ecological systems. The overview of the method is discussed through its evolution, suitability for application in environmental modeling, and examples in environmental research.

Participant Engagement: Fuzzy-logic cognitive mapping facilitates interactive stakeholder involvement throughout the entire modeling process, offering a possibility for discussion, negotiation, consensus building, and social learning.

Models/Outcomes: Final outcomes are semi-quantitative models representing most significant components and their relationships within a social-ecological system. These can be used to increase understanding of a particular issue, analyze possible changes to the system, and project future scenarios, or serve as a communication or perception analysis tool.

Challenges: The results of this semi-quantitative method can only be interpreted compared to other components in the model, and cannot be taken as absolute real-value outcomes. Also, to overcome possible biases, subjectivity, and difficulties in assigning weights to interconnections in the model, numerous stakeholders must be involved.

Ž. Malek (✉)
Environmental Geography Group, Faculty of Earth and Life Sciences, Vrije Universiteit Amsterdam, Amsterdam, The Netherlands

Risk, Policy and Vulnerability Program, International Institute for Applied Systems Analysis (IIASA), Laxenburg, Austria
e-mail: z.malek@vu.nl

S. Gray et al. (eds.), *Environmental Modeling with Stakeholders*,
DOI 10.1007/978-3-319-25053-3_7

7.1 Introduction

When addressing environmental issues, we often come across gaps in knowledge, limitations in data, and uncertainties in understanding. This lack of information to apply analytical models has resulted in a growing need for alternative models that build knowledge and generate solutions starting from stakeholders' perceptions (Hurtado 2010). Indeed, the use of both expert and local stakeholders' knowledge has been growing in environmental modeling (Özesmi and Özesmi 2004). Still, participatory environmental modeling can be challenging, due to several reasons. First, environmental issues involve numerous actors with different perspectives and conflicting interests, which are often characterized by intangible causes and key uncertainties (Mingers and Rosenhead 2004). Furthermore, participatory environmental modeling can be time-consuming, complicated, and incomprehensible to stakeholders otherwise unfamiliar with modeling. This is also demonstrated by the clear gap between the demands of researchers and their quantitative simulation models, and the stakeholders' needs for simple decision support tools (van Kouwen et al. 2008). Finally, information provided by experts can be unclear, incomplete, and subject to personal biases (Krueger et al. 2012; Page et al. 2012).

Fuzzy-logic cognitive mapping (FCM) can serve as a means for clear, transparent participatory modeling, and improving otherwise lengthy and complicated procedures of gathering expert-derived data. FCM is a soft-knowledge methodology, where a number of identified concepts and the relations between them are depicted in the form of a graph. This allows a semi-quantitative description of various interactions within a system, and enables visualizing causal reasoning. Thus, significant information about a system can be encoded and visualized, helping to reduce uncertainties and exceed limitations in knowledge and data (Hobbs et al. 2002). Since its emergence from cognitive mapping, FCM has evolved into a means to confront uncertainty, going beyond simple description and visualization to offer the potential to simulate complex systems. FCM can be used to develop simple qualitative models of a particular system, to quantify causal relationships of measureable physical variables, or to model abstract and complex theories. Therefore, it is suitable for modeling a variety of systems, and has been applied in several disciplines, including banking, information technology, and engineering.

This chapter introduces fuzzy-logic cognitive mapping as a participatory environmental modeling approach, able to overcome the unknowns in knowledge and data. The chapter describes FCM's evolution into a method for identifying key issues and modeling system structure. Moreover, being a transferable modeling approach applied to a number of environmental issues, its application in environmental modeling and decision support will be presented.

7.2 Description

Fuzzy-logic cognitive maps are semi-quantitative, mental models of a given system. They are graphical representations of the behavior of complex systems based on the modelers' expertise and understanding of a particular domain (Kosko 1986). Due to their ability to represent complex models, they are considered to be an alternative to other system modeling approaches. A fuzzy-logic cognitive map consists of numerous concepts representing components of a system, and the causal links between these concepts, describing how different concepts are thought to influence each other. The concepts and relationships are represented in graphical form, allowing easy visualization and control of the system.

The graphical representation of the system consists of a directed graph: nodes connected with edges in the form of arrows. The nodes represent concepts, which are the most significant components of the system as defined by the experts involved. They can be vague or abstract ideas—such as aesthetics or satisfaction—or measurable physical quantities—such as precipitation or percentage of vegetation cover (Özesmi and Özesmi 2003). Moreover, they can represent logical propositions (thresholds of a specific process), state variables (quality, abundance), rare events (weather extremes), and decisions (harvest quotas), and can thus describe the management of a particular system (Hobbs et al. 2002). Directed edges connecting the nodes represent the causal relationships between different concepts. The assigned weights of the edges quantify how the concept at the beginning of the edge influences the concept at the other end (McNeill and Thro 1994). Figure 7.1 shows a simplified FCM as a graph consisting of nodes and weighted connections. For modeling complex systems the edges can be defined as feedback loops, therefore FCM could be considered as a system dynamics approach (Kok 2009). The mathematical representation of a Fuzzy-logic Cognitive Map is represented by the numerical values of nodes and edges and the vector matrix calculation (van Vliet et al. 2010). The numerical values of nodes range between 0 and 1, and edges between −1 and 1, thus describing the value of concepts, and strength and direction of the causal relationships. A positive relationship means an increase (decrease) of the first concept leads to an increase (decrease) in the other, and the negative relationship means an increase (decrease) in the first concept leads to a decrease (increase) in the other. A weight with a value 0 indicates no relationship between the two concepts. Whereas numerical values for edges are defined by expert opinion or empirical data, the values of nodes can either be calculated by the model, or are fixed boundary conditions (Hobbs et al. 2002).

The mathematical representation in the form of a vector matrix enables a calculation of numerous structural metrics. The centrality of a variable shows the significance of the variable in a FCM (Özesmi and Özesmi 2003). It is the sum of the outdegree and the indegree of a variable. The outdegree indicates the effect of the variable on other variables and is the sum of all vectors exiting the variable. The indegree is the sum of all vectors entering the variable, and provides information on how the variable is affected by other variables. Another structural metric is complexity, which is the ratio between transmitter variables (variables that only affect other variables and are not affected by other variables) and receiver variables (variables only affected by other variables and not affecting other variables). Structural density is a metric comparing the number of all identified connections in a FCM to the number of all possible connections between variables. It demonstrates the perceived number of possible management options (Gray et al. 2014).

a

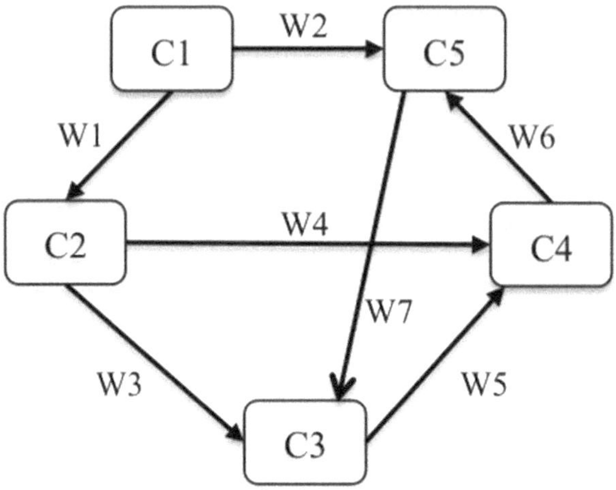

b

	C1	C2	C3	C4	C5
C1	0	W1	0	0	W2
C2	0	0	W3	W4	0
C3	0	0	0	W5	0
C4	0	0	0	0	W6
C5	0	0	W7	0	0

Fig. 7.1 An example of a fuzzy-logic cognitive map: (**a**) the graphical representation of FCM, (**b**) relationship vector matrix as a mathematical representation of FCM. Nodes Cx represent concepts with a state value. The weights and arrows of the causal relationships Wx represent the influence one concept has on another

Finally, the relative number of connections per concept indicates the degree of connectedness in the system (Özesmi and Özesmi 2004).

Constructing a fuzzy-logic cognitive map demands the involvement of experts or stakeholders, as the method takes advantage of their knowledge and experience. The first step in the construction process is the identification of appropriate concepts to include in the model by the involved stakeholders, either directly from their own expertise and experience or from a pre-populated list of potentially relevant concepts. Afterwards, the stakeholders identify causal relationships among these concepts and describe them as negative or positive, allowing the draft of the first versions of the directed graph. Finally, these relationships are estimated and ranked as numerical values, or defined as a set of linguistic variables than can later be transformed to values between −1 and 1 (from negatively very very strong, to positively very very strong). The constructed FCM can be modified later and altered at any time. The concepts and their relationships can be either constructed through interviews, group sessions, questionnaires, or document interpretation (Ülengin and Topçu 1997). The approach is semi-quantitative despite the possibility of quantifying the values of nodes and edges supported by empirical data. This is due to the fact that the quantification is performed based solely on the relationships between the concepts, and that the outcomes can only be compared within the system. Also, the concepts, their relationships, and assigned weights are subjective, purely reflecting the perspectives and opinions of participants. However, they are typically authoritative and not random due to the selection of modelers with substantial expertise and/or experiential knowledge in the domain of the system (Tan and Özesmi 2006). A single expert can construct a FCM. Involving a group of experts however improves the reliability of the FCM, as the approach allows a knowledge aggregation from multiple sources (Özesmi and Özesmi 2004; Stach et al. 2005; Gray et al. 2012). Constructing a FCM collectively through a group meeting, can minimize misunderstanding, improve knowledge, and accelerate the constructing process (Hobbs et al. 2002).

After a FCM has been developed, it can be used to model a system. In order to simulate the system, initial values of all concepts—indicated by the initial vector—must be defined. Using different initial vector states, FCM can be applied to analyze the outcome of different scenarios. Each model run begins with defining a scenario that is the initial situation of the system. The concepts then interact based on the relationship matrix defined in the FCM development phase. The FCM can be run for many steps, where in every step the values of the concepts are updated. The new value of each concept is calculated by summarizing the impact of all other concepts and by squashing the overall impact using a barrier function (Yesil et al. 2014). Squashing functions are used to limit the value of the updated concept usually in a range between 0 and 1. Besides squashing the overall impact of concepts, these activation functions also lead to nonlinearity of the model represented by an FCM.

7.3 Evolution of FCM

Fuzzy-logic cognitive mapping evolved from concept and cognitive mapping, studying structures, interconnections and causal relationships of a particular issue. The development of the approach went hand-in-hand with the recognized deficiency of other methods to deal with complex systems, related to model causal relationships and feedback loops. Originally meant to model social, economic, and political systems, FCM has developed into a tool for modeling systems and analyzing decisions and processes in a range of different scientific areas.

The concept of semi-quantitative representation of a system originates from graph theory, which was formulated by Euler in 1736 and has undergone significant development by mathematicians since (Biggs et al. 1999). The methods of graph theory are used to analyze the structural properties of a graph, such as a fuzzy-logic cognitive map (Özesmi and Özesmi 2003). With its help, we can understand the complexity of the modeled system, for example by describing the centrality and density of its graph, thus providing quantitative indicators to describe the graph's characteristics. The term cognitive mapping (CM) was originally coined in 1948 by Tolman and was later adopted by Axelrod to model decision-making processes using directed graphs where a set of nodes are connected by directed edges. The theory of directed graphs was developed in the 20th century for structural studies (e.g., in anthropology) (Hage and Harary 1983). The goal of CM was to construct a graphical representation of a person's system of belief about an issue on a conceptual, qualitative level (Axelrod 1976). Besides being a participatory technique, CM went beyond simple listing of ideas by organizing them into a map showing the interactions between these ideas, thus structuring them (Mendoza and Prabhu 2006). CM has been used to capture different mental models and deal with strategy making (Ackermann and Eden 2004). It can, however, result in large and incomprehensible models that are difficult to analyze, and do not take into account indirect variables and feedback loop (Jetter and Schweinfort 2011). To overcome these limitations, Kosko (1986) modified cognitive maps with fuzzy logic. Unlike Axelrod's CMs, where the relationships are described by the discrete values 0 or 1, the strengths of causal relationships in Kosko's FCM are fuzzy and range between −1 and 1. By being defined as positive or negative, they describe the direction and type of causality. Moreover, the causal links in FCM are cyclically dynamic, where the effect of altering one node affects other nodes in the path, allowing the study of feedback loops within a cycle (Mendoza and Prabhu 2006). The use of fuzzy-logic and consideration of causality proved to be useful when incorporating vague and qualitative knowledge (Rotmans 1998).

The flexibility of the tool is manifested through its diverse applications. In engineering, FCM has been widely used for controlling and supporting, as well as projecting future outcomes of changes to processes. It has been applied for supervision of manufacturing systems (Stylios and Groumpos 1999), human reliability in industrial facilities (Bertolini 2007), and safety evaluations (Enrique Peláez and Bowles 1996) to name a few. In information technology (IT), FCM has been applied mostly to support IT project management. Applications in IT range from evaluating investments in information systems (Irani et al. 2002), knowledge-based data mining of information from the internet (Hong and Han 2002), automatic generation of semantics for scientific e-documents (Zhuge and Luo 2006), modeling the success of IT projects (Rodriguez-Repiso et al. 2007), to predicting software reliability (Chytas et al. 2010). FCM has also been used in medicine (e.g., for aiding medical diagnosis) (Innocent and John 2004) and tumor grading (Papageorgiou et al. 2006). Due to its usefulness for decision-support, FCM has found its way into business, where it was used for analyzing market needs and potential, idea and concept development and evaluation, as steps in developing a new product (Jetter 2006). In social and political sciences FCM has served to model strategic issues and decision-support, and as well as complex social and economic systems. Among others, it has been applied to study political development (Taber 1991), and the influence of police presence on theft occurrence (Carvalho 2013). FCM has thus proved to be a suitable tool for system modeling and decision-making support, paving the way for its application in environmental issues.

7.4 Fuzzy-Logic Cognitive Mapping in the Environmental-Modeling Context

Having been previously applied in numerous applications throughout several disciplines, fuzzy-logic cognitive mapping has also been utilized in the research of social-ecological systems. First, the need for application of FCM in environmental modeling and decision-making was demonstrated by the growing demands for participatory approaches when addressing environmental issues. Secondly, FCM emerged as a means to incorporate expert knowledge when modeling complex systems facing uncertainties in data and knowledge. Moreover, due to their ability to study feedback loops and causal relationships, FCM has been applied numerous times to study the consequences of changes to the environment (e.g., under different conditions of the system or decisions) thus enabling scenario studies.

7.4.1 Facilitating Public Participation

Numerous actors such as experts, scientists, decision makers, and other stakeholders are involved when addressing environmental issues. In the past however, managing these issues has mostly been assigned to experts, with marginal involvement of local communities or a wider range of stakeholders. Due to the ineffectiveness of this traditional top-down approach to dealing with the challenges of sustainable environmental management, the need for participatory management has arisen (Mendoza and Prabhu 2006). However, instead of facilitating a "one-way" participation, more interactive processes providing opportunities for discussion, deliberation, negotiation, and consensus building have become acknowledged as a major component of dealing with environmental issues (Patel et al. 2007). Fuzzy-logic cognitive mapping enables the involvement of experts and the public throughout the entire modeling process. This is mainly due to its transparent development process, as the experts and other stakeholders need to be involved to construct the model from its beginning: identifying the concepts, their relationships and the strength of these relationships. This way, the acceptability of the final model is improved (Stach et al. 2005). Furthermore, FCM can promote cognitive learning (van Vliet et al. 2010).

FCM facilitates participation through supporting all four arguments for public participation: normative, substantive, instrumental, and social learning (von Korff 2007). Numerous experts and other stakeholders (e.g., members of the public) can be involved in FCM, which can lead to a wide variety of opinions on environmental issues. In this way, FCM reflects a broad spectrum of public and professional values, following normative reasons for participation. The substantive argument claims that the involvement of a wider group of people can offer detailed local information, uncover mistakes or lead to alternative clarifications (Beierle and Cayford 2002). Through FCM, we can thus gather more and improved information. For example, this is of high importance when identifying relationships in an ecosystem. Von Korff (2007) furthermore describes the instrumental argument, which states that participation can legitimize the final decision. Stakeholders can so consider the final social-ecological model as a more relevant and reliable. The concluding argument of public participation describes the idea of social learning. Active stakeholder involvement promotes learning for (and from) all involved parties, and leads to better knowledge about other participants' views and values. FCM has indeed proven to serve as a successful learning and communication tool (e.g., in the case of forest management) (Mendoza and Prabhu 2006). Besides supporting decision-making and problem solving, it also serves as a tool for learning and negotiation (Eden et al. 1992).

Despite its relatively comprehensible, visually guided method of model construction, FCM can still be difficult to understand by those not used to flow diagrams. Nevertheless, experts are normally familiar with conceptual models and thus are able to understand FCM (Vennix 1996; Pahl-Wostl and Hare 2004). There are several reasons for using FCM as a participatory tool ahead of other semi-quantitative methods (van Vliet et al. 2010). First, they are easy to teach and explain. Second, all stakeholders should be able to understand them, as the basics are comprehensible. Moreover, FCM has a high level of integration, which is particularly needed for complex environmental issues. Also, the construction of an FCM can be completed in a short time, leading to lower costs and less consumed time of the stakeholders (Kosko 1992). Lastly, the method results in a description of a system that provides sufficient complexity to explain a wide variety of environmental issues (Wainwright and Mulligan 2013).

7.4.2 Expert Knowledge to Deal with Data and Knowledge Limitations

Solving environmental issues is often especially difficult due to large uncertainties or incomplete data and knowledge. Scientific data might be unavailable for a particular case study, or its level of detail may be insufficient to perform an analysis. Usually, there are also limitations of a definite cost on obtaining information about the system—this can be demonstrated by the case of collecting field data, usually restricted both by time and money. Especially in environmental studies, there might, however, be abundant local knowledge of experts or the public, familiar with the environmental issue (e.g., a particular ecosystem).

It is still a big challenge to incorporate this local knowledge, as typical models have no means to achieve it (Özesmi and Özesmi 2004). FCM does not result merely in a list of ideas or perceptions: it results in a semi-quantitative model, based on people's knowledge. No hard data is needed to construct the model, however it can also be taken into account when identifying concepts and assigning weights to relationships. FCM can help to identify qualitative variables, and even relate them to quantitative variables. Additionally, important intangibles can be identified, thus leading to a possible incorporation of socio-economic driving forces and consequences. All this leads to more than improved data, as it can also be used as a means of communication between stakeholders and scientists, and more importantly, for support of further model development (van Vliet et al. 2010).

As mentioned before, fuzzy-logic cognitive maps (FCMs) can be constructed by using different approaches, from interviews and group discussions, to document analysis (Özesmi and Özesmi 2003). This flexibility allows the researcher to apply the suitable involvement approach, also depending on the availability and preferences of the stakeholders. During the process, FCMs can be reviewed, edited and compared to other fuzzy-logic cognitive maps. FCMs can be constructed jointly in a group session, or by aggregation of numerous individual FCMs, where the concepts have to be predefined. This goes for any number of maps, thus leading to integration also for a large set of individual FCMs (Jetter and Schweinfort 2011). The flexibility of the approach is also due to the fact that they can easily be edited or extended by adding new concepts and establishing new relationships between them at any time of the construction process. The graphic part can easily be translated into a vector matrix, containing all information about the relationships between the concepts. In this way, the slow, burdensome task of filling out a matrix can be avoided, requiring less time spent on the parameterization of the model (De Jouvenel 2000). Moreover, no particular modeling software is needed to calculate the output of a fuzzy-logic cognitive map, as traditional statistical and spreadsheet software can suffice.

A two-way communication with a group of experts can also serve as a possibility to validate FCMs. In spite of the difficulties of validation in terms of traditional historical data and statistical validation, it is possible to test the approach using other procedures. FCMs can be compared to other models representing the same or similar social-ecological issue. Secondly, experts can evaluate whether the model logic and its results are reasonable. Through evaluating whether the changes to concepts result in realistic changes in the results, a sensitivity analysis can be performed (Kok 2009). The symbolic representation of FCMs can also be matched to a real life issue (e.g., a decision process or workflow). Additionally, it is possible to test whether a model run over a certain number of iterations results in reasonable changes to the concepts. FCMs can be tested in several commonly employed validation procedures in order to provide information on the acceptance of a model (Rykiel 1996).

7.4.3 Simulating Changes to the System and Decision Outcomes

Social-ecological systems are dynamic, evolving through changes of their components or the relations between them. Feedback must be taken into account when updating the condition of the components, and propagation of causal relationships. Fuzzy-logic cognitive mapping allows feedback loops, thus being able to handle this complexity and help to understand short- and long-term dynamics (Kok 2009).

The basic concept of FCM is established as a semi-quantitative system dynamics approach involving feedback. A change to a single concept results in the changes to all concepts it directly affects. Through a network of causal relationships, other concepts are subsequently subject to change. Consequently, changes in other concepts can affect the concept initiating these changes (Kosko 1986). An example of a feedback loop is presented in Fig. 7.2, using an example of deforestation in a rural mountainous area. FCMs are therefore not static, and can be used to study changes to the social-ecological system, either in the form of changing conditions in the environment (e.g., precipitation), socio-economic driving forces (e.g., population, demand for resources), or management decisions (e.g., changes to harvest technique or quantity). Therefore, FCM offers much more than just an explanatory use, and can be applied to project and evaluate a possible future. This can be done either by identifying key future issues or guiding the exploration of plausible future scenarios (Probst and Gomez 1992; Ackermann and Eden 2004).

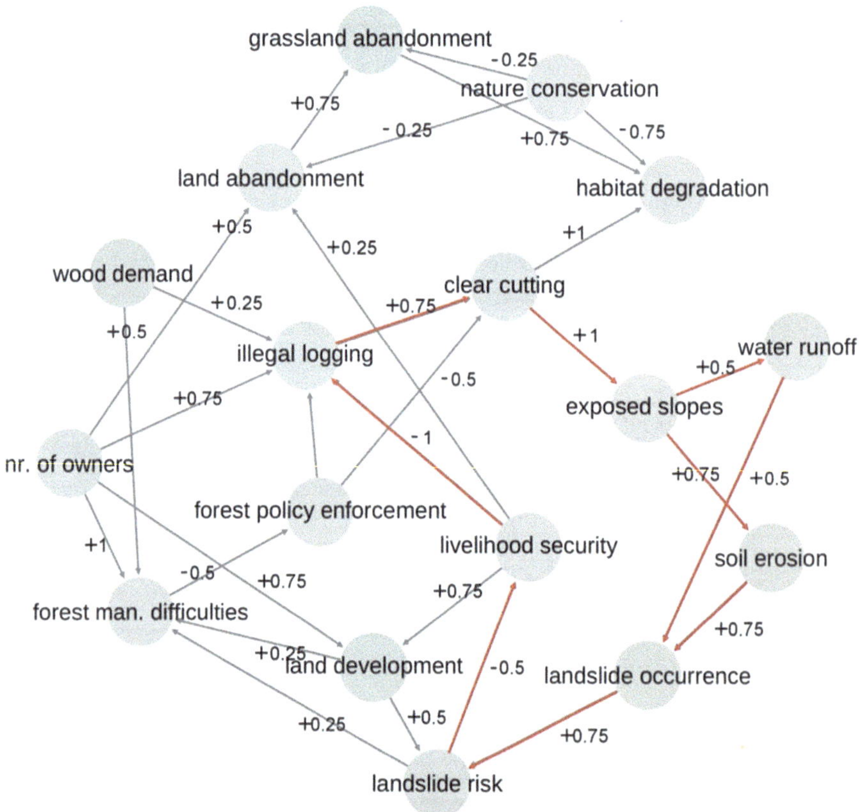

Fig. 7.2 Feedback loop on a simplified example of deforestation in a rural area in the Romanian Carpathians, modified from Malek et al. (2015). The feedback marked with red, depicts how a decrease in livelihood security triggered by the external fall of communism in the late 1980s affected illegal logging, with a consequent increase in landslide risk due to forest clear cutting on slopes. Finally, the increased landslide risk resulted as a negative feedback to livelihood security

Revealing key feedbacks is one of the strongest points of FCM, not only as it enables the study of "what if" scenarios, but also as it leads to aggregating information of simulation models to the level of decision-making (van Kouwen et al. 2008). People's difficulties in understanding complex systems are usually an obstacle when discussing the results of quantitative environmental simulation models. Among others, people tend to focus on a limited number of variables, ignoring feedbacks and overlooking the temporal dimension when thinking about future changes (Senge 1990; Acar and Druckenmiller 2006; Jetter and Schweinfort 2011). By involving both experts and stakeholders in constructing the model from the outset, the model and its simulation results are in their domain. Also, the simple and transparent construction method improves the trust of all the involved stakeholders in scenario analysis, impact assessments, and final decision evaluation and choice (Mendoza and Prabhu 2006).

7.4.4 *Examples of FCM in Environmental Research*

Fuzzy-logic cognitive mapping has been used to model how different social-ecological systems operate. One of the first applications of FCM in environmental sciences, were ecological models based on expert and stakeholders knowledge. Radomski and Goeman (1996) applied FCM to improve decision-making in sport fisheries by involving fisheries biologists and fisheries managers. Hobbs et al. (2002) used FCM to define management objectives for the complex ecosystem of Lake Erie. They involved numerous scientists, managers, and the public to construct a complex model of an ecosystems. The work of Özesmi and Özesmi includes applications of FCM to obtain the opinions of different stakeholders when establishing a national park, solving the conflict of population displacement due to construction of a hydro plant, facilitating participatory wetland management, comparing the perceptions of different stakeholder groups regarding a salt lake ecosystem, and identifying needs for ecosystem conservation strategies (Özesmi and Özesmi 2003, 2004; Tan and Özesmi 2006). Gras et al. (2009) have applied FCM to develop an individual-based predator model. In their model, the behavior of individual agents is modeled by FCM, allowing the evolution of the agent behavior. Kontogianni et al. (2012) analyzed the perception of Ukrainian stakeholders for risks to the marine environment of the Black Sea. They used FCM to generate a model for environment management based on laymen's perceptions of ecosystem resilience, risk management, and possible future scenarios. Gray et al. (2012) applied FCM on a case of fisheries management to integrate stakeholder knowledge. By collecting representations of stakeholders' mental models, they aimed to evaluate similarities and differences in their perceptions of the same social-ecological system. Another example is the application of FCM to support Long-Term Socio Ecological Research by Wildenberg et al. (2014). They applied FCM in five case studies to explore, analyze, and communicate the perceptions of key stakeholders affected by conservation management.

Fuzzy-logic cognitive mapping has proven to be successful in forest management, where decision-making is characterized by high uncertainty due to the variety of social-ecological interactions. Skov and Svenning (2003) combined FCM with GIS-based spatial operations to predict ground flora species richness. This approach, based on standard forestry maps together with expert knowledge, was shown to be an efficient way of predicting the spatial pattern of species diversity under a set of different forest management scenarios. Carvalho et al. (2006) have combined FCM with voronoi cellular automata to simulate the propagation of forest fires. They used rule-based FCM to model the dynamic behavior of individual forest fire cells. Mendoza and Prabhu (2006) used FCM for participatory forest management. They applied it to an Indonesian case study area, where a state-owned forest was subject to large pressures in the form of deforestation for urban and agricultural expansion and tourism. Ramsey et al. (2012) modeled forest response to deer control in New Zealand using a Bayesian algorithm to train their FCM. Their aim was to extract expert knowledge on the response of growth rates of tree seedlings to lower deer densities.

Besides forest management, fuzzy-logic cognitive mapping has been applied to management of other natural resources, such as water and soil, as well as to agriculture and conservation. Giordano et al. (2005) identified issues in water resources conflicts in southern Italy using FCM. Here, FCM was used to structure the issues of drought, and inform the involved participants about water management alternatives. Ramsey and Norbury (2009) developed a model to assist decision making on pest management relying on qualitative information. They used FCM to develop a complex food webs model and applied it to a dryland ecosystem in New Zealand. Papageorgiou et al. (2009) applied FCM for cotton yield management in precision farming. Their FCM modeled the behavior of cotton yield under a set of key factors in cotton crop production as recognized by the experts. Ortolani et al. (2010) analyzed the Belgian farmers' perceptions of agri-environmental measures with FCM. They extracted causal relationships between environmental management measures and numerous socio-economic and biophysical variables from questionnaires and interviews with farmers. Murungweni et al. (2011) applied FCM to analyze livelihood vulnerability in the Great Limpopo Conservation Area in Southern Africa. Their emphasis was on evaluating feedback mechanisms in social-ecological systems to reveal possible changes to a livelihood system under different scenarios. In the study by Văidianu et al. (2014), FCM was applied to examine stakeholders' perceptions for improving the management of the Danube Delta Biosphere Reserve in Romania. The key concepts were gathered for supporting future communication on sustainable development and biodiversity conservation of the area.

7.5 Limitations of Fuzzy-Logic Cognitive Mapping

Owing to its rather broad and semi-quantitative methodology, fuzzy-logic cognitive mapping is a flexible approach, transferable to basically any problem. On the other hand, its main weaknesses are also connected to the methodology. Whereas some drawbacks related to its subjective and qualitative nature can be improved easily by involving additional experts, other issues cannot be resolved and have to be taken into account when interpreting the results.

The minor drawbacks of FCM are related to graphical representation and stakeholder involvement. First, the simple and open structure of the symbolic representation of the system offers a suitable framework for participation of non-expert stakeholders. However, this vagueness can serve as a concern for more technical experts and researchers, especially as results gathered through participation can have a lower degree of accuracy (Mendoza and Prabhu 2006; Gray et al. 2012). Second, the stakeholders involved must have adequate knowledge of the topic under analysis to be able to estimate the strength of relationships between the concepts. This can result in the exclusion of some stakeholders, which could otherwise provide great value to the process (Kok 2009). Furthermore, all biases of involved stakeholders are encoded in the maps as well (Kosko 1992). Nevertheless, the subjectivity and robustness of the model generated through FCM can be improved by involving numerous experts and informed stakeholders.

The major limitations are related to the methodology of the approach itself. First, relationships in FCM are only semi-quantified, as they are not described by real-value parameter estimates (Craiger et al. 1996). This places a significant constraint on the interpretation of results. Second, despite providing information on the values of concepts after a defined number of iterations, these cannot be directly converted into time steps. The relationships between unrelated and often loosely defined concepts do not have a temporal dimension. This issue can be partially solved if the processes studied all operate at the same temporal scale (Kok 2009). Another weakness of the method lies in the process of defining the weights of the semi-quantified relationships. The methodology is based on gathering opinions and representing the belief system of numerous involved stakeholders. In this way, the final fuzzy-logic cognitive map represents an agreement between different opinions. Agreement can be achieved through combining multiple fuzzy-logic cognitive maps, or constructing one map in a workshop setting, both with drawbacks. Combining single FCMs into a final map can result in no identified relationships, if for example the model values are simply averaged. In a workshop setting on the other hand, issues may arise when involving oppositional stakeholders. The views of involved stakeholders can thus be diverse and potentially contradictory, with FCM not providing a solution for stakeholder disagreement. This limitation, however, can also be used to understand how different stakeholders view the important concepts and relationships of a system (Özesmi and Özesmi 2003). Furthermore, notable limitations also need to be taken into account when using FCMs for simulation of a system. Here, the role of initial vectors remains understudied. The starting values of concepts present a calibration step of a model, however, the values are based on subjective values—either on expert knowledge or agreement between stakeholders. Also, squashing functions limiting the value of a concept in every step of the simulation reduce the influence of the starting values of concepts. In this way the use of knowledge on initial concepts that are characterized with a higher degree of certainty or accuracy is limited (Motlagh et al. 2014). These methodological concerns constrain the application of FCM for simulation purposes.

7.6 Conclusion

Fuzzy-logic cognitive mapping has emerged as a useful participatory instrument for modeling complex social-ecological systems. Moreover, through successful applications in numerous domains, it has become established as an effective technique for decision-making support in environmental issues.

The rising demand for participatory approaches in environmental issues is well acknowledged, and FCM has proven to be an effective approach for discussing, planning, negotiating, and building consensus. FCM leads to a semi-quantitative, graphical representation of the behavior of a complex system. Its graphical and semi-quantitative nature allows effortless and quick visualization and control of the analyzed system. It can combine expertise from scientists, experts, decision-makers, and other stakeholders from different disciplines, thus including a broader spectrum of public and expert opinion. Therefore, it can help to bridge the gap between science and decision-making. It can offer more and improved information, available in a detail otherwise impossible to achieve with other techniques.

This is especially significant in environmental issues, where hard data is often unavailable or knowledge of a system is uncertain. Due to the complexity of social-ecological systems, it is sometimes difficult to identify important intangibles or establish relationships between socio-economic and physical variables. As the key stakeholders have been involved throughout the complete model construction process, FCM lead to more reliable and relevant model outcomes. Moreover, the method has proven to be a successful learning and communication tool, facilitating the exchange of ideas and opinions between different stakeholders. Due to its ability to model feedback loops, FCM has great potential in future environmental research, studying consequences of environmental changes or decisions regarding a particular social-ecological system.

References

Acar W, Druckenmiller D (2006) Endowing cognitive mapping with computational properties for strategic analysis. Futures 38:993–1009. doi:10.1016/j.futures.2005.12.018

Ackermann F, Eden C (2004) The practice of making strategy: a step-by-step guide. Sage, London

Axelrod RM (1976) Structure of decision: the cognitive maps of political elites, 1st Paperback Edition ed. Princeton University Press, Princeton

Beierle TC, Cayford J (2002) Democracy in practice: public participation in environmental decisions. Resources for the Future, Washington

Bertolini M (2007) Assessment of human reliability factors: a fuzzy cognitive maps approach. Int J Ind Ergon 37:405–413. doi:10.1016/j.ergon.2005.12.009

Biggs NL, Lloyd EK, Wilson RJ (1999) Graph theory 1736–1936. Oxford University Press, Oxford/New York

Carvalho JP (2013) On the semantics and the use of fuzzy cognitive maps and dynamic cognitive maps in social sciences. Fuzzy Sets Syst 214:6–19. doi:10.1016/j.fss.2011.12.009

Carvalho JP, Carola M, Tome JAB (2006) Forest fire modelling using rule-based fuzzy cognitive maps and voronoi based cellular automata. NAFIPS 2006:217–222. doi:10.1109/NAFIPS.2006.365411

Chytas P, Glykas M, Valiris G (2010) Software reliability modelling using fuzzy cognitive maps. In: Glykas M (ed) Fuzzy cognitive maps. Springer, Berlin/Heidelberg, pp 217–230

Craiger JP, Weiss RJ, Goodman DF, Butler AA (1996) Simulating organizational behavior with fuzzy cognitive maps. Int J Comput Intell Organ 1:120–123

De Jouvenel H (2000) A brief methodological guide to scenario building. Technol Forecast Soc Chang 65:37–48. doi:10.1016/S0040-1625(99)00123-7

Eden C, Ackermann F, Cropper S (1992) The analysis of cause maps. J Manag Stud 29:309–324. doi:10.1111/j.1467-6486.1992.tb00667.x

Enrique Peláez C, Bowles JB (1996) Using fuzzy cognitive maps as a system model for failure modes and effects analysis. Inf Sci 88:177–199. doi:10.1016/0020-0255(95)00161-1

Giordano R, Passarella G, Uricchio VF, Vurro M (2005) Fuzzy cognitive maps for issue identification in a water resources conflict resolution system. Phys Chem Earth Parts ABC 30:463–469. doi:10.1016/j.pce.2005.06.012

Gras R, Devaurs D, Wozniak A, Aspinall A (2009) An individual-based evolving predator-prey ecosystem simulation using a fuzzy cognitive map as the behavior model. Artif Life 15:423–463. doi:10.1162/artl.2009.Gras.012

Gray S, Chan A, Clark D, Jordan R (2012) Modeling the integration of stakeholder knowledge in social–ecological decision-making: benefits and limitations to knowledge diversity. Ecol Model 229:88–96. doi:10.1016/j.ecolmodel.2011.09.011

Gray SA, Zanre E, Gray SRJ (2014) Fuzzy cognitive maps as representations of mental models and group beliefs. In: Papageorgiou EI (ed) Fuzzy cognitive maps for applied sciences and engineering. Springer, Berlin/Heidelberg, pp 29–48

Hage P, Harary F (1983) Structural models in anthropology. Cambridge University Press, Cambridge

Hobbs BF, Ludsin SA, Knight RL et al (2002) Fuzzy cognitive mapping as a tool to define management objectives for complex ecosystems. Ecol Appl 12:1548–1565. doi:10.1890/1051-0761(2002)012[1548:FCMAAT]2.0.CO;2

Hong T, Han I (2002) Knowledge-based data mining of news information on the Internet using cognitive maps and neural networks. Expert Syst Appl 23:1–8. doi:10.1016/S0957-4174(02)00022-2

Hurtado SM (2010) Modeling of operative risk using fuzzy expert systems. In: Glykas M (ed) Fuzzy cognitive maps. Springer, Berlin/Heidelberg, pp 135–159

Innocent PR, John RI (2004) Computer aided fuzzy medical diagnosis. Inf Sci 162:81–104. doi:10.1016/j.ins.2004.03.003

Irani Z, Sharif A, Love PED, Kahraman C (2002) Applying concepts of fuzzy cognitive mapping to model: the IT/IS investment evaluation process. Int J Prod Econ 75:199–211. doi:10.1016/S0925-5273(01)00192-X

Jetter AJ (2006) Fuzzy cognitive maps for engineering and technology management: what works in practice? In: Technology Management for the Global Future 2006 PICMET 2006, pp 498–512

Jetter A, Schweinfort W (2011) Building scenarios with fuzzy cognitive maps: an exploratory study of solar energy. Futures 43:52–66. doi:10.1016/j.futures.2010.05.002

Kok K (2009) The potential of fuzzy cognitive maps for semi-quantitative scenario development, with an example from Brazil. Glob Environ Chang 19:122–133. doi:10.1016/j.gloenvcha.2008.08.003

Kontogianni A, Papageorgiou E, Salomatina L et al (2012) Risks for the Black Sea marine environment as perceived by Ukrainian stakeholders: a fuzzy cognitive mapping application. Ocean Coast Manag 62:34–42. doi:10.1016/j.ocecoaman.2012.03.006

Kosko B (1986) Fuzzy cognitive maps. Int J Man-Mach Stud 24:65–75. doi:10.1016/S0020-7373(86)80040-2

Kosko B (1992) Fuzzy associative memory systems. In: Kandel A (ed) Fuzzy expert systems. CRC press, Boca Raton

Krueger T, Page T, Hubacek K et al (2012) The role of expert opinion in environmental modelling. Environ Model Softw 36:4–18. doi:10.1016/j.envsoft.2012.01.011

Malek Ž, Boerboom LGJ, Glade T (2015) Future forest cover change scenarios with implications for landslide risk: an example from buzau subcarpathians, romania. Environ Manage 56: 1228–1243. doi:10.1007/s00267-015-0577-y

McNeill FM, Thro E (1994) Fuzzy logic: a practical approach. Academic Press Professional, Inc., San Diego

Mendoza GA, Prabhu R (2006) Participatory modeling and analysis for sustainable forest management: overview of soft system dynamics models and applications. Forest Policy Econ 9:179–196. doi:10.1016/j.forpol.2005.06.006

Mingers J, Rosenhead J (2004) Problem structuring methods in action. Eur J Oper Res 152:530–554. doi:10.1016/S0377-2217(03)00056-0

Motlagh O, Tang SH, Jafar FA, Khaksar W (2014) FCM relationship modeling for engineering systems. In: Papageorgiou EI (ed) Fuzzy cognitive maps for applied sciences and engineering. Springer, Berlin/Heidelberg, pp 49–64

Murungweni C, van Wijk MT, Andersson JA et al (2011) Application of Fuzzy Cognitive Mapping in Livelihood Vulnerability Analysis. Ecol Soc. doi:10.5751/ES-04393-160408

Ortolani L, McRoberts N, Dendoncker N, Rounsevell MD (2010) Analysis of farmers' concepts of environmental management measure: an application of cognitive maps and cluster analysis in pursuit of modelling agents' behaviour. In: Glykas M (ed) Fuzzy cognitive maps: advances in theory, methodologies, tools and applications. Springer, Berlin/Heidelberg, pp 363–381

Özesmi U, Özesmi S (2003) A participatory approach to ecosystem conservation: fuzzy cognitive maps and stakeholder group analysis in Uluabat Lake, Turkey. Environ Manag 31:0518–0531. doi:10.1007/s00267-002-2841-1

Özesmi U, Özesmi SL (2004) Ecological models based on people's knowledge: a multi-step fuzzy cognitive mapping approach. Ecol Model 176:43–64. doi:10.1016/j.ecolmodel.2003.10.027

Page T, Heathwaite AL, Thompson LJ et al (2012) Eliciting fuzzy distributions from experts for ranking conceptual risk model components. Environ Model Softw 36:19–34. doi:10.1016/j.envsoft.2011.03.001

Pahl-Wostl C, Hare M (2004) Processes of social learning in integrated resources management. J Community Appl Soc Psychol 14:193–206. doi:10.1002/casp.774

Papageorgiou EI, Spyridonos PP, Stylios CD et al (2006) Advanced soft computing diagnosis method for tumour grading. Artif Intell Med 36:59–70. doi:10.1016/j.artmed.2005.04.001

Papageorgiou EI, Markinos A, Gemptos T (2009) Application of fuzzy cognitive maps for cotton yield management in precision farming. Expert Syst Appl 36:12399–12413. doi:10.1016/j.eswa.2009.04.046

Patel M, Kok K, Rothman DS (2007) Participatory scenario construction in land use analysis: an insight into the experiences created by stakeholder involvement in the Northern Mediterranean. Land Use Policy 24:546–561. doi:10.1016/j.landusepol.2006.02.005

Probst GJB, Gomez P (1992) Thinking in networks to avoid pitfalls of managerial thinking. In: Maruyama M (ed) Context complex. Springer, New York, pp 91–108

Radomski PJ, Goeman TJ (1996) Decision making and modeling in freshwater sport-fisheries management. Fisheries 21:14–21. doi:10.1577/1548-8446(1996)021<0014:DMAMIF>2.0.CO;2

Ramsey DSL, Norbury GL (2009) Predicting the unexpected: using a qualitative model of a New Zealand dryland ecosystem to anticipate pest management outcomes. Aust Ecol 34:409–421. doi:10.1111/j.1442-9993.2009.01942.x

Ramsey DSL, Forsyth DM, Veltman CJ et al (2012) An approximate Bayesian algorithm for training fuzzy cognitive map models of forest responses to deer control in a New Zealand adaptive management experiment. Ecol Model 240:93–104. doi:10.1016/j.ecolmodel.2012.04.022

Rodriguez-Repiso L, Setchi R, Salmeron JL (2007) Modelling IT projects success with fuzzy cognitive maps. Expert Syst Appl 32:543–559. doi:10.1016/j.eswa.2006.01.032

Rotmans J (1998) Methods for IA: the challenges and opportunities ahead. Environ Model Assess 3:155–179. doi:10.1023/A:1019019024003

Rykiel EJ Jr (1996) Testing ecological models: the meaning of validation. Ecol Model 90:229–244. doi:10.1016/0304-3800(95)00152-2

Senge PM (1990) The fifth discipline: the art & practice of the learning organization. Doubleday, New York

Skov F, Svenning J-C (2003) Predicting plant species richness in a managed forest. For Ecol Manag 180:583–593. doi:10.1016/S0378-1127(02)00646-1

Stach W, Kurgan L, Pedrycz W, Reformat M (2005) Genetic learning of fuzzy cognitive maps. Fuzzy Sets Syst 153:371–401. doi:10.1016/j.fss.2005.01.009

Stylios CD, Groumpos PP (1999) A soft computing approach for modelling the supervisor of manufacturing systems. J Intell Robot Syst 26:389–403

Taber R (1991) Knowledge processing with fuzzy cognitive maps. Expert Syst Appl 2:83–87. doi:10.1016/0957-4174(91)90136-3

Tan CO, Özesmi U (2006) A generic shallow lake ecosystem model based on collective expert knowledge. Hydrobiologia 563:125–142. doi:10.1007/s10750-005-1397-5

Ülengin F, Topçu İ (1997) Cognitive map: KBDSS integration in transportation planning. J Oper Res Soc 48:1065–1075. doi:10.1057/palgrave.jors.2600444

Văidianu MN, Adamescu MC, Wildenberg M, Tetelea C (2014) Understanding public participation and perceptions of stakeholders for a better management in Danube Delta Biosphere Reserve (Romania). In: Papageorgiou EI (ed) Fuzzy cognitive maps for applied sciences and engineering. Springer, Berlin/Heidelberg, pp 355–374

Van Kouwen F, Schot PP, Wassen MJ (2008) A framework for linking advanced simulation models with interactive cognitive maps. Environ Model Softw 23:1133–1144. doi:10.1016/j.envsoft.2008.02.006

Van Vliet M, Kok K, Veldkamp T (2010) Linking stakeholders and modellers in scenario studies: the use of Fuzzy Cognitive Maps as a communication and learning tool. Futures 42:1–14. doi:10.1016/j.futures.2009.08.005

Vennix J (1996) Group model building: facilitating team learning using system dynamics, 1st edn. Wiley, Chichester/New York

Von Korff Y (2007) Re-focusing research and researchers in public participation. In: Proceedings of CAIWA, Basel

Wainwright J, Mulligan M (2013) Introduction. In: Wainwright J, Mulligan M (ed) Environmental modelling: finding simplicity in complexity. John Wiley & Sons, Chichester, pp 3–7

Wildenberg M, Bachhofer M, Isak KGQ, Skov F (2014) Use and evaluation of FCM as a tool for long term socio ecological research. In: Papageorgiou EI (ed) Fuzzy cognitive maps for applied sciences and engineering. Springer, Berlin/Heidelberg, pp 221–236

Yesil E, Urbas L, Demirsoy A (2014) FCM-GUI: a graphical user interface for big bang-big crunch learning of FCM. In: Papageorgiou EI (ed) Fuzzy cognitive maps for applied sciences and engineering. Springer, Berlin/Heidelberg, pp 177–198

Zhuge H, Luo X (2006) Automatic generation of document semantics for the e-science Knowledge Grid. J Syst Softw 79:969–983. doi:10.1016/j.jss.2005.08.022

Chapter 8
FCMs as a Common Base for Linking Participatory Products and Models

M. van Vliet, M. Flörke, C. Varela-Ortega, E.H. Çakmak, R. Khadra, P. Esteve, D. D'Agostino, H. Dudu, I. Bärlund, and K. Kok

8.1 Introduction

Social, economic, and biophysical systems are increasingly intertwined. The analysis of these complex systems necessitates multi-disciplinary approaches, including stakeholder participation (Website Mont Fleur 2011; Voinov and Bousquet 2010).

M. van Vliet (✉)
Public Administration and Policy Group, Wageningen University, P.O. Box 8130, 6700 EW Wageningen, The Netherlands

Water Consultancy Group, Movares B.V., Utrecht, The Netherlands

Soil Geography and Landscape Group, Wageningen University, P.O. Box 47, 6700 AA Wageningen, The Netherlands
e-mail: Mathijs.vliet@gmail.com

M. Flörke
Center for Environmental Systems Research, University of Kassel, Wilhelmshöher Allee 47, 34109 Kassel, Germany

C. Varela-Ortega • P. Esteve
Department of Agricultural Economics, Statistics and Management, Universidad Politécnica de Madrid, Madrid, Spain

E.H. Çakmak
Department of Economics, TED University, 06420 Ankara, Turkey

R. Khadra • D. D'Agostino
Mediterranean Agronomic Institute of Bari, Via Ceglie 9, 70010 Valenzano, Italy

H. Dudu
Institute for Prospective and Technological Studies, EC-JRC, Seville, Spain

I. Bärlund
Department Aquatic Ecosystem Analysis (ASAM), Helmholtz Centre for Environmental Research – UFZ, Brückstrasse. 3a, 39114 Magdeburg, Germany

K. Kok
Soil Geography and Landscape Group, Wageningen University, P.O. Box 47, 6700 AA Wageningen, The Netherlands

S. Gray et al. (eds.), *Environmental Modeling with Stakeholders*,
DOI 10.1007/978-3-319-25053-3_8

Local knowledge should, in particular, be incorporated when data is lacking, or when actions by stakeholders have a large influence on the system (Özesmi and Özesmi 2003). Furthermore, the involvement of stakeholders facilitates the uptake of research results in practice and contributes to the learning process of both stakeholders and scientists (Sterman 2004; Vennix 1999). Moreover, it can further increase the relevance and legitimacy of the research.

Recently, stakeholder participation has been increasingly used in more formalised efforts that include some type of modeling. There are multiple approaches to develop models in cooperation with stakeholders (Bousquet and Voinov 2010), often relying on the use of existing models. Incorporating the output of stakeholder consultations in quantitative models is, however, difficult (Cash et al. 2006; Martínez-Santos et al. 2010) as aspects of the stakeholder-generated models are either difficult to quantify by nature, not sufficiently defined, or the expected magnitudes of change are not specified. Consequently, stakeholders should not *only* be used to provide input for the mathematical model nor should it be their only task.

Additionally, most mathematical models are highly technical, which restricts the stakeholders' ability to fully understand the model's architecture or correctly interpret results. It is, therefore, crucial that stakeholders and other end-users understand the model, including how and when the model can be used (Refsgaard et al. 2005). Likewise, it is often difficult for mathematical, quantitative modelers to interpret stakeholder-generated results (Verburg et al. 2006). Thus, there is a need for methods and tools that can help to create a shared language and a common base for comparison of stakeholder products and mathematical models. Such tools should make assumptions explicit and provide a clear system description. Additionally, they should be able to integrate social, economic, and biophysical issues. Conceptual models have been frequently used to elicit knowledge from scientists from different fields (both social and natural sciences (e.g. Heemskerk et al. 2003)) and they have been used by both experts and stakeholders in participatory workshops (e.g. Hare et al. 2003; Simon and Etienne 2010; Magnuszewski et al. 2005). As such, this type of simpler (conceptual) models is potentially useful to fulfil the task at hand.

Fuzzy Cognitive Maps (FCMs; (Kosko 1986)) are such a conceptual model that can be used to develop system descriptions in a workshop setting. It also facilitates the analysis of the dynamic behavior of the system. FCMs have been applied successfully in a wide variety of cases, in which both social and biophysical aspects are often combined (e.g. Cole and Persichitte 2000; Kok 2009; Özesmi and Özesmi 2003; Gray et al. 2013). Importantly, FCMs have also been proposed in combination with other foresight methods (Jetter and Schweinfort 2011; Jetter and Kok 2014), thus enhancing the link between stakeholder-based qualitative scenarios and model-based quantitative scenarios. The latter is related to the often employed Story-And-Simulation approach (Alcamo 2008) in scenario development. Van Vliet et al. (2010) showed the potential for the use of FCMs to address the weak link between Story (stakeholder-based storylines) and Simulation (model-driven quantitative

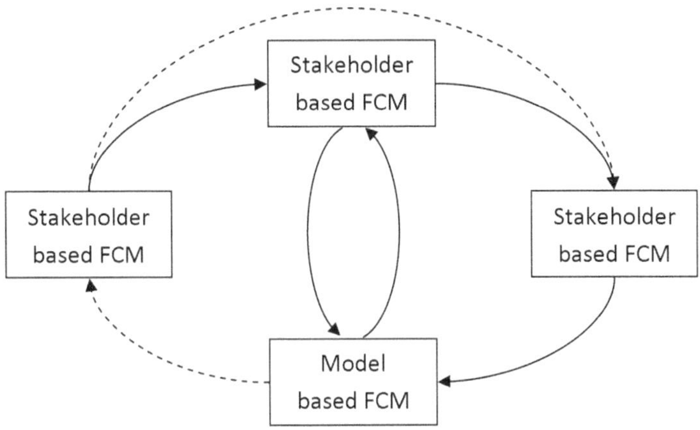

Fig. 8.1 Usage of FCMs in the communication between stakeholders and modellers. Based on van Vliet et al. (2010)

explorations). This potential has been partly substantiated in subsequent publications, where we have demonstrated a number of added values from using FCMs in stakeholder workshops (Kok and van Vliet 2011; van Vliet et al. 2012).

Yet, these publications do not address the most important assumed added value of FCMs, namely the potential to facilitate the translation of stakeholder-derived output to a mathematical model. As said, this potential has been hypothesised previously (Kok and van Delden 2009; Kok 2009; van Vliet et al. 2010) but remains untested.

Figure 8.1 illustrates the different approaches to use FCMs to facilitate communication between stakeholders and mathematical modelers. Other researchers have predominantly explored the potential of using stakeholder-based FCMs to inform models (top half of Fig. 8.1). In this chapter, we explore the potential of generating two FCMs that would allow a direct comparison between the system descriptions of stakeholders and of modelers.

8.1.1 Objectives

The main objective of this chapter is to use FCMs to compare stakeholders' system descriptions to a mathematical model. To study this we will:

- develop and analyze a Fuzzy Cognitive Map for the Mediterranean region based on the stakeholder-driven information from three case studies
- develop and analyze a Fuzzy Cognitive Map for the Mediterranean region based on the modeling architecture from a mathematical model
- compare both maps to identify crucial differences and similarities

8.2 Background

The following sections provide background by briefly introducing the SCENES project and the two main tools relevant to this chapter (FCM and WaterGAP), accompanied by an outline of how FCMs were developed by stakeholders and modelers.

8.2.1 SCENES

This study was part of SCENES, a 4-year EC FP6 research project, to develop and analyze a set of comprehensive scenarios of Europe's freshwater futures until 2050 (Kämäri et al. 2008). SCENES was a multi-scale project with scenarios being developed at pan-European scale, for ten river basins, and at an intermediate regional scale. One of the main goals of SCENES was to improve the Story-And-Simulation approach.

Within SCENES a novel qualitative, participatory scenario development methodology was designed and applied in the majority of the ten-river-basin-scale case studies (so-called Pilot Areas).

In this chapter, we focus on FCMs that were developed in three Mediterranean Pilot Areas, situated in the Guadiana (south-west Spain), Candelaro (south-east Italy), and Seyhan (south Turkey) river basins. These Pilot Areas all have large areas of irrigated agriculture but are different in, for instance, their type of irrigation systems, water users, and amount of natural areas.

The global water model WaterGAP (Water—Global Assessment and Prognosis) (Alcamo et al. 2003; Döll et al. 2003; Verzano 2009) was used to develop the quantitative pan-European scenarios. Data for the model was (partly) based on the quantification of stakeholder-generated scenarios developed at the pan-European level.

Because FCMs were developed at the Pilot Area scale and the WaterGAP model is on the pan-European scale, there was a need to upscale Pilot Area results and downscale WaterGAP results to the Mediterranean scale at which they were assumed to be valid. The three FCMs from the Pilot Areas were aggregated into one combined FCM.

8.2.2 Fuzzy Cognitive Maps

FCMs were first described in 1986 by Bart Kosko, who proposed them as a means to make qualitative cognitive maps computable. In contrast to other cognitive mapping approaches, FCMs enable an analysis of the dynamic properties of the system they represent and the identification of possible future system states and instabilities. After a slow start, between 2000 and 2015 the number of Fuzzy Cognitive Map-related publications saw a tenfold increase. Additionally, a growing number of publications are devoted to applications of FCMs across a variety of fields, such as business planning, medicine, and environmental management.

FCMs consist of concepts representing key factors (variables) of the system, joined by directional edges representing causal relationships (connections) between concepts. Each connection is assigned a weight that represents the strength of the causal relationship. FCMs can be represented graphically, in the form of boxes and arrows, and mathematically in the form of a vector and a matrix. The matrix consists of the weights of the connections. The vector shows the weight of the variables in the system.

Some factors are given a value, which forms the starting vector. The next state of the system can then be calculated by a matrix vector multiplication. Multiplication can be repeated as often as desired. If iterated, the system shows whether the value of a factor will increase, decrease, or stabilize.

Although the mathematical foundations underlying FCMs are simple, there is disagreement in literature on how to best interpret the quasi-dynamic behavior of the system by analyzing the changes in the values of the vector. A paper offering an overview of the polemic and new guidelines for interpretation is underway (Helfgott et al. in review). In anticipation, we follow methodological instructions as provided by Kok (2009) and Özesmi and Özesmi (2003).

8.2.3 WaterGAP Model

WaterGAP computes both water availability and water uses; thus, it computes the impact of climate change and other important driving forces on future water resources. The version of the model applied in this study, WaterGAP3, uses a five-by-five arc minutes grid (longitude and latitude, approximately 6×9 km in Central Europe). WaterGAP consists of two main components: a Global Hydrology Model to simulate the terrestrial water cycle, and a Global Water Use Model (Flörke and Alcamo 2004; Flörke et al. 2013) to estimate water withdrawals and water consumption in five water-use sectors. The aim of the Global Hydrology Model is to simulate the characteristic macro-scale behavior of the terrestrial water cycle in order to estimate water availability. Herein, water availability is defined as the total river discharge, which is the sum of surface runoff and groundwater recharge. The upstream/downstream relationship among the grid cells is defined by a global drainage direction map (DDM5) which indicates the drainage direction of surface water (Lehner et al. 2008). In a standard model run, river discharges in close to 20,000 river basins in Europe are simulated. The effect of a changing climate on runoff is taken into account via the impacts of temperature and precipitation on the vertical water balance.

River discharge is affected by water withdrawals and return flows. In WaterGAP, natural cell discharge is therefore reduced by the consumptive water use in a grid cell as calculated by the Global Water Use Model. This model consists of several modules that calculate water withdrawals and water consumption in the domestic, industry, irrigation, thermal electricity production, and livestock sectors. In this context, water withdrawals depict the total amount of water used in each sector while the consumptive water use indicates the part of withdrawn water that is consumed by industrial processes, human needs, or is lost by evapotranspiration. For most water-use sectors—except irrigation—only a small amount of water is consumed; whereas most of the water withdrawn is returned—probably with reduced quality or after being heated—to the environment for subsequent use. WaterGAP simulates water use for the agricultural and electricity-production sectors on a grid scale, but for the domestic and manufacturing sectors on a country scale. These country-scale estimates are downscaled to the grid size within the respective countries using demographic data. Grid cell outputs are then summed up to the river basin scale.

8.3 Methods

8.3.1 Development of Stakeholder-Based FCM

The development of the stakeholder-based FCM (FCM-SH) started at the Pilot Area level. Stakeholders were selected after a detailed stakeholder mapping exercise was conducted in all Pilot Areas so the participants reflected different types of views and levels of expertise in the water sector. Stakeholders included—among others—government officials, water authority personnel, farmers' associations, individual irrigators, and nature conservation groups. The broad range of stakeholders in the different Pilot Areas permitted the FCM-SH to represent the complexity and richness of the water and human systems. In two successive workshops, stakeholders developed an FCM that represented their perception of the current (water) system in their Pilot Area. This was done in two-to-three small groups, of six-to-ten people each. First, participants were asked to write down the most important aspects concerning the water system. Answers were clustered to form the FCM's variables. Stakeholders then established the relationships between variables and assigned polarity and weights to those relationships. In a second workshop, results from the first round were refined. After the second workshop, one combined FCM was developed for each Pilot Area (see Kämäri 2008; Cakmak et al. 2013; Khadra et al. 2011; Varela-Ortega et al. 2011).

These three Pilot Area FCMs were further aggregated into the FCM-SH presented here. There are a number of distinctly different methods to aggregate FCMs, ranging from a straightforward calculation of adding matrices without further involving stakeholders to organizing separate workshops in which stakeholders are invited to construct an aggregation version. Here, we opted to organize a workshop that involved experts at the Mediterranean level, rather than stakeholders from the three Pilot Areas, realizing that the interpretational step was not checked by the original, basin-scale, stakeholders. The aggregation started with merging identical variables. In all Pilot Areas, issues like water shortage, water demand, water price, and water quality were addressed. Most variables were addressed in at least two out of three Pilot Areas (see Table 8.1). Many Pilot Area variables related to similar issues but used different words, for instance water quality and water pollution. Other issues

Table 8.1 Overview of related variables in the three Pilot Area FCMs and how they are represented in the combined FCM-SH

FCM-SH	Seyhan (Turkey)	Guadiana (Spain)	Candelaro (Italy)
Present in three Pilot Areas			
Environmental policies	Sustainable water management	Common Agricultural Policies environmental requirements Protection of water courses	Water Framework Directive
Water quality	Water pollution	Water quality	Water quality
Good ecosystem condition	Soil degradation	Wetland conservation Biodiversity protection	Alteration of environment and of territory
Sustainable water management	Sustainable water management	Wetland conservation Culture of water use Water demand management	Sustainable rural development model Environmental awareness
Climate impact	Impacts of climate change	Drought impact	Climate and drought
Water saving methods	Use of water-saving methods	Improvement of water technologies	Technologic innovation Use of non conventional water
Groundwater exploitation	Use of groundwater	Imbalance demand/ supply	Groundwater exploitation
Population	Impact of increasing urbanization	Stabilization of rural population	Socio-economic dynamics
Water demand	Water demand	Imbalance demand/ supply	Water demand
Water price	Irrigation water price	Water price	Water cost
Water availability	Water supply Irrigation water use	Imbalance demand/ supply	Water scarcity
Irrigation efficiency	Irrigation efficiency	Water use efficiency	Technical assistance and efficiency
Agricultural support policies	Agricultural support policies	Common Agricultural Policy payments	Common Agricultural Policy
Infrastructure	Water delivery losses Irrigation infrastructure	Hydraulic infrastructure	Lack of infrastructure
Present in two Pilot Areas			
Rural development policies		Rural development programs	Sustainable rural development model Financial resources
Farm income		Farm income Socio-economic development	Socio-economic dynamics

(continued)

Table 8.1 (continued)

FCM-SH	Seyhan (Turkey)	Guadiana (Spain)	Candelaro (Italy)
Governance		Political will Policy enforcement Institutional coordination	Economic planning Local management policies Control and vigilance of territory
Present in one Pilot Area			
Water allotments		Water allotments	
Intensification of agriculture		Intensification of agriculture	

were represented in detail in one FCM and by just one variable in another. The experts merged the variables until 19 remained. See Table 8.1 for an overview of concepts in the final (FCM-SH) and the way they were represented in the three Pilot Area FCMs.

A similar procedure was followed to assign values to connections. A final step was the calibration of the (FCM-SH) to best represent the perception of the stakeholders and get a stable quasi-dynamic output, which facilitates the analysis of FCM output. As no squashing function was used, often the iteration results are first either chaotic, all values became zero, or all values became increasing larger. Because stable output is easier to interpret, the strengths of some connections were slightly adjusted. Often there are one or two connections that trigger a more stable output, often related to direct feedback loops (C1 impacting C2 and vice versa). This provides the modeler with additional insights into the workings of the system and often makes it possible to get a stable output, such as shown in Fig. 8.4 (Sect. 8.4.2). This output was discussed with the experts to make sure that it matched with their perceptions of the system. Where necessary, small changes in the weight of connections were made to better reflect the experts' and stakeholders' opinions.

The resulting final FCM-SH can be regarded as a product that represents the perspectives of stakeholders on a Mediterranean-wide system description. Also, it can be assumed that the product is representative of basins across the Mediterranean, with an emphasis on rather intensively managed, agriculturally dominated areas.

8.3.2 Development of a Model-Based FCM

The FCM-WG covers all aspects that WaterGAP deals with, yet it does not reflect all aspects in their full complexity. WaterGAP is driven by data generated by several other models, yet the FCM-WG only represents those components that are part of the WaterGAP model itself. For instance, results from the land-use model LandSHIFT (Schaldach and Koch 2009) are used in WaterGAP to establish the relationship between the number of livestock and the area required for crop production. As this relationship is not part of WaterGAP's main component, it was not represented in the FCM-WG, even though it is reflected in the WaterGAP output. As such, the FCM-WG does not reflect the diversity of issues that relate to WaterGAP, but rather WaterGAP itself. Strengths of connections were assigned to fit the Mediterranean region. Connections in the FCM were estimated by a small team of WaterGAP modelers, based on their experience with and their knowledge of the mathematical functions present in the model. Some of the WaterGAP parameters are differentiated according to regions.[1] From the modeling perspective, quantitative information from two regions—namely Southern Europe (Spain and Italy) and Western Asia (Turkey)—was considered. The connection strengths in the FCM therefore are an average for the two regions. Note that an FCM-WG for other regions would have had other weights and the focus of the FCM might also have been on other parts of the model if, for instance, the manufacturing sector is the largest water user in that other region.

8.3.3 Comparison of Both FCMs

There are many ways in which FCMs can be compared, ranging from a comparison of the list of variables that are included to a comparison of the iterative quasi-dynamic output. Below, an overview is given emphasising the two main ways in which the FCMs were compared. First, the system configuration of both FCMs was compared; second, the quasi-dynamic behaviors were assessed.

8.3.3.1 System Configuration

The list of variables taken into account in both FCMs can be used to compare the FCM system configurations. Which issues are taken into account in both FCMs versus those considered only in one? The next step is to compare pure transmitters—"drivers"—and pure receivers. Subsequently, several indicators can be calculated, including: centrality, density, number of receivers and transmitters, number of variables, and number of connections that shed light on the complexity of the graphic representation of the FCM.

[1] Within SCENES WaterGAP used seven regions following the composition of macro geographical (continental) regions and geographical sub-regions used by the UN (e.g., United Nations Statistics Division, http://unstats.un.org/unsd/methods/m49/m49regin.htm).

8.3.3.2 Quasi-Dynamic System Behavior

By running FCMs in which small changes are made to certain relationships, the effect of these changes on the rest of the system can be analyzed. This gives additional insights into the system depicted by the FCM. Four separate modifications have been made to each FCM. The effects of each of these changes on similar variables in both FCMs were studied. One modification changed the value of the starting vector of the pure transmitter "climate change"/"climate impact." Three other modifications were made in the connection strength between a pure transmitter and one variable. These changes were chosen because they could relatively easily be incorporated in both FCMs. As pure transmitters are never affected by changes in the system, they are ideal for use to manipulate the system. All modifications were of similar strength (changing the values from 0.5 to 0.9). For each modification, both FCMs were iterated 200 times, which in all cases was sufficient to stabilize the values. To mimic a decrease in water availability, the strength of the connection from "drought impact" on "water availability" (in the FCM-SH) and from "climate warming" on "fresh water resources" (in FCM-WG) was changed. An increase in irrigation efficiency was simulated by changing the strength of the connection from "agricultural support policies" on "irrigation efficiency" (FCM-SH) and the starting value of "project efficiency" (FCM-WG). To mimic a decrease in intensification of agriculture, the strength of the connections from "environmental policies" on "intensification of agriculture" (FCM-SH) and of "irrigated crop production" on "area required for crop production" (FCM-WG) were changed.

8.4 Results

8.4.1 System Configuration

The comparison of the system configurations of both FCMs shows that they are similar in a number of important aspects (see Fig. 8.2 for FCM-SH and Fig. 8.3 for FCM-WG). First, there are seven variables that are very similar in both FCMs (only small differences in wording, e.g., irrigation efficiency versus project efficiency, climate impact versus climate change, and water demand versus total water withdrawals), and one that is exactly the same (water quality; see Table 8.2).

Second, in both FCMs the "water availability"/"freshwater resources" form a connection between the water quantity and water quality. Third, in the FCM-SH, "water availability" has the highest centrality, similar to "fresh water resources" in the FCM-WG. This clearly illustrates the importance of water quantity issues in the Mediterranean. In both FCMs, "total water withdrawals"/"water demand" have a high centrality.

Finally, in both FCMs, the variables with most receiving connections are "water demand"/"total withdrawals." "Water availability" has most transmitting connections

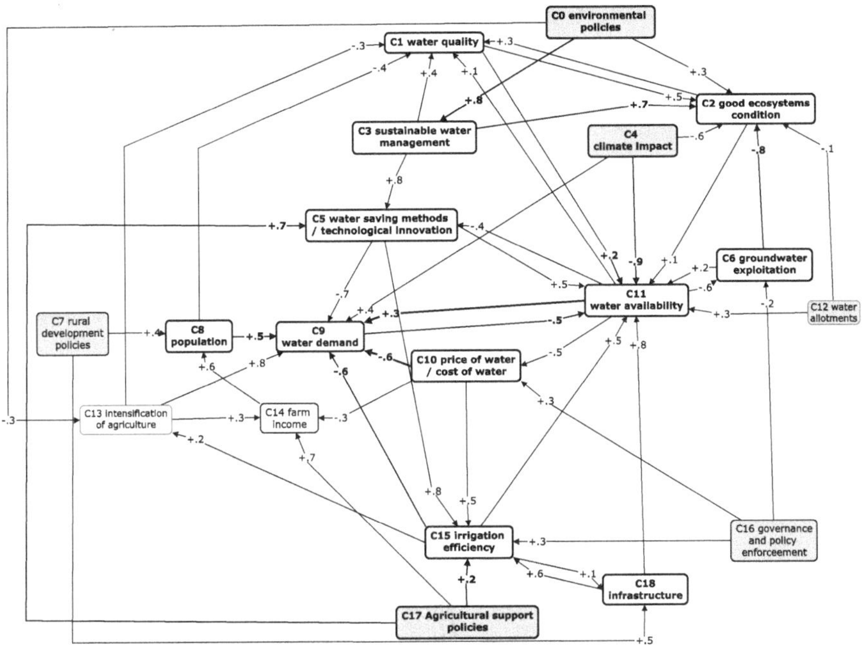

Fig. 8.2 Graphical representation of the stakeholder based FCM. *Grey* variables are pure transmitters that drive themselves and thus the system. *Bold* variables and thicker connections were represented in two or three Pilot Areas

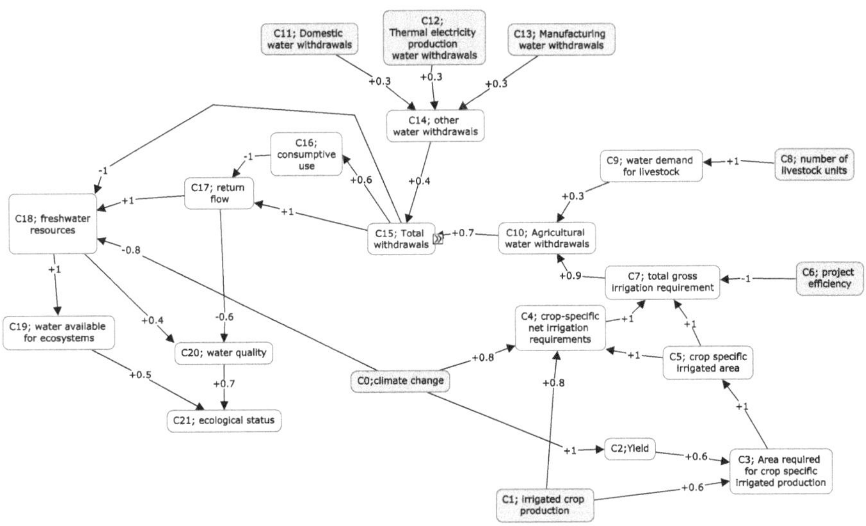

Fig. 8.3 Graphical representation of the WaterGAP based FCM. *Grey* variables are pure transmitters that drive themselves and thus the system

Table 8.2 Comparison of the stakeholder based FCM and WaterGAP based FCM

	Stakeholder based FCM	WaterGAP based FCM
Variables present in both FCMs	Water quality Irrigation efficiency Water availability Ecosystems condition Climate impact Water demand Population	Water quality Project efficiency Fresh water resources Ecological status Climate change Total withdrawals Domestic water withdrawals
Variables present in only one FCM	Intensification of agriculture Environmental policies Agricultural support policies Governance and policy enforcement Rural development policies Ground water exploitation Water price/cost of water Sustainable water management Effectiveness of control Water saving methods Water allotments Farm income	Crop specific irrigated area Number of livestock Water demand for livestock Agricultural water withdrawals Other water withdrawals Irrigated crop production Area required for crop specific irrigated production Crop specific net irrigation requirement Total gross irrigation requirement Manufacturing water withdrawals Thermal electricity production water withdrawals Return flow Consumptive use Water available for ecosystems Yield
Number of variables	19	22
Number of connections	49	29
Density (C/V^2)	0.14	0.06
Average value per connection	0.46	0.74
# pure transmitters	6	7
# pure receivers	0	1 (ecological status)
Highest centrality (*#connections and absolute value of connections*)	Water availability (*14 connections; abs value 5.9*)	Fresh water resources (*5 connections; abs value 4.2*)
Average centrality (*out + ingoing connections*)	5.16 connections (*abs value 2.37*)	2.64 connections (*abs value 1.96*)
Most receiving connections (*# of connections*)	Water availability (*9 connections*) Water demand (*7 connections*)	Crop specific net irrigation requirements, total gross irrigation requirement, other water withdrawals, fresh water resources (*all: 3 connections*)
Most transmitting connections (*# of connections*)	Water availability (*5 connections*)	Climate change, total water withdrawals (*both 3 connections*)

in FCM-SH, while climate change and total water withdrawals have most transmitting connections in the FCM-WG.

There are also differences: FCM-SH gives more weight to social aspects and policies compared to FCM-WG. The FCM-SH is focused on water quantity and irrigation, but water quality and social issues also play a strong role. All parts of the FCM are related to each other. The FCM-SH is more complex and denser, with many feedback loops, some of them consisting of loops between two variables. It has many more connections (49 versus 29) and fewer variables (19 versus 22) and, therefore, a density that is more than double that of the FCM-WG (0.14 versus 0.06; see Table 8.2). The most central variable in FCM-SH is water availability.

The FCM-WG is less complex and less dense and has one pure receiver, whereas the FCM-SH has no pure receivers. It has a strong focus on agriculture—specifically on irrigation—but also includes other water withdrawals such as for domestic, thermal electricity production and for the manufacturing sectors, some of which are missing in the FCM-SH. The FCM-WG focuses on water quantity, but also shows implications on water quality.

The starting values of the pure transmitters in the FCM-SH depended on the number of Pilot Area FCMs in which they were addressed; this served as an indicator of their importance. Those present in all three Pilot Areas (environmental policies, climate impact, and agricultural support policies) were assigned a starting value of 1, those in two (rural development policies and governance and policy enforcement) a starting value of 0.6, and those in one (water allotments) 0.3. All other variables got a starting value of zero.

Because the FCM-WG has no feedbacks, the WaterGAP modelers gave all pure transmitters the same starting value and deviated their effects on the rest of the system by giving different values to the relationships from these pure transmitters to other concepts.

8.4.2 *Quasi-Dynamic System Behavior*

Table 8.3 shows the relationships that were modified regarding the four modifications as explained in Sect. 8.3.3.2 and the effects of these changes on seven similar variables in both FCMs. An example of the quasi-dynamic output of both FCMs is given in Fig. 8.4.

In both FCMs, increasing climate impact directly affected the water availability and (irrigation) water demands. It further directly affected ecosystems conditions in FCM-SH and yields in FCM-WG. The climate-change-induced changes showed the same direction of change in both FCMs, except for water demand. The magnitude of change was, in most cases, larger in the FCM-WG. Water demand decreased in the FCM-SH. This is due to the mechanisms to mitigate climate impacts that are present in the FCM-SH. A decrease in water availability leads to more water-saving methods and a higher price of water; both of which, in turn, lower the water demand. Also,

Table 8.3 Comparison of system description of both FCMs for four different changes, with the effects on a number of similar variables in both FCMs

Changes made		Variables affected						
		Water quality	Water availability/ fresh water resources	Good ecosystems condition/ ecological status	Water demand/total withdrawals	Population/ domestic water withdrawals	Irrigation efficiency/ project efficiency	Intensification of/ agriculture crop specific irrigated area
Climate change impact increase	**Stakeholder based FCM**: starting vector climate impact (C4)	Decreases moderately	Decreases moderately	Decreases moderately	Decreases strongly	Decreases minimal	Increases slightly	Increases moderately
	WaterGAP based FCM: starting vector climate warming (C0)	Decreases strongly	Decreases strongly	Decreases strongly	Increases strongly	Not affected	Not affected	Increases moderately
Decrease in water availability	**Stakeholder based FCM**: drought impact on water availability (C46->C11)	Decreases slightly	Decreases slightly	Decreases slightly	Decreases strongly	Decreases minimal	Increases slightly	Increases moderately
	WaterGAP based FCM: climate warming on fresh water resources (C0->C18)	Decreases slightly	Decreases moderately	Decreases moderately	Not affected	Not affected	Not affected	Not affected

Increase in irrigation efficiency	Stakeholder based FCM: Agricultural support policies on irrigation efficiency (C17->C15)	Increases minimal	Increases moderately	Increases slightly	Not affected	Increases minimal	Increases moderately	Increases strongly
	WaterGAP based FCM: project efficiency (C6)	Increases slightly	Increases slightly	Increases slightly	Decreases slightly	Not affected	Increases strongly	Not affected
Decrease in intensification of agriculture	Stakeholder based FCM: environmental policies on intensification of agriculture (C0->C13)	Increases strongly	Increases slightly	Increases moderately	Decreases strongly	Decreases slightly	Decreases minimal	Decreases strongly
	WaterGAP based FCM: irrigated crop production on area required for crop production (C1->C3)	Increases moderately	Increases moderately	Increases moderately	Decreases moderately	Not affected	Not affected	Decreases strongly

Changes were made by changing the value of the connection or starting value from 0.5 to 0.9 (80 %) of the variable mentioned in the second column
Strongly >50 %; moderately 15–50 %; slightly 5–15 %; minimal <5 %; not affected <1 %

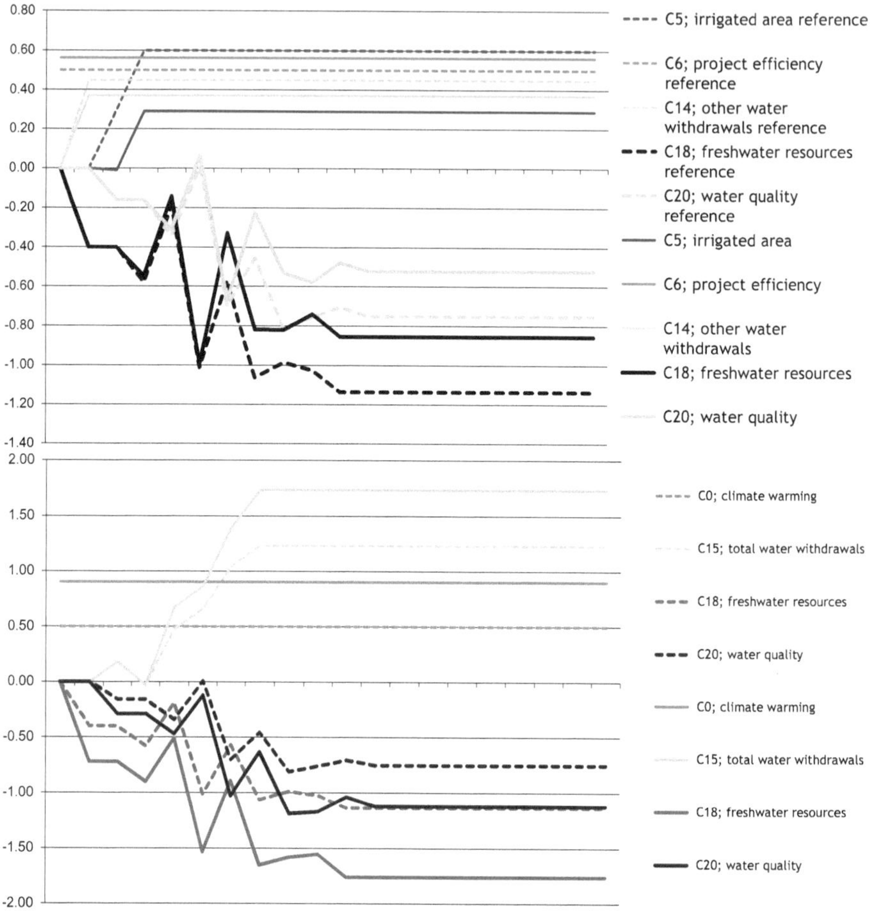

Fig. 8.4 (**a**) Iteration results of two FCM-SH runs with a change in climate impact (starting value 0.5 resp 0.9), showing the impact on water quality, water demand and water availability. *Dotted lines* show the reference iteration run (C4 = 0.5); *non-dotted lines* the iteration with a stronger climate impact. Y-axis showing the value of the variables, X-axis number of iterations. (**b**) Iteration results of two FCM-WG runs with a change in climate impact (starting value 0.5 resp 0.9), showing the impact on water quality, water demand and water availability. *Dotted lines* show the reference iteration run (C0 = 0.5); *non-dotted lines* the iteration with a stronger climate impact. Y-axis showing the value of the variables, X-axis number of iterations

irrigation efficiency is increased in reaction to increasing water prices. These mechanisms were not present in the FCM-WG, resulting in an increase of water demand.

In the FCM-WG, four out of the seven variables studied are not affected by a decrease in water availability as the model calculates the potential water demand of each sector; in the FCM-SH, almost all variables are affected. The directions of change are the same, while the magnitude of change for water availability and ecosystem conditions is larger in the FCM-WG. This shows the role of feedbacks in the FCM-SH; the water availability decrease is partly balanced by a decreasing demand and increasing irrigation efficiency, which are pure transmitters in the FCM-WG and therefore not affected.

In both FCMs, the directions of change are the same if irrigation efficiency is increased, but magnitudes differ for most variables. In both cases, water quality increases, but in the FCM-SH the change is smaller. FCM-SH describes a positive relationship from irrigation efficiency via water availability to water quality, but also includes a negative feedback. Increasing irrigation efficiency leads to an increase in intensification of agriculture, which has a negative effect on water quality. In the FCM-WG, the irrigated area is not affected by the change in irrigation efficiency. There is only a positive effect on water quality through lower water withdrawals.

Also, with a decrease in intensification of agriculture, the directions of change are the same, but magnitudes differ. In both cases, water demand decreases, which in turn increases water resources and water quality. The FCM-SH shows a stronger increase of water quality. The variables that are not affected in FCM-WG are slightly or minimally affected in the FCM-SH.

8.5 Discussion and Outlook

The objective of this chapter was to analyze the potential of using FCMs to link stakeholder information to mathematical models. Three sub-objectives were identified, all of which will be discussed below. Subsequently, the FCM tool is put in the context of other projects and tools. Both aspects combined lead to conclusions about the potential of FCMs to function as a common base for linking stakeholders and modelers.

8.5.1 Development of FCM-SH, Based on Information from Three Case Studies

In general, aggregating FCMs is not as straightforward as it might appear, and the aggregation of three local FCMs into one regional stakeholder-FCM in this paper was no exception. Particularly, the step of merging similar variables forced us to make choices to obtain a clear and relatively simple system description, which was needed to facilitate comparison with FCM-WG. Often, each Pilot Area used different variable descriptions to represent the same process. Once the names of these variables were harmonized, merging became easier. As anticipated, detailed insight on the knowledge and perceptions of the stakeholders in the Pilot Areas was essential in making these choices. Somewhat surprisingly to the experts from the Pilot

Areas, after this process it was concluded that the three FCMs were more similar than they first appeared. However, as a result, some of the Pilot Areas' specific details and the diversity among them were lost in FCM-SH.

In summary, the method we employed—including an expert meeting—to develop the FCM-SH was adequate and resulted in a product that reflected the main processes in the three Pilot Areas. We conclude that a separate meeting is a necessary element that cannot be replaced by using simple arithmetic to aggregate.

8.5.2 Development of FCM-WG, Based on a Mathematical Model

Developing a model-based FCM forced modelers to be precise on how the model works. To keep it simple, the modelers chose to show only WaterGAPs main component, leaving out many links that are used when calculating the input for that main component. They further opted to give it the same (agricultural) focus as the FCM-SH. Other water users (domestic, thermal energy production, and industry) were, therefore, represented by a single variable only. Each of these water uses could also have been represented by a more extensive part, similar to agricultural water use. Additionally, the model uses multiple crops, which could not be reflected in the FCM, as it would have led to multiple variables for concepts such as crop-specific net irrigation requirements. Therefore, one variable was used with an average value for the outgoing connections.

These choices resulted in an overly simplified representation of the model. As such, the FCM-WG is not an accurate representation of the complexity of the WaterGAP model. Yet, the FCM-SH similarly oversimplifies the complexity of case-specifics in the three Pilot Areas. In this type of exercises the communication of the model's architecture should mirror the level of detail of the system description from the stakeholders' side. Only in this way, are products comparable. Modelers should try to find a balance between being specific on how the model works and keeping it simple enough for communication and comparison.

As a last exploratory step, we compared the quasi-dynamic output of the FCMs with the results of the actual WaterGAP model to compare the system behavior of the model with that of both FCMs. The difference (percentage change) between two runs for a number of indicators was used for comparison. Both FCMs were first iterated with the connection strengths and starting values as shown in Figs. 8.2 and 8.3 (reference run). Subsequently, changes were applied to the starting vector of various variables—and in some cases to the strength of connections—to reflect the differences between the WaterGAP runs. Then, the quasi-dynamic output was compared with the output from the reference run. Percentage differences between variable values of the reference and second run were compared to the percentage changes between the two WaterGAP model runs. The quasi-dynamic output of

FCM-WG did not fit the model runs completely, but the direction and magnitude of change of the compared variables were similar to the changes in the model runs. Knowing that the starting values and connection strengths of the FCM-WG were not calibrated on the model beforehand, the results show that modelers are capable of mimicking the system behavior of the model in an FCM. However, there were also differences. Thus, in future exercises, we advise calibration of the model-based FCM before comparison with the stakeholders' FCM is undertaken. We refrain from including the full details of the comparison in this chapter because of the dissimilarities between the structure and, thus, functioning of the FCM-WG and the WaterGAP model and the current discussions on the interpretation of the quasi-dynamic output.

8.5.3 *Comparing Both FCMs to Identify Crucial Differences and Similarities*

The comparison between both FCMs showed that although the systems are pronouncedly different in terms of key variables and key indicators (density, number of connections, etc.), system dynamics were similar. This demonstrates the added value of FCMs above qualitative conceptual models—that cannot simulate (quasi-) dynamic output—in analyzing the influence of feedbacks (Kok 2009). Results for an increase in climate impacts, for example, immediately showed the impact of adaptation measures that were included in the FCM-SH and not in the FCM-WG. Although not impossible, the effects of multiple feedback loops are difficult to reason through without the dynamic output. The capability of FCMs to include these loops together with the consideration of policy and social issues by stakeholders, enabled the identification of mitigation and adaptation processes that are not reflected in WaterGAP simulations.

The comparison of the quasi-dynamic output of the two FCMs was slightly hampered by an almost total lack of identical variables. Similar variables can be used to compare, but they do not always match completely. For instance, intensification of agriculture and irrigated areas are related, but there are also ways to intensify agriculture other than by increasing the irrigated area. Part of this problem is caused by the process by which both FCMs were developed. Modelers must be specific when developing the model, and will therefore include specific variables in their FCM. Stakeholders used clusters of issues to derive variables, a process that leads to less well-defined variables. The FCM-SHs variables also do not necessarily have to be quantifiable, contrary to those derived from quantitative models.

In summary, by reproducing knowledge using the same tool, comparison is facilitated and leads to important insights in differences and communalities. Yet, a number of problems remain that hamper the analysis.

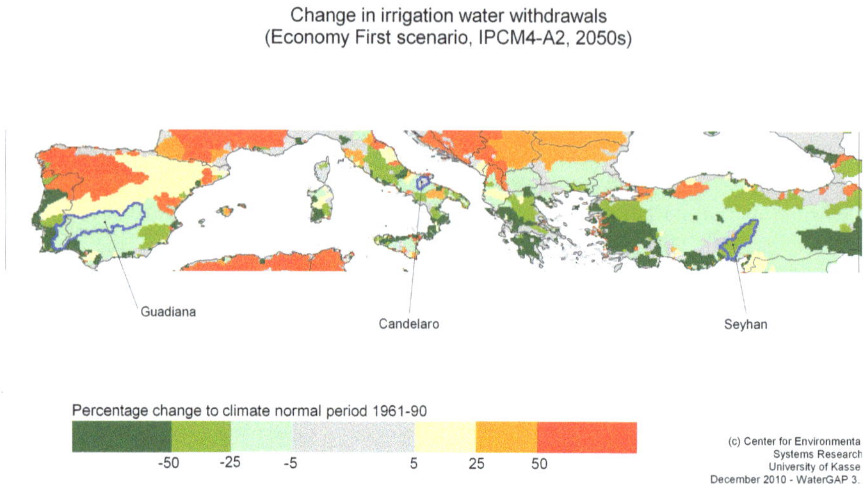

Fig. 8.5 WaterGAP output for a change in irrigation water withdrawals under the Economy First scenarios (IPCM4-A2, 2050), showing the spatial differences in irrigation water withdrawals between the three Pilot Areas and within the three countries

8.5.4 Comparing FCMs and Mathematical Models

FCMs and models complement each other. FCMs (like all conceptual models) are not well suited to be spatially explicit (Voinov and Bousquet 2010), but can capture feedbacks and include social processes. WaterGAP (like many mathematical models) is quantitative and spatially explicit but fails to include drivers for which no data is available. Figure 8.5 illustrates the specific strengths of WaterGAP with a typical example of its output. On the map, large spatial differences can be observed across the Mediterranean region and between the three Pilot Area watersheds. This serves to position the FCMs in the spatial heterogeneity that is not considered in this chapter, but that is part of water-related issues as discussed here.

Thus, the mathematical models are well-suited to show the temporal and spatial uncertainty of the issue at hand; whereas an FCM can show the systemic uncertainty in feedbacks and the effect of more social processes that are lacking in the mathematical model. This chapter has provided a first attempt at how this complementarity could be explored.

8.5.5 Recommendations to Better Match FCMs and Models to Bridge Between Qualitative and Quantitative Scenarios

Future studies are needed to shed more light on how the link between FCMs and mathematical models could be further strengthened. Ideally, the FCM-SH could also be used to directly provide input to the mathematical models. As said, there is current debate on how to interpret the semi-quantitative quasi-dynamic output of FCMs. Translation of the results to percentage changes that might be directly compared to mathematical model output is, therefore, a step too far (but might be undertaken when this debate is settled). Future studies should work on further improving the FCM and look for options to combine FCMs with other tools. Below are some possible options:

- There is ample literature on the mathematical properties of the matrix and the vector x matrix multiplication that define the Fuzzy Cognitive Map and its quasi-dynamic behavior. A paper is underway that will explore the mathematical properties of FCMs and—importantly—how to interpret these properties (Helfgott et al. in review). More work is needed to explore the domain of application, in particular. As the number of publications that link FCMs to scenario-based explorations grows, the quasi-dynamic output of FCMs will gain importance and, thus, peer-reviewed demonstrations of its interpretation.
- Comparison between the FCM-SH and FCM-WG was slightly hampered by an almost total lack of identical variables. A stronger involvement of modelers in the stakeholder workshops could help to increase the number of identical variables, for instance by suggesting the use of some of the key concepts from the FCM-WG. However, one should take care not to lower stakeholder buy-in in the developed FCM. Also, including concepts in the SH-FCM that cannot be quantified—and are thus harder to model—helps to show the impacts of such issues on the concepts in the model, thus showing the added value of combining qualitative and quantitative approaches.
- In many FCM-related publications, multiple FCMs (often based on individual interviews) are constructed and aggregated into one, summarizing system description. Currently, there is no consensus on how to best construct the aggregated Fuzzy Cognitive Map. This chapter has provided a method to do so by organizing an expert meeting, which has its pros and cons. More work is needed to explore the effect of different aggregation methods on key system properties of the aggregated product.
- In essence, Fuzzy Cognitive Mapping is a "quick and dirty" tool. When dealing with stakeholders, the speed with which FCMs can be developed is a large advantage. However, when aiming to improve the communication with mathematical models, enhancing the functionality of FCMs might take preference. FCMs could be improved in multiple ways, including allowing for non-linear relations, memory, and delays, or relations influencing relations rather than variables. Also, spatial and temporal explicitness could be included. By incorporating these aspects, FCMs could better mimic "the real world." Incorporating delays can tackle part of the problem that time is ill-defined. Memory in the form of an internal feedback of a variable on itself can represent stocks. There is, however, a risk that this will make them overly complex for some groups of stakeholders.
- Most system dynamic tools can deal with delays, stocks, and flows and non-linear relations and could, therefore, be used instead of FCMs. Either the whole

process could be done with other system dynamic models or as a next step after the use of FCMs. Depending on the stakeholders' knowledge, this could be done with the same stakeholder group, a smaller group, or with experts only. These system dynamic models could provide quantitative output that could be used directly in the mathematical model. The ability to use particular stocks and flows closely resembles the logic of most mathematical models and yields the same quantitative output.

- Parameter quantification by stakeholders (e.g., via Fuzzy Sets (Alcamo 2008; Eierdanz et al. 2008)) could be combined with FCMs to show the relative change in variables. Experiences taught us that stakeholders perceive their knowledge as insufficient to provide quantitative estimates and prefer to rely on mathematical models for this information (Kok et al. 2015). Yet, future studies could attempt to improve methods that engage stakeholders in providing quantitative information in combination with system dynamics tools.

8.6 Conclusions

This study set out to analyze the potential of FCMs to function as a common base for linking participatory products and models, by representing both stakeholder knowledge and a mathematical model as an FCM. This enabled a direct comparison of system perceptions, both in the system configuration and dynamics. The results show that both systems can indeed be directly compared and similarly analyzed. Yet, despite the comparability, results also show that fundamental differences in how stakeholders perceive the system (and how they are limited by the format that standard FCMs offer) and what is represented in a model-based description limits communication. Importantly, the direct use of FCMs to provide input to mathematical models remains problematic.

FCMs can nevertheless be useful in the process of quantification of stakeholder products by showing the direction and magnitude of change and making assumptions explicit. FCMs can show the implications of social aspects in stakeholder output that are hard to quantify and, thus, to deal with by mathematical models. Mathematical models, in turn, can show spatial and temporal details that are difficult to include in FCMs. Perhaps most importantly, FCMs are likely to aid the communication between modelers and stakeholders, rather than their products. An FCM of a mathematical model helps to open up the "black box" of that model, which will increase stakeholder understanding and acceptance of the model. In turn, stakeholders' underlying assumptions are made explicit and structured when using FCMs, which provides the—otherwise often vague—stakeholder output in a manner closer to that with which modelers usually work.

In conclusion, Fuzzy Cognitive Mapping is a very promising tool for linking stakeholders, modelers, and their respective products. The system dynamics of FCMs can play an important role in the quantification, communication, and dissemination processes. Yet, its "quick and dirty" nature leaves many issues unaddressed and future work is needed to substantiate the claims made here.

Acknowledgements We would like to acknowledge the work of the stakeholders in the Pilot Area workshops who developed the local FCMs. We would further like to thank all who helped in organizing the workshops and analyzing the workshop results, especially Gema Carmona and Irene Blanco, from the Guadiana Pilot Area, Nicola Lamaddalena and Alessandra Scardigno, from the Candelaro Pilot Area, and Ozlem Tonguc, from the Seyhan Pilot Area. We would like to thank Christof Schneider for helping with the model runs of WaterGAP. We further want acknowledge the funding of this research as part of SCENES, by the EC under contract 036822.

Disclaimer The views expressed are purely those of the authors and may not in any circumstances be regarded as stating an official position of the European Commission.

References

Alcamo J (2008) The SAS approach: combining qualitative and quantitative knowledge in environmental scenarios. In: Alcamo J (ed) Environmental futures: the practice of environmental scenario analysis . Elsevier, Amsterdam, pp 123–150

Alcamo J, Döll P, Henrichs T, Kaspar F, Lehner B, Rösch T, Siebert S (2003) Development and testing of the WaterGAP 2 global model of water use and availability. Hydrol Sci J 48(3):317–338. doi:10.1623/hysj.48.3.317.45290

Bousquet F, Voinov A (2010) Preface to this thematic issue (thematic issue - modelling with stakeholders). Environ Model Softw 25(11):1267

Cakmak EH, Dudu H, Eruygur O, Ger M, Onurlu S, Tonguc O (2013) Participatory fuzzy cognitive mapping analysis to evaluate the future of water in the Seyhan Basin, Turkey. J Water Clim Chang 4(2):131–145

Cash DW, Adger W, Berkes F, Garden P, Lebel L, Olsson P, Pritchard L, Young O (2006) Scale and cross-scale dynamics: governance and information in a multilevel world. Ecol Soc 11(2)

Cole JR, Persichitte KA (2000) Fuzzy cognitive mapping: applications in education. Int J Intell Syst 15(1):1–25

Döll P, Kaspar F, Lehner B (2003) A global hydrological model for deriving water availability indicators: model tuning and validation. J Hydrol 270:105–134

Eierdanz F, Alcamo J, Acosta-Michlik L, Kromker D, Tanzler D (2008) Using fuzzy set theory to address the uncertainty of susceptibility to drought. Reg Environ Chang 8(4):197–205. doi:10.1007/s10113-008-0069-1

Flörke M, Alcamo J (2004) European outlook on water use, final report, EEA/RNC/03/007. Center for Environmental Systems Research, University of Kassel, Kassel, Germany

Flörke M, Kynast E, Bärlund I, Eisner S, Wimmer F, Alcamo J (2013) Domestic and industrial water uses of the past 60 years as a mirror of socio-economic development: a global simulation study. Glob Environ Chang 23(1):144–156. doi:10.1016/j.gloenvcha.2012.10.018

Gray S, Gagnon A, Gray S, O'Dwyer B, O'Mahony C, Muir D, Devoy R, Falaleeva M, Gault J (2013) Are coastal managers detecting the problem? Assessing stakeholder perception of climate vulnerability using fuzzy cognitive mapping. Ocean Coast Manag 94:74–89

Hare M, Letcher RA, Jakeman AJ (2003) Participatory modelling in natural resource management: a comparison of four case studies. Integr Assess 4(2):62–72

Heemskerk M, Wilson K, Pavao-Zuckerman M (2003) Conceptual models as tools for communication across disciplines. Conserv Ecol 7(3)

Helfgott A, Lord S, Bean N, Wildeberg M, Gray S, Gray S, Vervoort J, Kok K, Ingram J (in review) When Fuzzy Cognitive Maps Aren't Fuzzy: fuzzy cognitive maps and system dynamics models in participatory social and environmental decision-aiding processes. Environmental Modelling & Software

Jetter A, Schweinfort W (2011) Building scenarios with fuzzy cognitive maps: an exploratory study of solar energy. Futures 43(1):52–66

Jetter AJ, Kok K (2014) Fuzzy cognitive maps for futures studies-a methodological assessment of concepts and methods. Futures 61:45

Kämäri J (2008) Results of first round of Pilot Area workshops, SCENES Deliverable IA2.2. SYKE, Helsinki

Kämäri J, Alcamo J, Bärlund I, Duel H, Farquharson G, Flörke M, Fry M, Houghton-Carr H, Kabat P, Kaljonen M, Kok K, Meijer KS, Rekolainen S, Sendzimir J, Varjopuro R, Villars N (2008) Envisioning the future of water in Europe – the SCENES project. E-WAter (2008/03):1–28

Khadra R, D'Agostino DR, Scardigno A, Lamaddalena N (2011) Down-scaling pan-European water scenarios to local visions in the Mediterranean: the Candelaro Basin case study in Italy. J Water Clim Chang 2(2-3):180–188. doi:10.2166/Wcc.2011.008

Kok K (2009) The potential of fuzzy cognitive maps for semi-quantitative scenario development, with an example from Brazil. Glob Environ Chang Hum Policy Dimens 19(1):122–133

Kok K, Bärlund I, Flörke M, Holman I, Gramberger M, Sendzimir J, Stuch B, Zellmer K (2015) European participatory scenario development: strengthening the link between stories and models. Climatic Change 128(3–4):187–200

Kok K, van Delden H (2009) Combining two approaches of integrated scenario development to combat desertification in the Guadalentin watershed, Spain. Environ Plan B Plann Des 36(1):49–66

Kok K, van Vliet M (2011) Using a participatory scenario development toolbox: added values and impact on quality of scenarios. J Water Clim Chang 2(2-3):87–105. doi:10.2166/Wcc.2011.032

Kosko B (1986) Fuzzy cognitive maps. Int J Man Mach Stud 24(1):65–75

Lehner B, Verdin K, Jarvis A (2008) New global hydrography derived from spaceborne elevation data. Eos Trans AGU 89(10):93

Magnuszewski P, Sendzimir J, Kronenberg J (2005) Conceptual modeling for adaptive environmental assessment and management in the Barycz Valley, Lower Silesia, Poland. Int J Environ Res Public Health 2(2):194–203

Martínez-Santos P, Henriksen HJ, Zorrilla P, Martínez-Alfaro PE (2010) Comparative reflections on the use of modelling tools in conflictive water management settings: the Mancha Occidental aquifer, Spain (thematic issue - modelling with stakeholders). Environ Model Softw 25(11):1439–1449

Özesmi U, Özesmi SL (2003) A participatory approach to ecosystem conservation: fuzzy cognitive maps and stakeholder group analysis in Uluabat Lake, Turkey. Environ Manag 31(4):518–531

Refsgaard JC, Henriksen HJ, Harrar WG, Scholten H, Kassahun A (2005) Quality assurance in model based water management - review of existing practice and outline of new approaches. Environ Model Softw 20(10):1201–1215

Schaldach R, Koch J (2009) Conceptual design and implementation of a model for the integrated simulation of large-scale land-use systems. In: Athanasiadis PA, Mitkas A, Rizzoli E, Marx-Gómez J (eds) Information technologies in environmental engineering. Springer, Berlin/Heidelberg, pp 425–438

Simon C, Etienne M (2010) A companion modelling approach applied to forest management planning (thematic issue - modelling with stakeholders). Environ Model Softw 25(11):1371–1384

Sterman JD (2004) Business dynamics: systems thinking and modelling for a complex world. Irwin/McGraw-Hill, Boston

van Vliet M, Kok K, Veldkamp T (2010) Linking stakeholders and modellers in scenario studies: the use of fuzzy cognitive maps as a communication and learning tool. Futures 42(1):1–14. doi:10.1016/j.futures.2009.08.005

van Vliet M, Kok K, Veldkamp T, Sarkki S (2012) Structure in creativity; effects of structuring tools on results of participatory scenario development workshops. Futures 44(8):746–760

Varela-Ortega C, Esteve P, Carmona G, Blanco I, Khadra R, D'Agostino D, Lamaddalena N, Cakmak E, Dudu H, Kok K, van Vliet M (2011) Integrating stakeholder-based fuzzy cognitive maps for understanding water systems in the Mediterranean. Paper presented at the Futures of European Waters, Budapest, Hungary

Vennix JAM (1999) Group model-building: tackling messy problems. Syst Dyn Rev 15(4):379–401
Verburg PH, Rounsevell MDA, Veldkamp A (2006) Scenario-based studies of future land use in Europe. Agric Ecosyst Environ 114(1):1–6
Verzano K (2009) Climate change impacts on flood related hydrological processes: further development and application of a global scale hydrological model. Reports on earth system science, vol 71-2009. Max Planck Institute for Meteorology, Hamburg, Germany
Voinov A, Bousquet F (2010) Modelling with stakeholders. Environ Model Softw 25(11):1268–1281
Website Mont Fleur (2011) The Mont Fleur Scenarios. Mont Fleur Conference Venue. http://www.montfleur.co.za/about/scenarios.html. Accessed Feb 2011

Chapter 9
Extending Participatory Fuzzy Cognitive Mapping with a Control Nodes Methodology: A Case Study of the Development of a Bio-based Economy in the Humber Region, UK

Alexandra S. Penn, Christopher J.K. Knight, Georgios Chalkias, Anne P.M. Velenturf, and David J.B. Lloyd

Chapter Highlights

Approach: We used fuzzy cognitive mapping combined with a novel analytic approach from network controllability to explore the causal structure of the development of a regional bio-based economy and to identify "control configurations"—subsets of factors which could theoretically be used to steer the system to any given state.

Participant Engagement: A variety of regional stakeholders engaged in three workshops to construct and verify the cognitive map, evaluate the controllability of factors, and, hence, the optimal control configurations for the system from their perspective.

Models/Outcomes: Six possible control configurations of the bio-based economy system map were calculated and stakeholders chose two as the most optimal. These were used to focus decision making and future modeling work.

Challenges: Control configurations are dependent on map structure generated in an intersubjective group context, hence the line between thinking tool and definitive model must be made clear and robustness testing performed. Different stakeholders perceive factors as differently controllable and the process must be adapted to take this into account.

A.S. Penn (✉)
Evolution and Resilience of Industrial Ecosystems Project, Department of Sociology, Centre for Environmental Strategy, University of Surrey, Guildford, Surrey GU2 7XH, UK
e-mail: a.penn@surrey.ac.uk

C.J.K. Knight • D.J.B. Lloyd
Evolution and Resilience of Industrial Ecosystems Project, Department of Mathematics, University of Surrey, Guildford, Surrey GU2 7XH, UK

S. Gray et al. (eds.), *Environmental Modeling with Stakeholders*,
DOI 10.1007/978-3-319-25053-3_9

9.1 Introduction

9.1.1 Tools for Steering Complex Systems

The work that we report in this chapter, including the development of the control nodes methodology, was carried out in the context of the Evolution and Resilience of Industrial Ecosystems (ERIE) Project at the University of Surrey (http://www.surrey.ac.uk/erie). The explicit aim of this interdisciplinary project (along with three other Engineering and Physical Sciences Research Council (EPSRC)-funded "Complexity Science for the Real World" projects) was to combine novel mathematical and computational tools and techniques with complexity science approaches to produce decision-making frameworks for stakeholders and policy makers in industrial networks. In our particular case study we worked alongside industrial, local-government, and NGO stakeholders in the Humber region of the UK, with the aim of creating management tools to contribute to regional development.

Ultimately, our motivation for gaining an understanding of the causal structure of this system is that the system stakeholders want to steer or influence it. We typically have multiple objectives: We may want to increase certain things—such as bio-based energy production, jobs, and sustainability—and, perhaps decrease others. Our difficulty is to work out how to do so. With this in mind, our work is framed within an adaptive management approach (Waltner-Toews and Kay 2005), in which we see steering complex systems as an ongoing, cyclic, participatory process within which both regional actors and scientists are explicitly seen as stakeholders. Such a process has a number of stages: collaboratively developing a system understanding through participatory work and modeling; using this understanding to list plausible system scenarios; choosing a vision for the system; design of interventions along with monitoring strategy and infrastructure to allow adaptation of interventions according to feedback from the system itself; and then implementation, monitoring, learning, and strategy adaptation. Designing context-appropriate system interventions that will allow us to manipulate the system toward our chosen goals is a challenging problem. It is usually impractical—or at least difficult and expensive—to attempt to control every factor within a system directly. Additionally, the network of interactions is complex. There are many interconnected factors and many connections; altering the value of one factor will affect the rest of the system. A more effective and less costly way to manage the system than struggling to impose control at every level is to exploit its causal structure. To aid in more rigorous approaches to this, we aim to contribute to an additional stage in the adaptive management process

---

G. Chalkias
Department of Sociology, Evolution and Resilience of Industrial Ecosystems Project, University of Surrey, Guildford, Surrey GU2 7XH, UK

A.P.M. Velenturf
Evolution and Resilience of Industrial Ecosystems Project, Centre for Environmental Strategy, University of Surrey, Guildford, Surrey GU2 7XH, UK

of finding effective system "levers" to enable design of efficient interventions. This, like all modeling processes within adaptive management, will consist of interplay between a variety of mathematical and computational tools and stakeholder knowledge and participation. The control nodes methodology presented in this chapter provides one such tool, but its utility will depend explicitly on its use in the context of a full participatory process.

9.1.2 The Humber Region Case Study

The Humber region faces significant new challenges and opportunities with the transition to a low-carbon economy. It is one of the UK's most important energy hubs, with strategic energy generation facilities and infrastructure based around fossil fuels, and new investment in large-scale renewable energy technologies. The development of a bio-based economy has been recognized as a key opportunity for regional economic growth by regional industrial fora (Hull Forward Limited 2009; Energy A, Group E 2007). This is due to the presence of required infrastructure and support industries, availability of feedstock from the substantial agricultural hinterland, and bulk imports via the large local ports. Numerous biodiesel and bioethanol facilities already exist or are under construction and the region expects to become the center of an emerging UK biofuel industry responsible for 50 % of UK production within the next 5 years. Significant investment is also underway in energy from biomass and biowaste facilities, alongside developments in biorefinery to produce high-value chemicals. The estuary is also of national and international biodiversity and conservation importance, and due to climate change presents increasing flood risk management issues—both of these issues can cause friction over proposed development. Furthermore, neighboring communities face significant socio-economic problems including unemployment and fuel poverty. Development of the region and its economy is thus affected by, and affects, linked biophysical, industrial, economic, social, and governance systems, populated by many diverse actors. Understanding and managing the interactions of the components of these systems as they develop will be crucial in addressing the balance between economic development, efficient use of resources, reduction in environmental impacts, and job creation on a regional and national scale. We aim to design model-based decision-support tools for the region to facilitate effective management of the transition to a bio-based economy.

9.1.3 Participatory Modeling in Our Methodology

In a multi-actor, industrial context such as this, data needed to construct a model may be sparse, commercially sensitive, or not centrally collected; the situation may be changing rapidly, influenced by many factors, and highly regionally-specific. We also require ways to enhance stakeholder engagement with modeling and complexity approaches to regional management so our tools are useful and usable. Hence, input and "buy-in" from expert stakeholders is vital to a successful outcome. Participatory modeling (PM), in which stakeholders in a system of study are actively involved in some aspect of the creation or evaluation of models of that system, is a particularly effective way to achieve both aims. Stakeholders can bring valuable first-hand knowledge and ideas to a research process (Ramanath and Gilbert 2004; Bousquet and Trebuil 2005; Batten 2009; Barreteau 2003) aiding, for instance, development of models and scenarios, interpretation of results, and formulation of collective strategies or policy alternatives. Their involvement in a model's construction can also provide a sense of ownership which makes it more likely to be used. As participatory modeling includes a broad spectrum of methods, we clarify our approach below.

We view participation as the *collective* and *active* involvement of various actors in the modeling process. By active, we mean that participants are not passive information receivers/transmitters but have the opportunity to understand, criticize, and reflect on relevant concepts and—at least to some degree—intervene in the modeling process. By collective, we mean that participants are involved simultaneously with at least some possibility for interaction. According to the participation typology proposed by Pretty (1995), this view would then be classified as interactive participation where stakeholders share the diagnostic and analytical methods and tools or results. In our process, stakeholders were involved throughout, from framing to constructing and evaluating models within a facilitated group context.

Models may be understood generally as "conceptual systems consisting of elements, relations, operations, and rules governing interactions… that are used to construct, describe, or explain the behavior of other systems." (Jonassen 2004). In this sense, models can either be classified as external and explicit or internal and implicit (Epstein 2008). Internal models reside in the mind and frame our perception of and thinking about a reference system. External, explicit models take the form of concept maps, equations, computer programs, etc. The relationship between internal and external models is dynamic and reciprocal: external models are imperfect manifestations of our internal models while the latter are changed or updated during the construction of the former.

From this perspective, participatory modeling is simultaneously a social process, a learning process, and a modeling process during which actors jointly develop external models and in the process review, challenge, and update their own and each other's internal models. As such, participatory modeling produces different outcomes at different levels: individual outcomes such as learning or conceptual change, group outcomes such as consensus building, and methodological outcomes such as modeling tools (Rouwette et al. 2002). (Note that these various outcomes of the modeling process tend to diverge and it is not possible to maximize all of them in the same process (Van den Belt et al. 2010).) Therefore, constructing a computational model may be only one, and perhaps not the most important, goal. In our process all outcomes are important, but we ultimately wish to construct external models that aid in a collective system-management process.

In participatory modeling, a polarization exists between the positivist paradigm—which assumes that objective truth exists and modeling must approach it as closely as possible in order to better inform decision-making—and the constructivist paradigm—according to which reality is socially constructed, thus to understand it and improve decision-making, it is necessary to refer to those who construct it (Voinov and Bousquet 2010). Using FCM, stakeholders construct a model of their perceptions of the system that is, in that sense, empirical. However, in terms of its intended role in a cyclical decision-making, intervention, and learning process of socio-technical-economic system management, we consider it to be both conceptual and empirical.

Ultimately, Voinov and Bousquet (2010) identify two objectives that may co-exist within participatory processes: to "(a) enhance the stakeholders' knowledge and understanding of a system and its dynamics under various conditions, as in collaborative learning, and (b) identify and clarify the impacts of solutions to a given problem, usually related to supporting decision making, policy, regulation or management." In the context of an adaptive management process we wish to use both these properties at different stages. Within this chapter we describe our progress in using a particular participatory modeling approach, Fuzzy Cognitive Mapping, combined with novel analytical tools in order to facilitate both these possible goals within the course of an ongoing participatory process.

9.2 Methodology: Expanding Fuzzy Cognitive Mapping with Network Controllability Analysis

9.2.1 Fuzzy Cognitive Mapping

Given the challenges of our particular system, the limited time that our stakeholders had available, and our goals of increasing stakeholder engagement with "whole systems" approaches, we chose to use Fuzzy Cognitive Mapping (FCM) (Kosko 1986). As described in this Chapter, FCM is a methodology that can capture qualitative knowledge from a variety of domains and in which the stakeholders themselves are able to construct the model and view results within the course of a one-day workshop. FCM is widely used for problem-solving in situations where numerous interdependencies are thought to exist between the important components of a system, but quantitative, empirically-tested information about the forms of these interdependencies is unavailable (Taber 1991; Craiger et al. 1996; Schneider et al. 1998; Hobbs et al. 2002; Fons et al. 2004; Mendoza and Prabhu 2006; Soler et al. 2012). The method aims to encapsulate the qualitative knowledge of expert participants or system stakeholders in order to rapidly construct a simple systems-dynamics model of a specified issue. It is considered particularly useful when behavior and decisions of stakeholders play an important role in determining the outcome of a system's development; when detailed local knowledge, but not scientific data, is available; and in problems where public or stakeholder participation is desirable or required

(Ozesmi and Ozesmi 2004). The model produced via an FCM process can be used for scenario testing and to facilitate further discussion and interaction within/with a stakeholder group. However, the values of factors and the links between them can only be interpreted in relative terms (Kok 2009).

The process of model construction consists of several stages: First, stakeholders generate and select key concepts/factors that are important influences on, or parts of, the system of interest. Factors can be from any domain (social, economic, physical, etc.) and may be qualitative or quantifiable. Second, causal influences—positive or negative links—between factors are discussed and decided on, which allows for construction of a directed graph. Finally, participants rank and verbally describe the strengths of these influences between factors, ultimately producing a directed graph with weighted links, which we refer to as the cognitive map or FCM. FCMs may be generated collaboratively by a group of stakeholders at a workshop (Kok 2009; Jetter and Kok 2014), or by individuals via questionnaires or interviews (Ozesmi and Ozesmi 2004; Mouratiadou and Moran 2007). Disparate maps of the same system from different sources can be combined and normalized (Mouratiadou and Moran 2007, Kosko 1992; Banini and Bearman 1998; Khan and Quaddus 2004). Alternatively, conflicting structures resulting from different expert opinions or future possibilities can be investigated as alternative scenarios (Jetter and Kok 2014; Kafetzis et al. 2010).

Graphs may then be used as the basis for simple dynamical models, with the weighted graph represented as an adjacency matrix used to update a vector of factor "values" (see Eq. 9.1). These are iterated forward to infer the possible, logical outcome of the system interconnections that participants have described, as well as the outcomes if links or their strengths are modified to represent alternative scenarios (Hobbs et al. 2002; Mendoza and Prabhu 2006; Soler et al. 2012; Papageorgiou and Groumpos 2005).

$$x_{n+1} = f\left(\boldsymbol{A}x_n\right)x_0 \quad \text{given} \tag{9.1}$$

Where $\boldsymbol{A}$ is the weighted connectivity matrix, f is the thresholding function or functional mapping (which may take a variety of forms), and n is the discrete time step. The state vector x_n contains real values for all the key factors identified by participants.

9.2.2 *Interpretation of Fuzzy Cognitive Maps*

It is clear that any graph that stakeholders produce will be a representation of their own opinions and expertise about their system and cannot be separated from the intersubjective group context. The strength of this technique is not, therefore, in obtaining a "definitive" model of a given human system, but in its ability to engage stakeholders, promote learning and discussion among disparate groups, enhance understanding of whole systems approaches, and extract a starting point for systems modeling where data on system structure is not available and where important variables are qualitative or hard to quantify (Ozesmi and Ozesmi 2004; Mouratiadou and Moran 2007; Kafetzis et al. 2010). In the context of much participatory work, the FCM is therefore primarily an organizational learning tool and an aid to engagement. It is highly valuable in making explicit, then clarifying, mental models and provoking discussion among stakeholders. The rapid construction of a simple mathematical model from such a cognitive map serves an important function in making explicit to stakeholders what the consequences of their beliefs about lower-level causal structure actually entails for the whole system. That is, to check the internal consistency of stakeholders' cognitive maps of the system.

Despite the utility of dynamical models in aiding discussion, we found that different functional mappings gave inconsistent results when applied to the same map structure: they often had more impact on model output than changes in the map itself (Penn et al. 2013; Knight et al 2014). Without any principled and straightforward way to choose between different mappings, this limits the use of these techniques for checking the internal consistency of proposed map structure. (For a description of the variety of mappings available and their use in model analysis see McNeil.)

One possible way to avoid this problem is to simply analyze the map as a network, using tools from network theory to aid in interpreting the structure. The representation of sets of interactions or relationships between interacting entities as a network or graph has become widespread in numerous fields (Borgatti et al. 2009; Proulx et al. 2005). Network analysis has proved to be a useful tool in understanding whether specific network structures are vulnerable to failure and which particular nodes in a given network exert a strong influence on its processes (May et al. 2008). A network analytic approach can be applied to the causal inter-relations between factors produced in an FCM process as long as care is taken in the interpretation of results.

9.2.3 Control Nodes Methodology

While keeping in mind the provisos that apply due to the intersubjective nature of our graph, network analysis offers various novel possibilities when using an FCM in decision-support processes. One is the application of a network controllability methodology to determine the network's potential "control nodes" (Liu et al. 2011). These are subsets of nodes within a network, the state of which that one would need to be able to control in order to steer the whole network to any state within finite time. These exist due to the structure of causal connections: Altering any factor will influence other factors in the network and altering some factors will have more influence on the network than others.

As discussed in Sect. 9.1.1, a central goal of ERIE is to combine modeling and participatory work to find effective levers or points of intervention within a given system to facilitate an adaptive management process. Finding control nodes offers a potentially useful approach.

For most networks there exist numerous subsets of nodes (factors, in our case) that we can use to control the state of the whole network—we call these "control configurations." Any particular network will usually possess many control configurations of different sizes, that is, containing different numbers of nodes. Liu et al. discovered a technique to calculate the minimum size of control configuration for a given network; however, their method does not identify which nodes these minimal configurations contain. Therefore, the information needed to discuss the potential use of these system levers is not available. In order to use this technique in our adaptive management process, we developed a method to identify the nodes contained in all the control configurations (of minimum size) for a given network. The specific computational details are given in Penn et al. (in preparation). In brief however, Liu et al. showed that there was a one-to-one relationship between control configurations (of minimum size) of a network and the "maximum matchings" of the network—that is, the maximum set of links that do not share start or end nodes, and that the control configurations could be directly generated from the maximum matchings. Finding a maximum matching is a "graph coloring" problem where the objective is to "color" the maximum number of links in the network under the constraint that there can be a maximum of one "colored" link entering, and one "colored" link leaving, any given node. Liu et al. showed that for a maximum matching, the set of nodes which do not have a "colored" link entering them form a minimal control configuration (of which there may be many). This reworking of the problem allows the application of the well-known polynomial Hopcroft-Karp algorithm (Hopcroft and Karp 1973) in the place of previous exponential time algorithms. This makes computation of the minimum control configurations for smaller networks feasible within a workshop scenario.

9.2.4 Incorporating Control Nodes into a Participatory FCM Workshop

This technique computes only the control configurations of minimum size of a given network, but arguably, it should be easier to manipulate a given system using the smallest number of points of intervention possible. Given this, the minimally-sized control configurations offer a set of plausible options for system intervention with adaptive management in mind. It is evident however, that the factors that are mathematically determined to be the most effective at controlling the network due to their position within its structure, may not be the factors that are most controllable from the point of view of a particular set of system stakeholders. Some factors are the product of the interaction of numerous large-scale effects, some are controlled by different sorts of actors or organizations at different scales. For this reason, to render this technique as useful as possible, the control configurations should each be evaluated according to their "real world" controllability. We designed a process of factor controllability scoring within a workshop context to allow ranking and evaluation of the control configurations in terms of total controllability according to the particular stakeholders present. Essentially, this involves stakeholders rating each factor as easy, medium, or hard to control during group work. This discussion is carried out without any prior presentation of the results of the control configuration analysis to avoid any possible bias of results. Scores for each factor are then displayed, discussed with the whole group, and consolidated as averaged numerical values. This allows us to rank the mathematically produced control configurations according to the stakeholders' perceptions of their controllability. We describe the results of applying the control nodes methodology to an FCM as part of our on-going participatory process in Sect. 9.3.

9.3 Case Study: Applying Control Nodes Methodology to Stakeholder-Produced Fuzzy Cognitive Maps

9.3.1 Producing a Cognitive Map of the Humber Bio-Based Energy System

As part of our on-going engagement with the Humber region, we facilitated FCM construction and verification workshops on development of a bio-based economy in the Humber region. FCM construction followed the standard form described in Sect. 9.2.1 and involved 11 participants representing industry, local authorities, and non-governmental organizations who collaboratively produced a single map. A verification and scenario-generation session was carried out 3 months later at a local environmental managers' meeting. The participant group had a similar composition and included both attendees of the original workshop and newcomers. Participants produced distinct, alternative structures for the map based on local or non-local feedstock production. The full details of workshop methodology and map output including transient dynamics and precise values of factors at the map fixed points (under both linear and sigmoidal mappings) are detailed in Penn et al. (2013). The map for the non-local feedstock supply scenario is shown in Fig. 9.1. In this scenario, feedstock is imported via the port rather than grown locally, meaning that there is no direct competition for land between feedstock production and industrial development. Availability of land for development is constrained by habitat regulations, however.

9.3.2 Humber Bio-Based Economy Control Nodes Workshop

In order to pilot the FCM control nodes methodology we ran a 3-hour workshop based around evaluation of a pre-existing FCM, the "non-local feedstock supply" scenario (Fig. 9.1). The workshop was designed to account for the fact that many of the participants had not been involved in construction of the original FCM. This

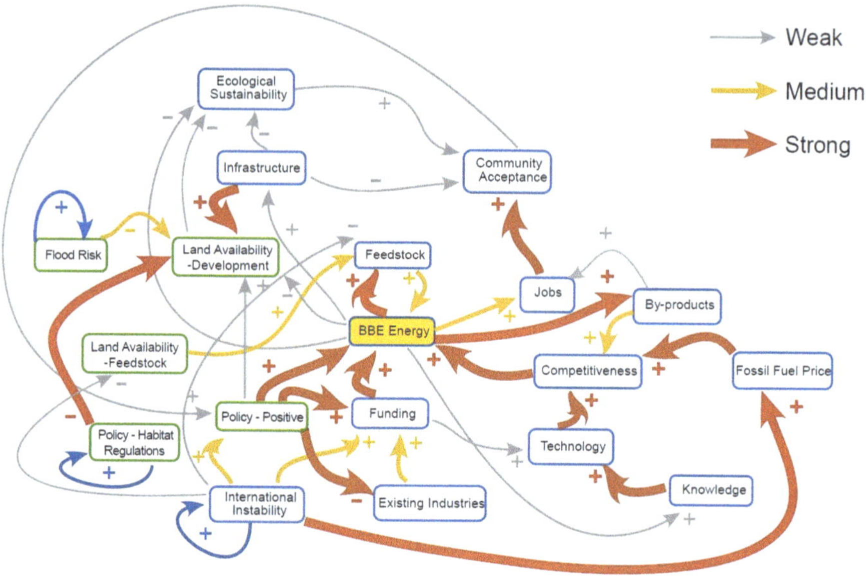

Fig. 9.1 Fuzzy Cognitive Map of non-local feedstock production scenario for Humber region bio-based economy. Factors outlined in *green* were added to the original FCM as a result of the verification exercise. International Instability (vs. UK stability), Flood Risk, and Habitat Regulations were identified as key external drivers of the regional system—a driver being defined as a factor with outgoing links only (these are denoted by self-reinforcing links which have strength 1 so that drivers are maintained at a constant value). The thickness of links denotes the strength of the influence (reproduced from Penn et al. 2013)

also allowed pre-calculation of the control configurations. Eleven regional stakeholders attended, again representing local authorities, industry, and NGOs. Five participants had previously attended one of the fuzzy cognitive mapping workshops.

After introducing the previous FCM and the network controllability concept, we began with a clarification of the non-local feedstock map. A table of previously agreed factor definitions was perused within brief small-group discussion. Unclear factors were fed back and clarified for the whole group. This led into the main body of the workshop, beginning with ranking "controllability" of the factors: First, small group discussion in which factors were grouped as easy, medium or hard to control; second, feedback and whole-group discussion of the differences between groups, allowing a rough consensus on factor controllability and preservation of key conflicts. Average factor controllability scores were calculated over results from all groups (as numerical values) and results for each factor identified as a control node in a given control configuration were simply added together to give a total score for that configuration. The top two control configurations were presented and contrasted with the lowest-ranked configuration. Results were followed up with discussion on the stakeholders' responses to the configurations and how they related to their perceptions or experience of the system, the nature and limitations of the nodes' controllability, and the utility of controllability ideas and methodology.

9.3.3 *Controllability Results*

9.3.3.1 Factor Controllability as Perceived by Stakeholders

As can be clearly seen, some factors achieved a consensus on their controllability, but a range of opinions existed on the controllability of the majority of factors. In particular, by-products, jobs, and bio-based energy production provoked a wide spread of views. Other factors such as land availability, development, or fossil fuel price were universally seen as easy or hard to control respectively (Table 9.1).

9.3.3.2 Control Configurations

We had previously calculated all the possible minimal control configurations for the network representing the "non-local feedstock supply" scenario shown in Fig. 9.1 using the methodology described in Section 9.2.3. One of these control configurations, configuration A, is shown in Fig. 9.2. The six factors highlighted in the diagram together form a minimal set of nodes that, if controlled independently, could be used to control the state of the entire network.

Table 9.1 Votes for controllability of all factors by the three groups and the averages used in calculation of total configuration controllability

	Easy	Medium	Hard	Average
Bio-based energy production	x		x	Medium
By-products	x	x	x	Medium
Community acceptance	x	xx		Medium
Competitiveness		xx	x	Medium
Ecological Sustainability		x	x	Medium
Existing industries		xx	x	Medium
Feedstock	x	xx		Medium
Flood Risk		xx	x	Medium
Fossil fuel price			xxx	Hard
Funding		xx	x	Medium
Infrastructure		xx	x	Medium
International instability			xxx	Hard
Jobs	x		xx	Medium
Knowledge	xx	x		Easy
Land availability: Development	xxx			Easy
Land availability: Feedstock		xx	x	Medium
Policy: Habitat Regulations		x	xx	Hard
Policy: Positive		x	xx	Hard
Technology	x	xx		Medium

Note: Some groups did not vote on all factors

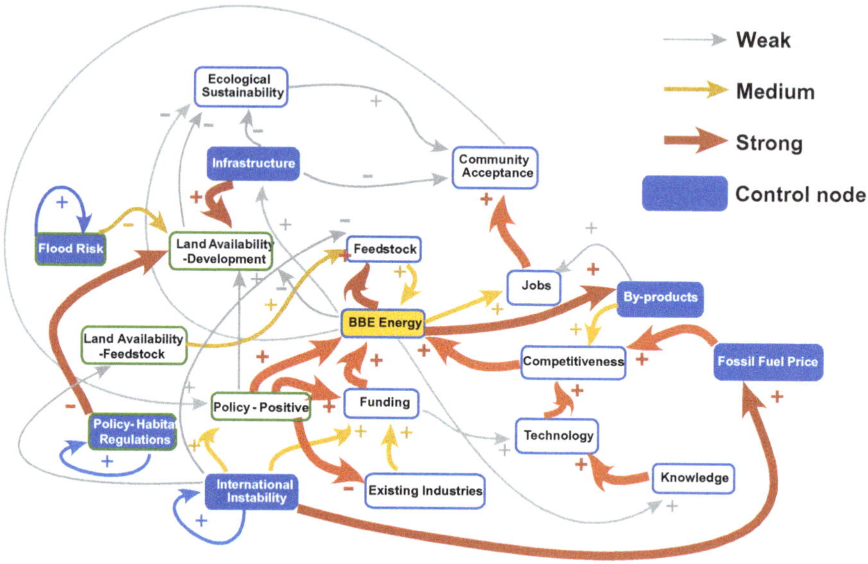

Fig. 9.2 Configuration A for the "non-local feedstock supply" scenario cognitive map. Factors highlighted in *dark blue* are control nodes

Table 9.2 Table of control nodes for control configurations A-F, each column represents a configuration and rows represent different factors

Factor	A	B	C	D	E	F
Flood Risk	x	x	x	x	x	x
Policy: Habitat Regulations	x	x	x	x	x	x
International Instability	x	x	x	x	x	x
Biological By-products	x		x		x	x
Knowledge		x		x	x	x
Infrastructure	x	x	x	x		
Land Availability: development			x	x		x
Fossil Fuel Price	x	x			x	

Factors are present in a configuration if the corresponding grid square contains an X

This particular network has six different minimal control configurations which overlap to some degree. All control configurations must by definition contain the driver nodes of the FCM (flood risk, policy: habitat regulations and international instability), as they have no incoming connections and hence cannot be indirectly controlled by any other node. ("Drivers" are external factors which influence, but are not influenced by the system.) All the control configurations contain six nodes which, excluding the three drivers, are all drawn from a subset of five nodes. The factors in each control configuration are summarized in Table 9.2.

Table 9.3 Control configurations ranked by sum of controllability score of all control nodes in configuration

Configuration	Controllability
A	9
B	10
E	10
C	11
D	12
F	12

Configurations are listed in order of ascending controllability

9.3.3.3 Stakeholder Ranking of Control Configurations

Compiling the results of stakeholders' estimates of factor controllability, we ranked the control configurations as described above. Total scores for each configuration are given in Table 9.3. According to stakeholders then, the most controllable configurations are D and F and the least controllable is A (as shown in Fig. 9.3). The principle factor in the low ranking of configurations A, B, and E is that fossil fuel price, which is considered hard to control, is a control node in all three. Configurations D and F, on the other hand, are the only two containing two control nodes ranked as easy to control by stakeholders: knowledge and land availability: development. It is interesting to note that "by-products," one of the factors that caused the most disagreement regarding controllability, was found to be a control node in four of the six configurations. Altering the controllability of the by-products factor to reflect the diversity of participant opinion would change which configurations were judged to be most controllable.

9.4 Discussion

9.4.1 Stakeholder Response to the Process

Compiled workshop feedback makes it evident that stakeholders had an overall very positive response to this activity and found it thought-provoking and useful in discussing the function of the system and their interactions with it. They did, however, find it challenging to define factor controllability in a simple way. Many of the factors in the FCM are broadly defined and/or composites and, hence, different examples or elements of these factors may be quite different in terms of their controllability. This may imply that the map itself might need to be restructured and factor definitions revisited for a controllability analysis if differences are too large. Varying types of stakeholders also perceived the controllability of factors very differently depending on their relationships to them. Local authorities, for example, felt they could take steps to increase skills and knowledge while industry did not; industrialists considered the nature of their by-products to be highly controllable while others assumed this was not the case. For this reason, when evaluating factor controllability the question of "controllable by who?" arose rapidly.

A general observation was made that scale was crucial in terms of controllability. Larger scale factors or factors that originated or were affected by dynamics outside the immediate region were perceived to be harder to control. Political factors determined by national government were seen to be hard to control by many stakeholders. However for others, such as local authorities—who are accustomed to working with and around such factors by negotiating with related agencies—these factors were seen as less difficult to control. Complex factors with multilateral influences, such as international instability and fossil fuel price, were universally seen as hardest to control.

We might expect those factors that were identified as external drivers to be labeled as hard to control since, by definition, they have no incoming connections from the regional "bio-based economy system." Interestingly however, this was not universally the case. Although international instability was seen as effectively impossible to control, policy on habitat regulations was seen by some stakeholders as controllable or manageable. Flood risk, the third external driver, was seen as being of medium controllability by the majority. More in-depth discussion revealed that this was because the risk could be mitigated by straightforward land management or infrastructure changes, some of which fell within the agency of local institutions or organizations. These differences could represent either a boundary issue with the mapping (that is, factors that control flood risk are not generally connected to the rest of the bio-based economy) or a difference in opinion about the existing system's causal structure due to the different stakeholders present. It is clear that viewing factors and causal structure through the lens of system control, might well lead to a realization of numerous indirect effects on the focal system that require additional factors and links to be considered.

When comparing the different control configurations, there was a generally positive and engaged reaction with stakeholders highly interested to see which nodes were considered crucial. There was particular interest in nodes that were unexpectedly found to be potentially important for network control, such as by-products and knowledge, and feedback that the results had challenged their thinking about system function. However, stakeholders also found it difficult to understand the differences between the various control configurations and why particular nodes were seen as control factors. This, of course, has no easy answer as given the nature of the complex system structure it is not usually immediately obvious why one given factor rather than another is a control node (excluding the straightforwardly explainable driver nodes). It would be helpful to develop further intuitive "hooks" to aid understanding, but ultimately this is a problem for any complex system analysis.

9.4.2 Who Is the Appropriate Audience?

Many participants felt that the method seemed most appropriate to actors perceived to be more powerful—national policy makers for example. This is perhaps inevitable given that factors that are structurally system drivers must always be control nodes. As discussed above however, drivers were not universally seen as being hard to control and apart from driver nodes, several configurations contained only nodes that the groups had labeled as of easy or medium controllability. It may well be important in future iterations of this exercise to explicitly ask stakeholders for controllability scores from their own perspectives or to consider more extensively which actors control which factors more easily depending on the purpose of the exercise. Despite this however, the exercise was seen by stakeholders to be useful as a risk-analysis tool for business. Determining control nodes gives an indication of potential vulnerabilities within the system. If a control node or configuration is known to be subject to external or internal shocks and stresses, then the whole system may be driven in an unexpected direction by change in this node. This may provide an indication of areas or interactions against which companies should consider buffering themselves. Ultimately, we might consider extending this methodology to answer questions about how the network could be rewired to reduce such vulnerability. Or, in a similar vein, to consider how the network would need to be restructured to better match what is controllable from a given actor's point of view with what constitutes an effective control node.

Domains of influence of different types of stakeholders clearly make a large difference to results. Some felt that the exercise would be more effective with groups of stakeholders of one type giving their input, with a subsequent consolidation across a broad range of views. Others mentioned the value of being exposed to different perspectives and it was clear that learning took place about how individuals in different types of organizations operated. Ultimately, the structure of the group would depend on its purpose. For a broad regional collaborative effort, stakeholders from different types of organizations would be more appropriate. An in-house activity for a particular type of organization might wish to discuss only what was controllable from their perspective and not require a range of external views. In terms of policy making however, the learning and discussion about controllability of different nodes from different perspectives might prove extremely useful in effective policy design. It is likely to be more effective to incentivize industry to manipulate a factor that is easily controllable for them for example.

9.4.3 Methodological Limitations and Further Work

There are several fundamental limitations of this approach. The first is shared by any method of FCM analysis, namely the sensitivity of the output to changes in map structure. When performing a cognitive mapping exercise we are of course recording the ideas of a particular group of stakeholders about how their system works. The map will be only one of numerous possible framings of the system, strongly dependent on the experience, bias, and perspectives of the individuals present in any given session. The changes in structure of the map that would be expected from different stakeholder groups, workshops, or map iterations could have a strong impact on which factors are calculated to be control nodes. To help this method progress, it will be important to perform a robustness analysis to allow a greater understanding of the extent of this effect. Even more important however, is to manage stakeholder expectations to make it clear that the control nodes technique is a thinking tool and that the product of this analysis is a way to focus further ideas on systems management, prompt evaluation of the conclusions, and provoke further exploration.

The second limitation of control nodes methodology in particular is that—even assuming that our map is an accurate representation of system causal structure—while the algorithm can calculate the nodes that must be controlled in order to drive the system to any given state, it gives no indication of *how* these nodes should be controlled to achieve a particular goal. That is, even if we are able to develop means to control these nodes, we would not know in which direction to steer them. This again, highlights the method's role as a thinking tool and a starting point for further investigation and modeling work rather than as the end point of a decision-making process about management options.

9.5 Conclusions

Returning to the two uses of participatory modeling described by Voinov and Bousquet (2010)—first as a tool for collaborative learning and developing stakeholders' knowledge and understanding, and secondly to support system management and decision-making—we believe our method helps bridge the gap between the first and second in FCM processes. FCM as used in participatory modeling has thus far been principally applied to support the first objective by providing a basis for discussion/thinking with no direct guidance for decision-making (although, see for example Ozesmi and Ozesmi 2004, for an exception). However, if the aim of our process is steering a system, this constitutes only half of the work that must be done. At the end of the process we have a complex network of factors ranked by a particular mapping function, but no direct idea of what this might mean for the governance of, or decision-making in, the system.

The control node methodology attempts to resolve this. It synthesizes network analysis and stakeholders' judgment and reduces the complexity of the system to a small set of cognitively manageable factors. Therefore, it becomes possible to provide several different reduced sets of factors on which the stakeholders should target their efforts if they wish to maximize the effectiveness of their decision-making. Further work exploring the intersection between different stakeholders' perceptions of controllability and the influence that particular factors actually have in a given causal structure could provide more nuanced and targeted tools.

As explicitly discussed in Sect 9.1.3, we consider the best use of this tool to be as part of an adaptive management process. In particular, as part of an additional step before the design of management plans that is focused on principled ways of finding effective system interventions. As such, the tool would be embedded in a cyclic process in which we treat our plans for intervention as hypotheses and their implementations as experiments to be monitored and learned from. It is to be expected that as the result of this learning we would revisit not just our management plans, but also our models as our understanding of the system is deepened by interaction with it.

Acknowledgements The authors wish to thank the numerous participants from the Humber region who gave their time and shared their knowledge in the course of this work. The partial support of the UK Engineering and Physical Sciences Research Council for programme grant EP/H021779/1 (Evolution and Resilience of Industrial Ecosystems (ERIE)) is gratefully acknowledged.

References

Banini GA, Bearman RA (1998) Application of fuzzy cognitive maps to factors affecting slurry rheology. Int J Miner Process 52:233–244

Barreteau O (2003) The joint use of role-playing games and models regarding negotiation processes: characterization of associations. J Artif Soc Soc Simul 6. Available at: http://jasss.soc.surrey.ac.uk/6/2/3.html

Batten DF (2009) Fostering industrial symbiosis with agent-based simulation and participatory modeling. J Ind Ecol 13:197–213

Borgatti SP, Mehra A, Brass DJ, Labianca G (2009) Network analysis in the social sciences. Science 323:892–895

Bousquet F, Trebuil G (2005) Companion modeling and multi-agent systems for collective learning and resource management in Asian rice ecosystems. In: Toriyama K (ed) Rice is life scientific perspectives for the 21st century. International Rice Research Institute, Los Banos, pp 401–404

Craiger PJ, Weiss JR, Goodman FD, Butler AA (1996) Modeling organizational behaviour with fuzzy cognitive maps. Int J Comput Intell Organ 1:120–133

Energy A, Group E (2007) The status of biofuels in Yorkshire and the Humber. Local Government Yorkshire and Humber Website. Available: http://www.lgyh.gov.uk/dnlds/BiofuelsFinalReport.pdf. Accessed 7 Oct 2013

Epstein JM (2008) Why model? J Artif Soc Soc Simul 11(4)

Fons S, Achari G, Ross T (2004) A fuzzy cognitive mapping analysis of the impacts of an ecoindustrial park. J Intell Fuzzy Syst 15:75–88

Hobbs BF, Ludsin SA, Knight RL, Ryan PA, Biberhofer J et al (2002) Fuzzy cognitive mapping as a tool to define management objectives for complex ecosystems. Ecol Appl 12:1548–1565

Hopcroft JE, Karp RM (1973) An algorithm for maximum matchings in bipartite graphs. SIAM J Comput 2:225–231

Hull Forward Limited (2009) Renewable energy opportunities in the Humber area. Hello from Hull and East Yorkshire Website. Available: http://www.hull.co.uk/websitefiles/RenewableEnergyBrochure.pdf. Accessed 7 Oct 2013

Jetter A, Kok K (2014) Fuzzy cognitive maps for futures studies - a methodological assessment of concepts and methods. Futures. doi:10.1016/j.futures.2014.05.002

Jonassen D (2004) Model building for conceptual change: using computers as cognitive tools. In: Grigoriadou M, Raptis A, Vosniadou S, Kynigos X (eds) Proceedings of the 4th Panellenic Conference with International Participations: Information and Communication Technologies in Education (ETPE2004), Athens, pp 4–17

Kafetzis A, McRoberts N, Mouratiadou I (2010) Using fuzzy cognitive maps to support the analysis of stakeholders' views of water resources use and water quality policy. In: Glykas M (ed) Fuzzy cognitive maps: advances in theory, methodologies, tools and applications. Springer, Berlin/Heidelberg, pp 401–404
Khan M, Quaddus M (2004) Group decision support using fuzzy cognitive maps for causal reasoning. Group Decis Negot 13:463–480
Knight CJK, Lloyd DJB, Penn AS (2014) Linear and sigmoidal fuzzy cognitive maps: an analysis of fixed points. Appl Soft Comput 15:193–202
Kok K (2009) The potential of fuzzy cognitive maps for semi-quantitative scenario development, with an example from Brazil. Glob Environ Chang 19:122–133
Kosko B (1986) Fuzzy cognitive maps. Int J Man Mach Stud 24:65–75
Kosko B (1992) Neural networks and fuzzy systems. Prentice-Hall International Editions, Englewood
Liu YY, Slotine JJ, Barabasi AL (2011) Controllability of complex networks. Nature 473:167–173
May RM, Levin SA, Sugihara G (2008) Complex systems: ecology for bankers. Nature 451:893–895
McNeil FM (2003) A rich collection of squashing functions. Fuzzy Systems Engineering Website. Available: http://www.fuzzysys.com/squash2.pdf. Accessed 9 Oct 2014
Mendoza GA, Prabhu R (2006) Participatory modeling and analysis for sustainable forest management: overview of soft system dynamics models and applications. Forest Policy Econ 9:179–196
Mouratiadou I, Moran D (2007) Mapping public participation in the water framework directive: a case study of the Pinios river basin, Greece. Ecol Econ 62:66–76
Ozesmi U, Ozesmi LS (2004) Ecological models based on people's knowledge: a multi-step fuzzy cognitive mapping approach. Ecol Model 15:43–64
Papageorgiou EI, Groumpos PP (2005) A new hybrid method using evolutionary algorithms to train fuzzy cognitive maps. Appl Soft Comput 5:409–431
Penn AS, Knight CJK, Lloyd DJB, Avitabile D, Kok K, Schiller F, Woodward A, Druckman A, Basson L (2013) Participatory development and analysis of a fuzzy cognitive map of the establishment of a bio-based economy in the Humber region. PLoS One 8(11), e78319. doi:10.1371/journal.pone.0078319
Pretty J (1995) Participatory learning for sustainable agriculture. World Dev 23(8):1247–1263
Proulx SR, Promislow DEL, Phillips PC (2005) Network thinking in ecology and evolution. Trends Ecol Evol 20:345–353
Ramanath MA, Gilbert N (2004) The design of participatory agent-based social simulations. J Artif Soc Soc Simul 7:13
Rouwette EAJA, Vennix JAM, Van Mullekom T (2002) Group model building effectiveness: a review of assessment studies. Syst Dyn Rev 18(1):5–45
Schneider M, Shnaider E, Kandel A, Chew G (1998) Automatic construction of FCMs. Fuzzy Sets Syst 93:161–172
Soler LS, Kok K, Câmara G, Veldkamp A (2012) Using fuzzy cognitive maps to describe current system dynamics and develop land cover scenarios: a case study in the Brazilian Amazon. J Land Use Sc 7:149–175
Taber RW (1991) Knowledge processing with fuzzy cognitive maps. Exp Syst Appl 2:83–87
van den Belt M, Kenyan JR, Krueger E, Maynard A, Roy MG, Raphael I (2010) Public sector administration of ecological economics systems using mediated modeling. Ann N Y Acad Sci 1185:196–210
Voinov A, Bousquet F (2010) Modelling with stakeholders. Environ Model Softw 25:1268–1281
Waltner-Toews D, Kay J (2005) The evolution of an ecosystem approach: the diamond schematic and an adaptive methodology for ecosystem sustainability and health. Ecol Soc 10(1):38

Chapter 10
Effects of Livelihood-Diversification on Sustainability of Natural Resources in the Rangelands of East Africa: Participatory Field Studies and Results of an Agent-Based Model Using the Knowledge of Indigenous Maasai Pastoralists of Kenya

Margaret Mwangi

Chapter Highlights

Approach: An integrated approach using statistical methods incorporated in Agent-Based Modeling (ABM) was used to examine data about the attitudes and behavioral responses of the Maasai pastoralists and their livelihood-diversification in the rangelands of East Africa.

Participant Engagement: A two-phase cross-sectional participatory survey was conducted to gather primary data which included (1) semi-structured interviews with households and key-informants and (2) discussions with focus groups were conducted.

Key Outcomes*:* Outcomes from agent-based livelihood-diversification modeling included new understanding about (a) current and future livelihood-diversification among the Maasai (b) current and future sustainability of natural resources under these new livelihood-diversification conditions and (c) the resulting drivers and impacts of these changes across the landscape.

Potentials and Shortcomings*:* The use of an integrated approach allowed for the capture and integration of socioeconomic, climatic, and environmen-

M. Mwangi (✉)
Department of Geography & African Studies Program,
The Pennsylvania State University, University Park, PA 16802, USA
e-mail: mmwangi@lycos.com

S. Gray et al. (eds.), *Environmental Modeling with Stakeholders*,
DOI 10.1007/978-3-319-25053-3_10

tal factors; and the statistical ABM models provided capacity to explain core feedbacks between the social and ecological systems. There were limitations in the approach in terms of scaling-up or scaling-out of attitudes and behavioral response of agents captured at the household level, given natural variations in response. Additionally, since most decision-making in Maasai-pastoralism vis-à-vis governance of natural resources is embedded in various social-networks, institutions, and power relationships, the accounting of the same in statistical methods incorporated in the ABM situation is challenging and there is a lack of standardized approach for analyzing attitudes and behavioral responses or attributes of an ever-evolving socioecological system like Maasai-pastoralism.

10.1 Introduction

Maasai-pastoralism is a coupled social-ecological system that relies on livestock; and by extension, critical rangeland resources (CRR) including water and natural pastures (Mwangi 2012, 2014). The spatiotemporal availability of these CRR is key to the functioning and subsistence of Maasai-pastoralism, and hence, the Maasai's reliance on this production system for generations.

Besides CRR, the operation of Maasai-pastoralism is defined by interlinked cross- scale social and biophysical factors, which influence the quality of pastoralism and therefore livelihoods. Because of this interconnectedness, pressures influencing either social or ecological facets of Maasai-pastoralism affect the other in complex ways; however, these feedbacks are difficult to understand. This behavior reverberates across the structural components contained within each facet of Maasai-pastoralism further affecting one or more components. Therefore, all of these factors must be integrated to address issues related to the future sustainability of local natural resources, and in particular, the relationship between changing environmental and social conditions, across Maasai rangelands of East Africa so that the future of this system can be evaluated to understand its current trajectory in relation to sustainability goals.

In the system of Maasai-pastoralism, spatiotemporal mobility is regularly employed as an adaptation strategy to accommodate fluctuations in access to, and availability of, CRR. CRR are highly sensitive to rainfall variability that characterizes the Maasai rangelands of East Africa. Consequently, the Maasai track them across spatiotemporal scales, and thus it is important to understand how these pastoralists select and utilize resources across the landscape. Developing such an understanding entails assessing Maasai pastoralists' migration, or daily movements, with their livestock in search of CRR under changing climatic and environmental circumstances Migration as a mechanism for tracking of CRR is not unique to the Maasai or to pastoralists of East Africa. Past studies have documented that pastoralists across

Africa have traditionally moved with their livestock across the vast arid and semiarid lands (ASALs) that characterize much of Africa (Behnke and Scoones 1993; Bassett 1988; Baker 1974; Mwangi 2005, 2007). The extraction of CRR by way of pastoralists' migration/daily-movement with their livestock with climate and environmental variability translates to selecting the most advantageous use of the variable pastures and water that characterize the ASALs. In fact, this undertaking ensures efficient use of dryland's dynamic forage resources (Behnke and Scoones 1993; Baker 1974); and explains why nomadism and transhumance have been vouched as the most efficient production system in these ASALs (Scoones 1995a, b; Niamir-Fuller 1999; Ellis and Galvin 1994; Smith et al. 2000).

Besides water and pastures, other natural rangeland resources such as saltlicks and wild biota contribute greatly to the movement of Maasai-pastoralism across the landscape. These resources are central to the livestock production system and understanding these dynamics requires asking two main questions: (1) what currently is required to ensure stable availability of these natural resources; and (2) what is the trajectory of sustainability of CRR—and indeed other natural resources—across Maasai rangelands as Maasai-pastoralism diversifies given anticipated socioeconomic and climate change. Recent studies have indicated that over 70 % of Maasai households have already diversified their livelihoods (Mwangi 2012) in ways that diverge from traditional practices that have ensured sustainable management of natural resources within the landscape that they inhabit. In this chapter, I propose a participatory modeling-based approach to explore the interactions of livelihood-diversification and sustainability of natural resources in the rangelands of East Africa using the case of Maasai-pastoralism. This study is built on the premise that the structural components and operation of primary production systems, such as Maasai-pastoralism, have a significant influence on the trajectory of sustainability of natural resources in these rangelands and is mediated by social and ecological factors that must be integrated to understand the future of this coupled system. The chapter is organized in three major sections. First, I provide an overview of the approach used including how model was development based on the agent-types and social structures on the landscape. This is followed by a section on historical context, social structures and dynamics, and how they relate to sustainability of natural resource in the Maasai's rangelands. Lastly, the output from model scenarios are used to help understand the influence of Maasai's attitudes and behaviors under specific climatic, environmental, and socioeconomic conditions on households' diversification.

10.2 Research Approach and Methods

This study uses an *integrated model-based approach* to understand social and ecological change in East Africa by combining climatic, social, economic, and environmental factors. The approach is integrated in that diverse methods and tools were utilized in data acquisition and analyses: mixed methods of analyses are used, diverse sources of data are utilized, and multiple agents are incorporated in the explanation of sustainability of natural resources in the Maasai-inhabited rangelands. Furthermore, the study is participatory in that the participants (the Maasai people) are engaged in the design of the survey and in data acquisition that was used in the current study.

The use of an integrated approach in data acquisition and analyses coupled with my long-term experience and robust contextual knowledge of the subject, particularly in the community's and biophysical contexts; and my established strong rapport with the Maasai community in the region, rendered a rich understanding of the problem context. Therefore specific agent types and agent attributes (e.g., rich/poor) were developed based on long term research engagement with the community and later validated based on focus groups.

Maasai pastoralists often diversify livelihood due to occurrences of drought, increased rainfall variability, land-shortages, loss of livestock to disease/pests, and unfavorable livestock-marketing institutions (Mwangi 2012, 2016). For this study, I used data and information derived from the county-wide project on *Adaptations of the Maasai*, to investigate the influence of the Maasai's attitudes and behaviors under specific climatic, environmental, and socioeconomic conditions on household's diversification. I examined dominant drivers for diversification beyond drought alone at the household-level.

The study acquired data from participatory interviews with household and key-informants, discussions with focus groups, participant observation, personal field—observations, retrieval from archives, and from review of the literature. These field studies availed household-level agent's attributes and behavioral responses associated with diversification. Data collection took place concurrently. Participants in the surveys included both females and males aged 18 years and older; this ensured representation of views across age and gender. Respondents liberally identified various livelihood-diversifications; analyses were done in SAS (SAS Institute Inc.).

A two-phase cross-sectional participatory survey was conducted to gather primary socioeconomic, climatic, and environmental data; and was based on integrated methods of FAO (1990), Friis-Hansen and Sthapit (2000), Smith et al. (2000) and Quinn et al. (2003). Specifics for the study area are in detailed in Mwangi (2012).

The first phase was a pilot study with households and focus-group discussions with members of the Maasai community and key-informants; and utilized a semi-structured survey interview. The pilot study helped with the design of an efficient survey tool, to derive core model parameters, and to provide baseline information and data for subsequent analyses and discussions where applicable, for the work that follows. These initial interviews provided the historical context, social structures, and dynamics; and how they relate to environmental variables.

The second phase was an in-depth participatory study with households, key-informants, and focus groups; from which the social and biophysical variables for this study were acquired. Any gathered data and information that is beyond the scope of this research is not reported in this study; rather it appears in other relevant publications (see Mwangi 2005, 2007, 2012). During this phase, randomized households were interviewed until the desired sample size was achieved (n = 120). Respondents liberally identified various adaptation strategies related to drought and non-drought situations—livelihood- diversification is an adaptation to both conditions—using a pre-prepared questionnaire whose structure was guided by recommended formats and information derived from established protocols detailed in works on similar or closely related subjects (e.g., Stevens 1986; FAO 1990; Smith et al. 2000; Quinn et al. 2003). The sample size (n = 120) was sufficient vis-à-vis ensuring the robustness of statistical analyses in a mixed model used in this study, particularly with regard to the independent variables used (e.g. diversification type, gender, age, and rainfall) . A reasonable sample size is 20 observations per independent variable (Stevens 1986). The interviews were conducted with the household-head and elicited responses regarding climatic, environmental, and household socioeconomic conditions.

Proportion value for the citing livelihood-diversification strategies was computed (0 = unmentioned, 1 = mentioned by all, see Table 10.1 for results). The usage of *Proportion value* in this work is guided by its application to closely related socio-ecological systems in the same region by Smith et al. (2000) who conducted a participatory study with the pastoral communities inhabiting the rangelands of southern Ethiopia and northern Kenya to explore livelihood risks. The parameters in the ABM model were determined by fitting the model to data/information derived from phase-1 survey and from long-term data from different archival sources. I used the initial survey (phase-1) for the study of *Adaptations of the Maasai Pastoralists* to parameterize the *wealth-submodel* and *climate-submodel*—only relevant variables in the context of this study are used (see Mwangi 2012). The most influential parameters on the probability of a household to diversify were determined using multiple regression techniques, and whenever a dependent variable was dichotomous, a logistic regression was utilized. To characterize the model, I relied on existing knowledge from my previous field studies (e.g., Mwangi 2005, 2007, 2012), empirical evidence, and long-term experiences. The model intent is to capture attitudes and behavioral factors that influence diversification.

The effects of various socioeconomic, environmental, and climatic conditions were investigated via logit and probit regression models respectively for time dependent (e.g., rainfall) and non-time dependent (e.g., gender) predictor variables; the choice of the predictor variables utilized in generating scenarios (see Table 10.2 for output) is guided by other studies conducted in the region (*ibid.*; Smith et al. 2000).

After analyzing the *Proportion value (Pv)* for livelihood-diversifications, I developed an ABM using SAS—and ran model scenario using the generalized estimating equations (GEEs) methods based on Liang and Zeger (1986)—which relies on determining agent-types and social structures in the landscape (see Sect. 10.3 for details on agent-types and social structures). Three main agent-based sub-models (*Wealth, Climate,* and *Household-demographic submodels*) are used with the parameters, description and valid values: (1) *Wealth-submodel* (based on median wealth): rich (household ≥ 14 members, livestock ≥ 20 heads, and/or land-accessed ≥ 50 hectares) = 1 else poor=0. (2) *Climate-submodel*: rainfall (reliable rainfall using all available data for this rangeland); drought (occurrence) = 1 else 0 (no drought). (3) *Household-demographics-submodel*: Education = household-head's highest level of

Table 10.1 Livelihood-diversification strategies among the Maasai

Livelihood-diversifications among the Maasai (n = 120, p < 0.05)				
	Diversification-citing[a] and associated contrast T (W)[b]			
Livelihood-diversification strategy (type)	<0.45	0.45–0.54	>0.54	EDA[c]
Employed in arable farming (small-scale)			***(***)	47
Employed in arable farming (large-scale)[d]	*(**)			34
Arable farming (own)			***(***)	100
Urban small-trader[e]	***(*)			38
Rural/home small-trader (dairy/produce)	***(*)			28
Trade (livestock & livestock-products)			***(**)	100
Lease out land (agricultural)	**(*)			25
Keep other peoples' livestock	***(*)			14
Herding hireling	***(*)			43
Milk-delivery	***(*)			16
Security-man/watchman/guard	***(***)			52
Transport (public)	*(*)			7
Government job[f]	***(*)			61
NGO/CBOs jobs	***(**)			56
Tourism-based enterprises		***(**)		34
Research (visitors & locals)	***(*)			37
Fuelwood[g] extraction & sales		***(***)		41
Harvesting & loading sand	***(*)			14
Other (multiple)			n.a (n.a)	n.a

Underlined denote mean estimate as follows: Wealth: poor > rich, Temporal: variable > fixed
ns not significant, *n.a* not analyzed
[a]Livelihood strategy-citing <0.45, 0.45–0.54, & >0.54, respectively denote least, moderately-, and frequently-mentioned diversifications; T=temporal-contrast (fixed vs. variable), W=wealth-contrast (herd-size)
[b]Asterisks (*) are p-values for contrasts; in & outside bracket are p-values for T & W respectively: *p < 0.05, **p < 0.01, ***p < 0.0001
[c]Exploratory Data Analysis (mean %)
[d]92% = irrigated agribusiness
[e]Shop, bar, restaurant tailor, open-air
[f]74% = general-election positions
[g]Charcoal & Firewood

education completed, his/her age; household with school-children = 1 else 0; Working member gender (1 = male, 0 = female), age (1 = youth, 0 = old). In addition to these three models, other sub-models for example livestock-productivity and market, location of household, were used (see Table 10.2 for details).

I employed the generalized estimating equations (GEEs) methods (Liang and Zeger 1986) to explain the influence of time-dependent critical predictor variables (and correlated data) on the Maasai's diversification; here, logistic regression models with repeated measures using the GENMODE procedure in SAS were used. Besides their capacity to analyze repeated measurements—especially categorical repeated measurements—the GEEs method has the advantage of accounting for correlations

Table 10.2 Agent-based model of livelihood-diversifications among the Maasai

Variable	Estimate	Effect	Rank[a]
Livestock productivity (1 = high, 0 = low)	0.9100	1.1588	**
Livestock loss to drought (1 = yes, 0 = no)	2.0014	38.0040	***
Livestock prices (1 = good, 0 = poor)	−1.0010	1.1174	*
Gender of working member (1 = male, 0 = female)	0.0020	3.4007	***
Age of working member (1 = youth, 0 = old)	0.2200	20.9000	***
Drought (1 = with drought, 0 = without drought)	0.6091	1.8388	***
Rainfall[b]	−1.7100	5.5290	*
Pasture available locally (1 = yes, 0 = no)	0.4003	1.5005	*
Water available locally (1 = yes, 0 = no)	1.0150	2.8001	***
Wealth (1 = rich, 0 = poor):			
Household-livestock-land	−1.5050	0.6936	***
Household-livestock	0.0650	3.0052	**
Household	0.1146	28.8213	***
Livestock	−1.8100	9.1510	**
Land accessed	0.9940	3.1598	**
Livestock-land	−4.0001	7.2228	***
Land tenure (1 = private, 0 = communal)	1.0004	19.7158	***
Cattle management labor (1 = yes, 0 = no)	1.1080	6.9290	ns
School-children (1 = yes, 0 = no)	0.9998	4.6822	*
Education (household head):			
None	−0.2009	1.7741	ns
Primary-school	0.1841	4.0084	**
Secondary-school	1.9072	11.6304	***
College	1.3907	14.0400	***
Age of household head	0.2220	22.4111	***
Household near local job-market (1 = yes, 0 = no)	1.0009	16.0371	***
Public transport reliable (1 = yes, 0 = no)	0.6400	9.2730	**

Value/DF = 1.0001
ns not significant
[a]Rank: *$p<0.05$,**$p<0.01$, ***$p<0.0001$
[b]Effect: Logit else Probit

among observations within each simulation (Liang and Zeger 1986; Diggle et al. 1994). In the GEEs method using logit transformation in SAS, if the estimate of drought = 0 (no drought) is −2.0389 (which is negative), it implies that when drought occurs (drought = 1), the ratio of possibility to diversify versus not to diversify becomes 7.6891 times larger than without drought. Likewise, using probit transformation in SAS, and the case of one's size of livestock-herd (a measure of household's wealth; (1 = rich, 0 = poor)), if livestock = 0.49, Odds = 3.01, which means that when livestock is 0.494, the probability of diversify is about three times that of the probability of not to diversify. The use of a bottom-up and statistical methods incorporated in ABM is not unique to my study. For example similar approaches have been used elsewhere (e.g., Lansing and Kremer 1993) to understand how sociocultural behaviors influence water-resource management in an agricultural socioecological system in Bali.

10.3 Structure and Operation of Maasai-Pastoralism and Resources Management

This section reports the results of the multiple participatory interviews with the Maasai of Kenya; particular emphasis is placed on resource management under the various levels of social organization that characterizes Maasai-pastoralism. Therefore, I explore the historical context, social structures, and dynamics in Maasai-pastoralism, and examine how they relate to environmental factors across Maasai rangelands. All quotations presented in the text are excerpts from the interviews or discussions with households, focus-groups, and/or with key-informants during data collection.

In Maasai-pastoralism, the management of natural rangeland resources occurs within and across levels of social organizations. For example, the calf-pastures (*Olopololi*) are managed at the household (*enkang*) level. In traditional terms, *Olopololi* are reserved for calves' grazing; are located near the *enkang,* or collection of such, under which it is controlled. Presently, mature cattle, small-stocks (mainly sheep and goats), and donkeys regularly utilize these *Olopololi*. The salt-licks and dry-season grazing fallbacks are managed at the level of *Maasai-Section* (Maa: *Iloshon,* plainly called *Section*, and is the sociocultural level below tribe, that is, the Maasai tribe). However, where convenient, *Olopololi* and dry-season grazing zones can be managed under a collection of adjoining clusters of *enkang*. Traditionally, dry-season grazing zones were limited to dry-season grazing and only to a brief duration in any day; these zones were often located away from the homestead.

The management of natural resources is a male responsibility and Maasai society is a patriarchal system. This is not to say that Maasai-women are not active managers of natural resources—they are. For example, they gather firewood from dry woody-species (Wild olives, Maa: *Oloirien*, are preferred for this purpose) that have died out from drought, pest/disease infestation, lightning strikes, or were felled by elephants; traditionally, standing/green trees were rarely used for firewood. Maasai-women track these dead woody-species across the landscapes they inhabit. Wood from standing/green trees are extracted for building houses (*enkaji*) by women (*Juniperus procera*, Maa: *Oltarakwai,* is the most preferred for house poles), and for constructing homesteads' enclosures that define the spatial extent of one's *enkang* by men (thorny acacias and *Tarchonanthus* spp., Maa: *Oleleshua* are predominantly used for this purpose).

Generally, natural resources, particularly CRR are managed by men operating at various sociocultural scales. For example, senior-elders advise the location of herds' grazing, while warriors (plur: *Il-Moran,* sing: *Ol-Moran*) scout for natural resources, particularly pastures. Members from the same clan (patriarchal ancestry, Maa: *Olgilata*) often share resources that are managed at the *enkang* or *Iloshon* level.

Iloshon boundaries are crossed in times of distress, for example during periods of drought. In fact, during periods of droughts or major catastrophes, protocol governing inter-tribal boundaries are disregarded; this is particularly so between the Maasai and Kikuyu communities. For example, during the prolonged devastating drought that occurred in the mid-nineteenth century (Waller 1988) some key informants indicated that the event—known as the *Emutai* among the Maasai (translated "to wipe out everything")— triggered a severe famine and most families (Maasai family: *olmarei*) sought refuge among the Kikuyu. Therefore, due to unspoken tribal-relations, reciprocity has been common between the Maasai and the Kikuyu. For example, the Kikuyu people cultivated the fertile and well-watered patches within the historically Maasai-dominated territories (also Maasai rangelands); while other communities such as the Ogiek gathered honey along riverine and forested areas and hunted across the greater Maasai rangelands. Suffice that, within the territories dominated by the Maasai in the past, resource use, access, and control was shared by multiple users whose spaces of extraction often overlapped. In fact, diverse users, tribes, and/or production systems co-existed and harmoniously shared land and rangeland-resources; the sharing was collaborative and socio-politically flexible, and ethnic identities were blurred.

Although male elders (Maa: *ilpaiyiani*) are the key decision-makers among the Maasai, all community members are custodians of the community's natural resources. In Maasai-pastoralism, CRR are communally shared and collectively managed; with several users enjoying independent rights of use. Land, one of the key natural resource-bases, historically was purely communal. However, this is no longer the case, and a new management system has been imposed including tenure land holding, particularly private and trustlands that is now common across Maasai rangelands (e.g., Sindiga 1984; Kimani and Pickard 1998; Waller 1988; Mwangi 2012; Kameri-Mbote 2002). Today, land management often occurs at the household level, on an individual basis, and households regulate access to the communal lands and reserve the right to exclude non-members

In the Maasai-pastoralism system, certain collective traditional protocols govern the access and rights to common pool natural resources (mainly land and land-resources), including access to CRR. Protocols that govern natural resources are embedded in the Maasai's social networks and power relations. These protocols designate spatial and temporal access to grazing lands, a divergence from the strategic communal governance structures of the past. This strategic management of natural resources was effected under the aegis of the collective-holding of land and rangeland-resources that traditionally characterized Maasai-pastoralism.

10.4 Maasai Knowledge, Values, and Preferences in Natural Resource Management

This section draws from the participatory surveys conducted with the Maasai of Kenya. The Maasai have specific knowledge systems, values, and preferences that allow informed management of the natural resources, and particularly the CRR. Specific indicators, such as behavior or presence of wildlife biota and or ecosystem conditions (e.g., Fig. 10.1), are used to foretell the status of a given natural resource. For example, with regard to behavior or presence of wildlife biota, heightened emigration of wildebeest and zebra signal looming shortages of CRR. Increased southward movements of these seasonally migratory-herbivores signify looming shortages of palatable graminoids, and denote the potential location of abundant pastures for the Maasai's livestock. In fact, most of the "*...wild animals that resemble cattle are living clocks that show the time of grass* ..." across the region (Maasai Respondent, personal communication, 2007).

High incidences of hyenas indicate that abundance of CRR is in the offing (Maasai-elder, personal communication, 2007), because "*... hyenas always follow food* ..." in this case, prey. Hyenas persistently howl at night during periods of drought, because "*... they are just so full of scavenged meat .... (ibid.)*" Frequent flocking of swallows and increased presence of noisy woodpeckers portend abundance of CRR. Frequent swarming of butterflies portend the same. Incidents of emaciated buffaloes or impalas—both resident-herbivores—denote dire shortage of CRR. Increased sightings of puff adders indicate that there is drought. Higher incidences of calving among the wild animals portend abundance of CRR, particularly pastures. Increased movement of wild animals and occurrences of these animals outside their normal habitats signifies that shortage of these CRR is at hand.

Concerning ecosystem conditions, indicators of shortage of CRR include changes in water and/or vegetation conditions. For example, reduction of levels of water in reservoirs including reduced flow of rivers and lowering levels of water in the boreholes and wells indicate shortage of this CRR. Whenever high levels of water are observed in the boreholes and wells during periods of drought, it signifies abundant rainfall (and hence pastures) in the neighboring places. During the dry-season and or periods of drought, browning and high flow of water in rivers, indicates high rainfall conditions in the upstream. With regard to vegetation, greening and flowering of certain trees and shrubs, for example the *Acacias tortilis*, indicate that rainfall is looming, and consequently heralds abundance in pasture and water.

Fig. 10.1 Indicators of changing natural resources and environmental conditions in the Maasai rangelands of East Africa: (**a**) reduced water volume and river flow, and (**b**) swarming butterflies. Photo Credit: Margaret Mwangi

Maasai elders, or a spiritual leader (*Oloibon*), forecast shortage of CRR for example by observing changes in the quantity and/or quality of local biota. Traditional knowledge about natural resources has been handed down over the generations. However, in general, "... *you have to be perceptive ... it is all about being very conscious that things* [plants, animals, and ecosystems] *are always talking ...*" narrates a Maasai elder during one of the interviews. Under conditions of predicted CRR shortages, Maasai elders often convene an open, dialogue-based meeting (Maa: *Entumo*) where mature women and young men often join the meeting—to make management decisions and allocate resources. Following this meeting, the *Il-Moran* are advised to scout for migratory destinations and individual Olmarei may also cull some of their cattle. In times of plenty, and/or when abundance in CRR is expected, livestock-holding is increased. Decisions concerning increase or culling of livestock occurs at the *Olmarei*- or *enkang*-level.

With regard to preference in environment and natural resource management, a myriad of strategies are employed. For example, patches of high potential ecozones are often reserved for dry-season grazing for a brief duration in any day; and so are other rangeland-resources. A detailed grazing sequence is developed to allow sustainability of wet- and dry-season grazing and *Olopololi* sites, for example. Concerning the rarely done extraction of wild fauna, the Maasai only kill animals on an as-needed basis. For example, they might consume wild animals, such as the eland, particularly in times of drought or other famine-causing catastrophes. In fact, they consider wild herbivores that resemble cattle (e.g., kudu, kongoni, and the like) as their second cattle that are provided by the land and used as appropriate given fluctuations in environmental or social conditions. The *Il-Moran* would hunt a kudu, which is consumed by a number of households during times of drought; sharing is an unspoken and strongly held virtuous norm among the Maasai. The kudu's skin and horns would be conserved for other uses—for example, the latter is blown during *eunoto* ceremony to call the attention of the *Il-Moran;* the former is used to make ropes for tying a bundle of firewood that is ferried on one's (female) back, restraining cow's legs when milking, and other uses. Among the Maasai, "... *you don't kill a wild animal unless it is perilously crucial ...*," a Maasai-elder informs during one of the participatory interviews.

In addition to preserving CRR for livestock during certain time periods, other age-old strategies have historically allowed the Maasai to manage CRR and, indeed, other natural resources across the rangelands of East Africa. For example, select senior elders and the *Oloiboni* judiciously extracted (and some still extract) medicinal parts (bark, root, leaves, or twigs) of certain plants as remedies for various ailments (e.g., *Warburgia* spp. (Maa: *Osokonoi*), is used for calming toothache). The extraction of these components has historically been done in such a way that the life of the tree/shrub remained unthreatened. As examples: the majority of the woody-species remained untouched; a woody-species was never ring-barked—rather a continuous strip of bark was left from the ground up, especially for rare species. Medicinal or food plants (mainly fruits and for brewing beer) and minerals are also extracted on an as needed basis. With regard to minerals—salt, for example—the Maasai traditionally took their livestock to salt-lick sites for only a brief duration in any day (usually once per week); these were often located away from the homestead, and were under the control of a collection of *enkang* and/or *Iloshon*.

Based on the examples above, it is clear that Maasai-pastoralism hinges on the exploitation of local CRR across spatial and temporal scales and formal and informal rules for using these resources have evolved to sustain both the Maasai people and their environment. This livelihood is primarily a better steward of natural resources because of the Maasai's lived experiences and judicious management of natural resources that is entrenched into their daily livelihood. Land and land-resources—and by extension the associated natural resources therein—are government-controlled, a situation that was initially appropriated without consultation with these traditional users. Simultaneously, this was an outcome and a cause of sociopolitical marginalization of the Maasai people (also see conceptual model Fig. 10.4 for a plausible implication of this). Marginalization, especially in making decisions and policies regarding issues affecting their physical landscapes, and by extension their livelihood (Mwangi 2012), contribute to erosion of traditional institutions that ensured the informed management of natural resources by the Maasai.

Given that land and land-resources are under differing tenure systems, the implication is that differing governance systems for natural resources that alter these traditional management practices are emerging. The manifestation of differing governance systems implies that the Maasai's institutions are no longer the *de facto* management practices in these rangelands. This does not mean that these pastoralists have shelved their social institutions of natural resource management. They have not. Rather, these disparate institutions (Maasai and non-Maasai, local and non-local) operate simultaneously. However, since the Maasai are highly marginalized, and heightened social and biophysical changes are occurring across the rangeland of East Africa, their institutions have been rendered considerably less effective (also see conceptual model Fig. 10.4 for a plausible implication of this). In fact, multiple policies that are unappraised by the Maasai have been implemented to manage land and land-resources, and the natural resources therein (e.g., ALDEV 1962; Kameri-Mbote 2002; Kimani and Pickard 1998; Sindiga 1984).

Complicating these mismatches in management approaches are issues associated with climate and other environmental changes. In fact, numerous studies have asserted that these rangelands are also experiencing dramatic variation in evaporation, desertification, and rainfall (IPCC 2001, 2007; McSweeney et al. 2007). These changes are expected to continue fluctuating lowering overall ecosystem stability in the region as some places will become wetter and others drier (IPCC 2001, 2007). Notwithstanding these projections, droughts across the Maasai's rangelands have been more frequent recently (Mwangi 2007, 2012) and temperatures seem to have increased as indicated by the declining glaciers on Mount Kilimanjaro (Kaser et al. 2004). Shifts in these climatic variables will alter growing seasons (including those of natural plants) (IPCC 2001, 2007). Rangeland, and indeed other types of ASALs across Africa, will suffer the most as the climate continues to change (IPCC 2007). In fact, even slight changes in climate could trigger cascading events, and result in significantly altered social and ecological states and are distinct in comparison to historical norms across the ASALs of Africa (Fig. 10.2) (Sivakumar et al. 2005).

Fig. 10.2 Portraits of livelihood-diversification and natural resource interactions in the Maasai rangelands of East Africa. (**a**) Acacia-based charcoal bagged for sale and (**b**) irrigated commercial floricultural greenhouses. Photo Credit: Margaret Mwangi

In addition to Maasai's livelihood diversifications in response to these changing ecological conditions, the landscape is also undergoing social change. Non-Maasai—both locals and non-locals practicing various types of livelihoods, particularly arable-farming—have immigrated to these historically Maasai-dominated rangelands. Natural resources-based diversifications in East Africa rangelands traverse the spectrum to include non-Maasai users. Consequently, the changing community dynamics also play a significant role in determining the sustainability of natural resources in the region. For example, the projected increase in human population for East Africa (IPCC 2001, 2007) translates to more demand for food and energy resources, and by extension, intensified extraction of natural resources. Note that, besides influences from climatic factors, the state of traditional and emergent CRR within which the dominant diversifications depend is continuously affected by persistent and intense anthropogenic pressures as human populations encroach.

In summary, these changing institutional arrangements, changing ecological conditions, and shifts in the community composition all interact and will have significant impacts on the future social-ecological health and sustainability in the region.

10.5 Livelihood-Diversification and Sustainability of Natural Resources in Indigenous Maasai-Pastoralism

This section provides empirical results from the participatory surveys conducted with the Maasai of Kenya and is organized in two parts. The first part details the types and attributes of livelihood-diversification among the Maasai to deal with the changing conditions listed above; the second investigates and explains the influence of the Maasai's attitudes and behaviors under specific climatic, environmental, and socioeconomic conditions on households' diversification using statistical methods incorporated in an agent-based model (ABM).

Table 10.1 shows types and attributes of livelihood-diversification strategies presently utilized by the Maasai. Strategy citing (*Pv*) represents the proportion of mentions of specific livelihood-diversification (0 = unmentioned, 1 = mentioned by all). Temporal (T)- and wealth (W)-contrast are shown, and they respectively represent a fixed/variable and rich/poor dichotomy.

Notably, most of the frequently-mentioned ($Pv > 0.54$) strategies, *namely*, employed in arable-farming (small-scale), arable-farming (own), and trade (livestock and livestock-products) have highly significant contrasts ($p < 0.01$). Others, for example, tourism-based enterprises and fuelwood extraction and sales have moderate-mention, and highly significant p-values. Some, like employed in arable-farming (large-scale) and herding hireling, were scantily-mentioned and have significant temporal-contrast and wealth-contrast ($p < 0.05$). Overall, an estimated 84.21 and 36.84 % livelihood- diversification strategies are derived from natural resources and non-natural resources respectively. In general, diverse, variously mentioned, and of various significance-values livelihood-diversifications are utilized by the Maasai people.

Table 10.2 shows the results from the statistical ABM of livelihood-diversification. The model goodness of fit has a value that is closer to 1 [(value/DF) = 1.0001], which indicates that the model is robust and is therefore reliable (Liang and Zeger 1986; SAS 1999). Livestock-loss to drought has a positive and highly significant effect on the likelihood of livelihood-diversification (38 %, $p < 0.001$). Rainfall has a negative impact on the likelihood of diversifying one's livelihood: specifically, it implies that when there is ample local rainfall, the ratio of the possibility to diversify versus not to diversify becomes 5.52 times larger than otherwise. Factors regarding the head of the household (e.g., attainment of education) are positive and highly significant; with increasing effects the higher the household-head progressed in his education from primary-school (4.01 %) through secondary (11.63 %) to college (14.04 %)—the numbers in parentheses are percent increase in the probability of livelihood-diversification with a unit (1 %) change in the indicated factor. The age of household-head (22.41 %) is positive and very highly significant in terms of its influence on the likelihood of households diversifying.

The estimate for gender (3.40) and age of the working member (20.90) of the household is positive and very highly significant. Similarly, nearness of a household to a local job-market (16.04) and access to a reliable public transportation (9.27) is positive and very highly significant. The implication is that, a 1 % increase in these variables will increase the probability of livelihood-diversification by 3.40, 20.90, 16.04, and 9.27 % respectively.

In Maasai-pastoralism, the key indicators of one's socioeconomic wealth include numbers of livestock, size of household, and land accessed (Mwangi 2012). Wealth based on the size of one's household-livestock (3.01 %), household (28.82 %), and land accessed (3.16 %) had significant positive impact on the likelihood of one's diversifying into other forms of livelihoods. Conversely, wealth based on household-livestock-land (0.69 %) and livestock (9.15 %), and livestock-land (7.22 %) has a strong negative effect on likelihood of livelihood-diversification. Overall, factors that directly emanate from the household-level (>95 %), and particularly related to characteristics of the household-dead and wealth (>50 %) have positive impacts on the likelihood of livelihood-diversification.

10.6 Scenarios for Change: Livelihood-Diversification and Scalar Environmental, Socioeconomic, and Climatic Changes

Current empirical evidence reveals that the Maasai people utilize multiple and different livelihood-diversification strategies, indicating multifaceted divergence from their traditional indigenous pastoralism. In addition to their multiplicity, and with regard to natural rangeland resources, most of these diversifications (>80 %) are predominately based on extraction of/from natural resources. This implies that the extraction of natural resources constitutes the dominant type of diversification in the rangelands of East Africa and is the focus of the following explication vis-à-vis sustainability of natural resources and associated environments.

Within the rangelands of East Africa—and with regard to the dominant natural resource-based diversifications—the core types of natural resources are water, pastures (graminoids, herbaceous, and woody-species), wild fauna, soil, and land. The first two types—essentially the CRR—dominate extraction of natural resources in Maasai-pastoralism, while all of them are widely extracted in the dominant livelihood-diversifications. Therefore, it implies that the sustainability of natural resources is buttressed on factors/processes emanating from these dominant diversifications, and therefore, serve as the livelihood bases.

Specific types of dominant natural resource-based diversifications were frequently-mentioned and statistically significant, for example arable-farming and trade in livestock and livestock-products (EDA, 100 %; $Pv > 54$; $p < 0.01$). The high frequency with which these dominant diversifications were mentioned, denotes an increasing rate of extraction of rangeland natural resources (CRR and land) relative to the past when CRR was sustainably extracted by the Maasai (see the Sect. 10.4). This, coupled with the frequent occurrences of drought in the region (Mwangi 2012), suggest that these natural resources are facing increased pressures. These results accord well with earlier observations of, for example, heightened water-shortages in the arid and semi-arid lands of the greater region (GoK-UNEP 2001; IRIN 2005; UNEP-GoK 2000.) in general, and in the Maasai rangelands in particular (Mwangi 2012). The estimate for local availability of water is positive and highly significant ($p < 0.0001$) implying that households with water available locally face higher odds of diversifying than those accessing this resource from distant places. This could be explained by the fact that water is a crucial resource in arable-farming, and its nearness in this drought-prone region translates to enhanced prospects for cultivation. Furthermore, permanent water reservoirs in these rangelands are often located far apart—a day's walk to a permanent water source is not uncommon. Therefore, where water is available locally—a river, swamp, or pipeline, for example—households opt for cultivation. By contrast, a significant negative estimate for rainfall is evident, which indicates that, when it rains locally, the likelihood of households diversifying diminishes. Specifically, a 1 % increase in rainfall reduces the chances of diversifying by approximately 5.53 %. This is explained by the fact that rainfall replenishes the amounts of CRR for the Maasai's livestock, which coupled with similar effects in livestock prices—a 1.12 % reduction of the likelihood to diversify into non-pastoral livelihood for any 1 % increase in livestock prices—and the Maasai pastoralists' steadfastness (Mwangi 2012) translates to a reduced propensity to engage in arable-farming. Taken together, these interpretations indicate that households actively engaged in livestock husbandry are likely to practice irrigated agriculture under conditions of reliable local water sources than otherwise. This coupled with that fact that these semi-arid areas are characterized by water-shortage implies that Maasais who predominantly engage in arable farming are actively contributing to water-stress in the rangelands.

In addition to water-stress, intensified arable-farming implies that soil resources are also being widely extracted, which presents a considerably new social and ecological context. That situation rarely occurred in historically Maasai-dominated rangelands—only a few patches were cultivated by the Kikuyu people, and it was predominantly done on a shifting basis. In fact, documented evidence shows that from 1973 to 2000, the area under rain-fed agriculture alone increased from 7213 to 24911 hectares in the current study area (Maitima and Olson 2006); much of this cultivation is intensive and practiced in ecologically superior zones (GoK 2002). In socioeconomic terms, soil is a new type of natural resource vis-à-vis the Maasai's production system: an emergent CRR. The following crucial questions emerge; and they should be the focus of future research and/or future management contexts related to the ability of the Maasai to adapt sustainably to future conditions under the model-based environmental or social conditions.

1. Are Maasai pastoralists adequately skilled in dealing with challenges associated with sustainable management of soil (Fig. 10.3a), for example maintenance of soil fertility, and indeed other *emergent* CRR (Fig. 10.3b, c)?
2. Can the same dexterity that Maasai pastoralists have achieved for natural pastures and water resources be transferred toward ensuring sustainable management of *emergent* CRR?
3. Can the observed Maasai's local and indigenous knowledge, values, and preferences inform sustainable management of *emergent* CRR across the rangelands of East Africa?

With regard to changes to CRR, the moderate-mentioning (see Table 10.1) coupled with high statistical significance of fuelwood (firewood and charcoal) extraction and sales—a dominant natural resource based diversification—has the same implication. Apropos this last point, the amplified utilization of woody-species to support these energy resources highlights manifestations of *emergent* CRR, *namely* trees and shrubs. Commercialization of these energy resources to meet the growing demands necessitates heightened extraction, particularly as population growth continues across scales.

Fig. 10.3 Portraits of arising impacts driven by, and drivers of, changes in use of critical rangeland resources of East Africa. (**a**) Gulley erosion; (**b**) active charcoal kiln; and (**c**) a bundle of firewood displayed for sale. Photo Credit: Margaret Mwangi

Notably, factors at the household-level have central impact by way of the likelihood to diversify (Table 10.2). Among the household-level factors, the most significant estimates of diversifications include age of working member (20.90 %), wealth based on the size of household (28.82 %), and household-head's attainment of college education (14.04 %) or his age (22.41 %). Their positive (+) effect implies that they amplify households' odds of diversifying. More specifically, a unit increase in these household-level factors would increase the possibility to diversify by at least 14 %. Most likelihood factors are under the control of the household—an indication that the Maasai are active participants in shaping livelihood-diversifications, and by extension the trajectory of natural resources sustainability in the rangeland spaces from which they extract the same.

That Maasai have subsisted on pastoralism in these rangelands since time immemorial, and given that natural resources contained therein have persistently remained diverse and productive, is an indication of their endurance, implying that the CRR have been sustainably managed. This latter stance can be attributed to judicious extraction, which has afforded the Maasais' lived and informed social-nature interaction, as revealed through the observed knowledge, values, and preferences that characterize Maasai-pastoralism (see Sect. 10.4). This Maasais' lived approach to natural resource management simultaneously incorporates sociocultural and socioeconomic sectors of livelihood as well as environmental and climatic factors. This interpretation accords well with the work of Magee et al. (2013) concerning factors that dictate sustainability of cultural, economic, ecological, and political natural resources. This incorporation of multifaceted factors in efforts that anticipate informing management of natural resources ensures sustainability of the same (Pahl-Wost 2007).

The fact that the estimate for land tenure was positive and highly significant implies that households living on privately owned land are likely to diversify compared to those inhabiting communal lands. This supports the study by Coast (2002) who found that the majority of private landowners (Maasai) tended to engage in non-livestock livelihoods. In addition, access to land alone (3.16%) as a form of wealth, was a significant determinant of diversifications: its effects are almost seven times less than private tenure. The results imply that increasing private land rights would increase the likelihood of households diversifying by a much higher margin than access to available land alone.

It is worth emphasizing that, because households have varied levels of diversifications, they have disparate influence on natural resources and sustainability of the same. Traditionally, Maasai pastoralists maintained strong informal social networks and institutions upon which access to and use of productive natural resources is governed (see Sect. 10.3; Mwangi 2007, 2012, 2016). Management of natural resources occurred across various scales that incorporated family, household, age, gender, and *Maasai-Section* sociocultural dimensions. Presently, natural resources are no longer entirely governed under Maasai systems, but rather under various governance systems and institutions. For example, the numerous wildlife sanctuaries that have been carved out of Maasai rangelands fall under government control. Conservation of the biota therein, particularly fauna, is under the aegis of various policies predominantly enacted by international entities such as the IUCN and CITES. Other resources, such as land, are under various forms of holding that range from private to communal to trust lands (Sindiga 1984; Kimani and Pickard 1998). With non-Maasai management of natural resources, additional parallel and sometimes overlapping institutional and governance structures emerge; a mismatch in scalar and collective management of rangeland resources is highly likely.

These results indicate that a major challenge confronting sustainability of natural resources in the Maasai rangelands of East Africa includes increased land use/tenure pressure (Mwangi 2012, 2016), impacts of the changing climate, and permeation of the effects of global socioeconomic and sociopolitical change (Brooks et al. 2005; IPCC 2001, 2007; McSweeney et al. 2007; Müller et al. 2014; Sivakumar et al. 2005). It is, therefore, crucial that we begin to understand potential trajectories of sustainability of natural resources amid these factors. The influence of the changing climate and/or the heightened permeation of socioeconomic pressures on the sustainability of environments and natural resources in the Maasai's rangelands can be linked to the effects of the various diversifications (Fig. 10.4), especially the dominant ones. From this conceptual model, various interlinked social, economic, political, and cultural factors and/or processes and their interactions shape the trajectory of sustainability of natural resources.

For the Maasai's rangelands, the trajectory of natural resource sustainability lies at the confluence of dominant natural resource-based livelihood-diversifications (Table 10.1), agent-based factors driving these diversifications (Table 10.2), and impacts of climate change and socioeconomic permeations (Fig. 10.4). From Fig. 10.4, factors such as the variable and historically-contingent socioeconomic and sociopolitical marginalization of the Maasai (see Mwangi 2012 for details) make important contributions to the trajectory of natural resource sustainability, particularly when they occasioned unequal access to land and land-resources.

The aforementioned unapprised policies (see Sect. 10.3) indicates both stress on—and exposure of—the rangelands to various deleterious effects including those from the changing climate, socioeconomic and sociopolitical globalization, and agent-based dominant diversification. Increased unsustainable extraction of natural resources will likely follow. The key concern pertains to how this unsustainable extraction will continue to evolve. Using the projections of effects of climate and socioeconomic change (e.g., Brooks et al. 2005; IPCC 2001, 2007; McSweeney et al. 2007; Sivakumar et al. 2005), diverse scenarios of sustainability of natural resources across the Maasai rangelands can be envisioned (Fig. 10.4).

Potential impacts on the traditional and emergent CRR are numerous. For example, projected water-stress and increased temperatures translate to amplified water shortages and impaired productivity of plants, respectively. Consequently, wild animals would suffer food- and water-shortages triggering a decline and/or alteration in species composition and structure. Apropos this last point, grazers, browsers, mixed-feeders and omnivores (and by extension, carnivores) will suffer differential

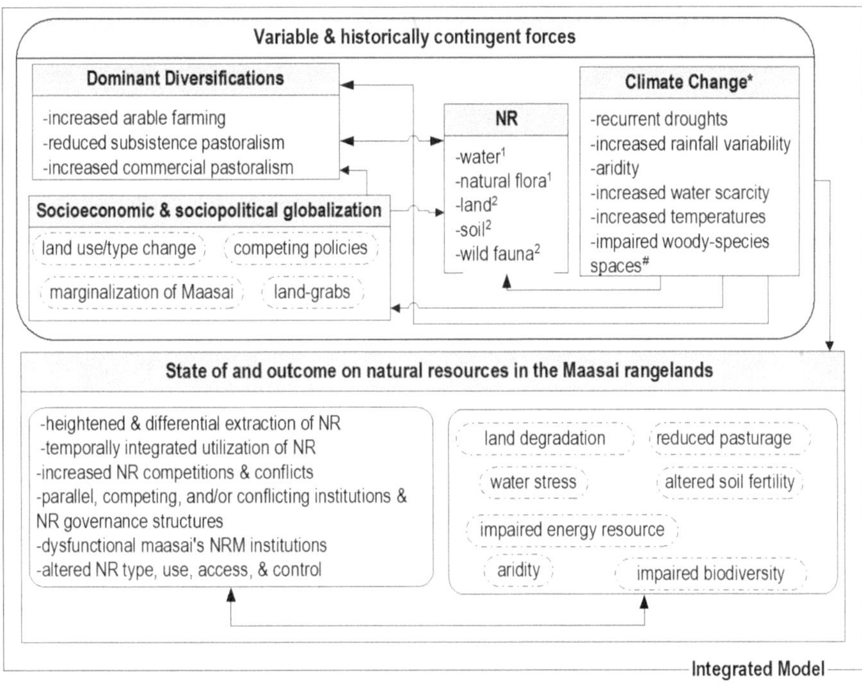

Fig. 10.4 Changing types and sustainability of natural resources (NR) in Maasai rangeland of East Africa in a conceptual model. [1]Traditional CRR, [2]emergent CRR. *Climate change projections (McSweeney et al. 2007; Müller et al. 2014; Sivakumar et al. 2005; IPCC 2001, 2007). Socioeconomic/sociopolitical factors (Brooks et al. 2005; Mwangi 2012, 2016) #Deforestation, woodland degradation, loss of forest quality. Porous borders denote interconnectedness

impacts due to their varied food sources and resultant feedbacks. Recurrent droughts have similar impacts. Recurrent droughts and increased aridity will hinder the proper operation of the presently intensive rain-fed arable-farming, triggering encroachment of irrigated arable-farming, thereby intensifying water-stress. Within these rangelands, irrigation is also practiced regardless of occurrence of drought or manifestation of dry-seasons. For example, much of the floriculture and horticulture is irrigated using water from rivers and ground-aquifers year-round. Besides irrigated arable-farming, forest woodlots, particularly of eucalyptus—a high-water-extracting woody-species of tree with deep roots that draw water from ground aquifers and is able to withstand drought conditions—have become common as demand for telecommunication poles, especially in South Africa, continues to grow. The encroachment of irrigated arable-farming and high-water-extracting plants (e.g., Fig. 10.2) translates to additional pressure on land, wildlife, soil, and other natural resources and highlights the interconnectedness of the impacts of climate change and anthropogenic activities on these CRR.

10.7 Conclusions and Emerging Themes

Maasai pastoralists have widespread knowledge about the environment and natural resource across the rangelands of East Africa. Among the Maasai, management of natural resources is inherently entrenched within their social organization, and it occurs at various interlinked scales that range from individual to community and beyond.

Traditionally, sustainability of natural resources in Maasai-pastoralism was ensured through informed use of rangeland resources and management arrangements that have evolved over time. More specifically, Maasai pastoralists tracked and monitored CRR across spatiotemporal scales, leading to collective action However, as the number of pressures on the landscape continue to evolve and shape the community and the environment, Maasai pastoralists have had to diversify their livelihoods.

The dominant livelihood-diversifications include—in order of increasing ascendancy—individual practices of arable-farming, trade in livestock and livestock-products, and employment in small-scale arable-farming. Livelihood-diversifications among the Maasai have occasioned the emergence of different types of CRR—from water and pastures to incorporate wildlife, forests, land, and soils—with consequent changes in the geography of sustainability of natural resources across the rangelands of East Africa.

Natural resources-based diversifications in East African rangelands traverse the spectrum to include non-Maasai users. Consequently, the trajectories of sustainability of natural resources are more diverse than previously thought.

Maasai-pastoralism is primarily a better steward of natural resources due to the Maasai's lived-experiences and their judicious management of natural resources that is entrenched in their daily livelihood. This, and given that factors at the household-level are the central drivers vis-à-vis the likelihood to diversify, it is, therefore, imperative that effective sustainable management of natural resources start at that scale. Overall, sustainability (or the lack of sustainability) of natural resources, particularly of emergent CRR, is buttressed on factors/processes emanating from these dominant diversifications. This study reveals that the Maasai's continuous adoption of natural resource-based diversifications, coupled with other tribes predominantly subsisting from similar production systems, implies that the natural resources are under persistent and intense pressure. Developing new sustainable management arrangements that address these issues will require understanding the interplay of these pressures, rather than evaluating any one pressure in isolation. The need for rethinking and/or redesigning approaches toward sustainability of natural resources cannot be overemphasized.

Acknowledgments Special thanks to the Maasai community of East Africa for their continued willingness to share their wealth of knowledge with me. Thanks to three geographers—Drs. Alan Taylor, Karl Zimmerer and the late C. Gregory Knight—for valuable discussions in the early framing of research arguments. Dr. Zimmerer, thank you for challenging me to *unpack* the system of nomadic Maasai-pastoralism: this article is one among the various *unpacked* outputs. Special thanks to The International Foundation for Sciences, United Nations University, PennState's Earth and Environmental Systems Institute, and Miombo Network for providing partial financial and logistic support for the broader projects that inform this chapter. The comments by Dr. Steven Gray and an anonymous reviewer helped improve this chapter, and are greatly appreciated.

References

ALDEV (1962) African Land Development Board. Land development in Kenya. Report for 1946–1962. Ministry of Agriculture and Animal Husbandry, Nairobi, Kenya

Baker SJK (1974) A background to the study of drought in East Africa. Afr Aff 73:170–177

Bassett TJ (1988) The political ecology of peasant-herder conflicts in the Northern Ivory Coast. Ann Assoc Am Geogr 78(3):453–472

Behnke RH, Scoones I (1993) Rethinking range ecology: implications for rangeland management in Africa. In: Behnke RH, Scoones, Kerven C (eds) Range ecology at disequilibrium. Overseas Development Institute, London, pp 1–30

Brooks N, Adger WN, Kelly PM (2005) The determinants of vulnerability and adaptive capacity at the national level and the implications for adaptation. Glob Environ Chang 15:151–163

Coast E (2002) Maasai socioeconomic conditions: a cross-border comparison. Hum Ecol 30:79–105

Diggle PJ, Liang KY, Zeger SL (1994) Analysis of longitudinal data. Clarendon, Oxford

Ellis J, Galvin K (1994) Climate patterns and land-use practices in the dry zones of Africa. BioScience 44(5):340–349

FAO (1990) Food and Agriculture Organization. The community's toolbox: the idea, methods and tools for participatory assessment, monitoring and evaluation in community forestry. FAO

Friis-Hansen E, Sthapit B (eds) (2000) Participatory approaches to the conservation and use of plant genetic resources. International Plant Genetic Resources Institute, Rome, Italy, p 214

GoK (2002) Government of Kenya. District development plan. Government Printer, Nairobi, Kajiado

GoK-UNEP (2001) Government of Kenya and UNEP 2001. Devastating drought in Kenya: environmental impacts and responses

IPCC (2001) Climate Change 2001: impacts, adaptation and vulnerability. intergovernmental panel on climate change-third assessment report. Cambridge University Press, Cambridge

IPCC (2007) Climate Change 2007: impacts, adaptation and vulnerability. intergovernmental panel on climate change-fourth assessment report. Cambridge University Press, Cambridge

IRIN (2005) United Nations Integrated Regional Information Networks. UN Office for the Coordination of Humanitarian Affairs, Kenya

Kameri-Mbote P (2002) Property rights and biodiversity management in Kenya. ACTS Press, Nairobi

Kaser G, Hardy DR, Mölg T, Bradley RS, Hyera TM (2004) Modern glacier retreat on Kilimanjaro as evidence of climate change: observation and facts. Int J Climatol 24:329–339

Kimani K, Pickard J (1998) Recent trends and implications of group ranch sub- division and fragmentation in Kajiado District, Kenya. Geogr J 164(2):202–213
Lansing JS, Kremer JN (1993) Emergent properties of Balinese water temples. Am Anthropol 95:97–114
Liang KY, Zeger SL (1986) Longitudinal data analysis using generalized linear models. Biometrika 73:13–22
Magee L, Scerri A, James P, Thom JA, Padgham L, Hickmott S, Deng H, Cahill F (2013) Reframing social sustainability reporting: Towards an engaged approach. Environ Dev Sustain 15(1):225–243
Maitima JM, Olson JM (2006) Arid and Semi-Arid agro-pastoral systems in transition. Lucid Working Policy Brief 2
McSweeney C, New M, Lizcano G (2007) UNDP Climate change country profiles: Kenya. University of Oxford, Oxford, p 26
Müller C, Waha K, Bondeau A, Heinke J (2014) Hotspots of climate change impacts in sub-Saharan Africa and implications for adaptation and development. Glob Chang Biol. doi:10.1111/gcb.12586
Mwangi M (2005) Effects of drought on nomadic pastoralism: impacts and adaptation among the Maasai of Kenya. Global environmental change, globalization and international security: new challenges for the 21st century. IHDP, Bonn, pp 378
Mwangi M (2007) Gender and drought hazards in the rangelands of the Great Horn of Africa: is gender equity the only solution? Women Environ Int 74/75:21–24
Mwangi M (2012) Effects of drought on nomadic pastoralism: impacts and adaptation among the Maasai of Kajiado District, Kenya. Unpublished PhD Dissertation, The Pennsylvania State University
Mwangi M (2016) Globalization, climate change, and evolution of an indigenous livelihood: the scalar transformation of Maasai-Pastoralism in the Rangelands of East Africa (in prep)
Niamir-Fuller M (1999) Conflict management and mobility among pastoralists in Karamoja, Uganda. In: Niamir M (ed) Managing mobility in African rangelands: the legitimization of transhumance. Intermediate Technology Publication, London, pp 18–46
Pahl-Wost C (2007) The implications of complexity for integrated resource management. Environ Model Softw 22(5):561–569
Quinn CH, Huby M, Kiwasila H, Lovett JC (2003) Local perceptions of risk to livelihood in semi-arid Tanzania. J Environ Manage 68:111–119
SAS (1999) SAS/STAT User's guide. SAS Institute Inc., Cary
Scoones I (1995a) Exploiting heterogeneity: habitat use by cattle in the communal areas of Zimbabwe. J Arid Environ 29:221–237
Scoones I (1995b) New directions in pastoral development in Africa. In: Scoones I (ed) Living with uncertainty: new directions in pastoral development in Africa. Intermediate Technology Publications, London
Sindiga I (1984) Land and population problems in Kajiado and Narok, Kenya. Afr Stud Rev 27(1):23–39
Sivakumar MVK, Das HP, Brunini O (2005) Impacts of present and future climate variability and change on agriculture and forestry in the arid and semi-arid tropics. Clim Chang 70(1–2):31–72
Smith K, Barrett CB, Box PW (2000) Participatory risk mapping for targeting research and assistance: with an example from East African pastoralist. World Dev 28(11):1945–1959
Stevens J (1986) Applied multivariate statistics for the social sciences. Lawrence Erlbaum Associates, Hillsdale
UNEP-GoK. United Nations Environment Programme (UNEP) and Government of Kenya (GoK) (2000) Devastating Drought in Kenya. Environmental impacts and responses. UNEP/GoK, Nairobi
Waller R (1988) Emutai: crisis and response in Maasailand, 1883–1902. In: Johnson D, Anderson DM (eds) Ecology of survival. Boulder, Westview, pp 73–113

Chapter 11
Level of Sustainable Activity: A Framework for Integrating Stakeholders into the Simulation Modeling and Management of Mixed-Use Waterways

Robert M. Itami, Randy Gimblett, and Aaron Poe

Chapter Highlights

Approach: We have developed a simulation/stakeholder engagement processes called "Level of Sustainable Activity (LSA)" to engage the boating community in the data collection, issue identification, and definition of management options to understand the relationship between the type and intensity of vessel traffic and impacts on quality of experience, safety, and environmental impact on urban and wilderness waterways.

Stakeholder Engagement: The boating community is segmented by vessel type and whether they are commercial or recreational users. These form the basis for face-to-face interviews and focus groups during the processes of issue identification, data collection, simulation model verification, and developing management options.

Models/Outcomes: The vessel simulation is used to characterize existing boating traffic and project changes in volume and density over 5- and 10-year time frames. The simulation outputs hourly traffic for weekends and weekdays during the peak boating season. Results are summarized by management zone for each vessel type.

R.M. Itami (✉)
GeoDimensions Pty Ltd, 16 Tullyvallin Crescent, Sorrento, VIC 3943, Australia
e-mail: Bob.Itami@geodimensions.com.au

R. Gimblett
School of Natural Resources and Environment, University of Arizona, Tucson, AZ, USA

A. Poe
Aleutian and Bering Sea Islands LCC, U.S. Fish and Wildlife Service,
1011 East Tudor Road, Anchorage, AK 99503, USA

S. Gray et al. (eds.), *Environmental Modeling with Stakeholders*,
DOI 10.1007/978-3-319-25053-3_11

Challenges: The quantitative outputs from simulation models are difficult for managers and the public to interpret especially where there is a complex mix of vessel types and boating schedules. The combination of simulation modeling with the Level of Sustainable Activity framework provides a means for the stakeholders to understand the management process, ensure important issues are addressed, and develop a broad set of management options. LSA has proven to be useful in both urban and wilderness waterways.

11.1 Introduction

Visitor simulation is a useful technology for describing and understanding complex human behavior in recreational environments. The development of Agent-Based and probabilistic models for the purposes of simulating outdoor recreation behavior over the last 20 years has yielded a wealth of experience in the technical and practical aspects of applying simulation to outdoor recreation environments including theoretical issues (Gimblett et al. 1996, 1998) software architecture (Itami 2002; Itami et al. 2003), field methods for collecting reliable input data (Itami 2008; Xia and Arrowsmith 2008), statistical techniques for analysing outputs of the simulation (Kiser et al. 2008), and using simulation outputs to improve applied visitor management (Itami 2002; Itami 2005; Manning et al. 2005).

Simulation models are notorious for being complex; requiring large data sets, an understanding of statistical methods, and producing quantitative outputs that can be hard to understand by decision makers and citizens. Yet, simulation models are often the only method of understanding complex systems with many interacting components. This is exactly the context for using simulation models for vessel traffic management. With the increased popularity of water-based recreation and recreational boating in particular; rivers, lakes and bays worldwide are becoming more crowded. Increased boating traffic raises safety issues, increases conflicts between recreational and commercial users, can create shoreline erosion and turbidity, and taxes infrastructure such as boat ramps, boat storage facilities and parking areas, and boat maintenance facilities.

Waterway managers have a general understanding of the concept of waterway capacity and are enthusiastic about the idea of using simulation to capture the spatial and temporal nature of the problem but are often perplexed as to how to interpret the outputs in a meaningful way that addresses the concerns of a diverse array of recreational and commercial waterway users. Without this understanding, management responses to river traffic volumes and densities may not address the needs of the boating community.

The Level of Sustainable Activity (LSA) framework was developed to better integrate stakeholders into the vessel management decision-making process, to provide better interpretation of vessel simulation outputs from a waterway user's perspective, and to involve them in the generation of management alternatives. The LSA framework uses a focus group approach to segment the boating users to understand—from their perspective—the objectives of users, the type of experience they are seeking, and the impact of varying levels of vessel traffic on these objectives and experiences. This is done with the use of visualizations representing a range of traffic volumes derived from the vessel simulation. These visualisations are used to confirm existing use levels, and to elicit responses from users to use levels projected by simulation models.

11.2 Background: Carrying Capacity for Water-Based Recreation

Carrying capacity is defined by Shelby and Heberlein (1986) as "the level of use beyond which impacts exceed levels specified by evaluative standards." In a recent monograph on carrying capacity, Whittaker et al. (2010) agree that capacity is the amount and type of use that is compatible with the management prescription for an area and is measured on a use level scale which includes (1) units of use, (2) timing, and (3) location components. Capacity can vary across an area for different uses, facilities, seasons or other "management-relevant situations."

Bosley (2005) reviewed seven studies of boating carrying capacity for lakes and reservoirs in the United States. She found most of these studies measured recreational boating capacity in terms of: boat density (acres per boat) derived by different methods for defining the useable or navigable water area by removing shallow water, water near shoreline facilities, and environmentally sensitive areas; determining the number, type, and speed of water craft (existing use), and defining users' perceptions of crowding (social carrying capacity). Environmental and safety issues are generally handled by assigning buffers around sensitive vegetation or erodible shorelines and facilities such as boating docks or swimming areas. A variety of recommended boating densities for single and mixed vessel types have been proposed in various studies as reported by Bosley (2005) in Table 11.1.

Most of these studies do not document how the standards are developed or how they relate to recreation satisfaction or "Social Carrying Capacity." One study by Environmental Resources Management (ERM) (2004) explicitly linked boating densities at Deep Creek Lake, Maryland, to Social Carrying Capacity. Boating densities for three lake zones were calculated for the five boating types in the Warren and Rea (1989) study. The useable acreage for each lake zone was divided by the recommended area for each boat type giving maximum number of boats for each type. A weighted average number of boats for each lake zone (referred to as estimated or calculated carrying capacity) were calculated by multiplying the proportion of each boat type based on summer boat counts in each zone yielding an overall estimated carrying capacity of 8.71 acres/boat.

The problem with the physical carrying capacity standards as shown in Table 11.2, is that they are generally set for lakes and reservoirs and do not adequately address user expectations in different settings (urban to wilderness). The ERM example at Deep Creek Lake improves on these standards by adjusting them to local perceptions of crowding. However, it fails to explore perceptions between different user groups and, therefore, the practice of averaging the values for different user groups into a

Table 11.1 Summary of boat density recommendations from recreational carrying capacity studies

Source	Suggested density	Boating uses
Ashton (1971)	5–9 acres/boat	All uses combined in Cass Lake
	4–9 acres/boat	All uses combined in Orchard Lake
	6–11 acres/boat	All uses combined in Union Lake
Kusler (1972)	40 acres/boat	Water skiing—All uses combined
	20 acres/boat	Water skiing
Jaakson et al. (1989)	20 acres/boat	Water skiing and motorboat cruising
	10 acres/boat	Fishing
	8 acres/boat	Canoeing, kayaking, sailing
	10 acres/boat	All uses combined
Wagner (1991)	25 acres/boat	All recreational activities
Warbach et al. (1994)	30 acres/boat	All motorized (>5 HP) uses
Warren and Rea (1989)	9 acres/boat	Motor boats
	1.3 acres/boat	Fishing from boat
	4.3 acres/boat	Sailing
	1.3 acres/boat	Canoeing, kayaking
	12 acres per boat	Water skiing

Source: Bosley (2005)

Table 11.2 "Reasonable" Boat Density Coefficients for WROS classes (Haas et al. 2004, p. 94)

WROS class range of boating coefficients		
	Low end of range	High end of range
Urban	1 acre/boat	10 acres/boat
Suburban	10 acres/boat	20 acres/boat
Rural developed	20 acres/boat	50 acres/boat
Rural natural	50 acres/boat	110 acres/boat (1/4 sq. mi.)
Semi primitive	110 acres/boat	480 acres/boat (3/4 sq. mi.)
Primitive	480 acres/boat	3200 acres/boat (5 sq. mi.)

single boating density may produce a result that does not satisfy key user groups and may not give management enough detailed guidance to manage use in a targeted way.

11.2.1 WROS: Water Recreational Opportunity Spectrum

The Water Recreational Opportunity Spectrum (WROS) (Haas et al. 2004) is an adaptation of the Recreational Opportunity Spectrum (ROS) for the USDI Bureau of Reclamation first described by Clark and Stankey (1979). This is demonstrated in Table 11.2 showing the low and high range of boating densities for six ROS classes ranging from urban environments to primitive environments. As can be seen, the densities decrease in the less urbanized settings. The WROS guidelines are aimed toward lakes and reservoirs and recognize that appropriate density levels are influenced by a number of interacting factors that need to be considered by recreation managers. These include the number and type of activities and the physical, social, and managerial attributes of the recreation setting. Together, these factors contribute to the quality of recreation experiences resulting in a set of recreation benefits. The WROS classes provide a decision-making context for determining the appropriateness of different recreation activities, facilities, boating densities, and managerial responses. The WROS decision process largely focuses on collecting data to determine the appropriate WROS class or classes for a water body, documenting setting attributes, and judging the compatibility of these attributes to the WROS Class.

The advantage of WROS is that it gives guidance to a broad range of decision makers and stakeholders about the fundamental concepts behind recreation planning and management and goes into some detail framing the issues that influence final capacity decisions within the range suggested for each WROS Class shown in Table 11.2. However, it is not clear how stakeholders should be consulted and what role they play in the decision-making process. Also there is no way to determine if the outcome of a prescribed capacity will result in the desired experience outcomes for visitors.

11.2.2 Level of Service: Capacity for Roadways

The Transportation Research Board (2000) has published the *Highway Capacity Manual 2000* (HCM 2000) that details a framework and methodology for examining facility (roadways, walkways, and bike trails) capacity for vehicular, public transport, pedestrians, and bicycles. The framework is a systems approach to analysis of traffic capacity and includes consideration of both qualitative and quantitative aspects of traffic management. HCM 2000 recognizes that the aim of traffic management is ultimately to achieve qualitative outcomes for the traveling public. The qualitative component is referred to as "Quality of Service" (QOS) which is defined as "a performance indicator of a traveler's perceived satisfaction with the trip." The quantitative side of traffic capacity analysis is the "Level of Service" (LOS) which is defined as "a quantitative measure describing operational conditions within a traffic stream, based on service measures such as speed and travel time, freedom to manoeuvre, traffic interruptions, and comfort and convenience."

Six LOS are defined for each type of facility with available analysis procedures. Letters designate each level, from A to F, with LOS A representing the best operating conditions (highest quality of service) and LOS F the worst. Each Level of Service represents a range of operating conditions and the drivers' perception of those conditions.

The value of the HCM 2000 methodology is the explicit relationship between human qualitative factors and quantitative measures of facility capacity for different travel modes and characteristics of the facility. Table 11.3 shows the relationships between Level of Service, Quality of Service, and Traffic Density. Quality of Service is highest at low traffic densities and lowest at high traffic densities. Level

Table 11.3 Relationship between level of service, quality of service and traffic density

Level of service	Quality of service	Traffic density/capacity
A	High	Low
B	↑	↑
C	|	|
D	|	|
E	↓	↓
F	Low	High

of Service encapsulates both of these concepts by providing a quantitative measure of traffic capacity at each level.

Service Levels (A through F) can be calculated for local conditions depending on vehicle mix, number of lanes, traffic control measures, and observed traffic flows. This allows flexibility in the implementation of the LOS framework to a wide range of local conditions which is also typical of outdoor recreation environments. Methodologies have been developed for vehicles, public transport, pedestrians, and bicycles, but not for either commercial or recreational vessels. However the overall concept should be adaptable to mixed-use waterways.

11.3 Level of Sustainable Activity (LSA) Framework for Waterways

Many of the fundamental characteristics of existing frameworks are useful, including:

- setting the environmental context (WROS),
- using stakeholder preferences to define capacity (ERM 2004) and,
- defining different levels of service for traffic management (Transportation Research Board 2000) linking human qualitative factors to traffic density levels.

However, the following issues remain unaddressed:

- establishing the quality of experience factors for different boating users (recreational and commercial),
- how different traffic densities impact boating safety and the user experience,
- how changing use levels will impact the experience within and between user groups, and,
- the effectiveness of different management strategies in managing traffic densities and maintaining desirable levels of user experience within each waterway zone.

Because of the great degree of uncertainty about all these aspects, a user-based approach was designed to address the fundamental problem of defining traffic capacity. The LSA approach requires engagement and participation from waterway users as well as data collection to characterize the boating environment and establish a baseline for vessel traffic monitoring and management.

11.4 The LSA Approach

The LSA approach is an integrated approach that provides a systematic method for:

- characterizing the waterway,
- identifying stakeholders,
- defining current and future levels of vessel traffic,
- defining environmental, social, and safety issues,
- differentiating the temporal and behavioral patterns of use between different stakeholder groups,
- drawing the relationships between traffic density and user perceptions and boating safety, and,
- defining a range of management responses to address the major issues.

The LSA approach is comprised of ten steps as follows:

11.4.1 Waterway Classification

In complex or large waterways it is strategically important to subdivide the waterway into homogenous zones that are characterized by similarity in use, adjacent land use or vegetation, bank condition, waterway width or depth (navigability), or other natural or social factors that my differentiate management response. The purpose of this type of waterway classification is to set the environmental and social context for management and provide the opportunity for a differentiated set of management responses where environmental and/or social conditions may warrant special management considerations.

11.4.2 Waterway Inventory

The waterway inventory is an objective description of the characteristics that are important for users and management of the waterway. Measurements of waterway area, river width and depth are important inputs for vessel capacity estimates. Other factors also impact capacity such as: shoreline bank conditions including erodibility, sensitive or protected vegetation, beaches, marinas, boat clubs, boat ramps, jetties and other boating infrastructure, as well as adjacent land uses that may be an attraction or hindrance to different boating activities. This inventory is reported for each waterway management zone determined in the previous step.

11.4.3 Selection of Stakeholder Groups

Stakeholder groups are selected not only by common interests but also by common vessel types. Typical groups might include:

- Independent recreational boaters either, motorized or non-motorized
- Boating or fishing clubs
- Commercial ferry and tour operators
- Commercial fishing
- Commercial shipping
- Waterway managers

Waterway managers are included because their perceptions and insight into issues can contribute positively to the understanding of the problem. Also, by including waterway managers in the process it is often possible to highlight miss-matches between the perceptions and attitudes of managers versus waterway users.

11.4.4 Define Issues

Waterway managers will have a good understanding of issues from their point of view, such as legal and legislative issues, boating safety, environmental impacts, and conflicts between specific user groups. However, they may not have a complete understanding of the issues important to different members of the boating community. For this reason, it is important to interview or survey waterway users to gain a complete understanding of issues. This is especially important where there may be entrenched, long-term social conflicts between user groups. Face-to-face interviews with feedback to managers and other users often provide insight into the origins and causes of these conflicts and solutions can be negotiated through the management process. By identifying a complete set of issues early, waterway users gain confidence in the process, and there is less chance of the process being sabotaged if "unforeseen issues" arise late in the management process. During the interview or survey process information relating to pattern of use and forecasted use levels can also be gathered—see below.

11.4.5 Pattern of Use Analysis

For each boating stakeholder group: define when they are on the water (season, time, and duration), the type of vessel they use, the number of vessels, the speed of the vessel, the conditions required (wind, depth, safety, etc.), and the typical path or area of the waterway used. Patterns of use can be defined from traffic observations or interviews with the boating community. Generally, both methods are used. The only systematic monitoring of boating traffic in our studies has been with commercial shipping; other sources of information can be gathered though counts at boat ramps and systematic boat traffic surveys. Results can be presented either in tabular form, charts, aerial photos, or maps (see GeoDimensions 2008) In more complex environments, pattern of use may be modeled in the form of a vessel traffic simulation (see GeoDimensions 2006, 2011; Itami 2008).

Pattern of use analysis is a key to understanding the temporal and spatial dynamics of boating use and provides a detailed insight into the requirements and habits of different boating users (especially boating clubs and commercial operators) and potential means of managing conflicts through scheduled use or spatial redistribution. It also provides the data for setting baseline or current conditions.

11.4.6 Forecast Use Trends

Management plans typically require a forecast of future vessel traffic. If there is no vessel traffic monitoring in place, which is usually the case, it is necessary to forecast growth by compiling secondary data and trying to fill holes in data through stakeholders. Forecasts can be based on existing statistics in tourism growth, boating license records, and growth trends for boating clubs, commercial operations and records for boat ramps, marinas, and boating businesses. Boating and fishing clubs generally keep good records and often have a clear idea of trends in club membership and changes in number of boats and scheduled uses. Commercial operators may be reluctant to divulge business plans, and independent recreational boaters can only be captured through direct observation, intercept surveys, or estimates from boat sales data or boat ramps. The key to getting "best guess" forecasts is to disaggregate the boating population by stakeholder group and get estimates for each group independently. In our experience, this gives remarkably good forecasts and is well suited as input into boating simulations.

In complex environments where there may be a large number of vessels with many different user groups, boating simulations may be the only means of getting accurate forecasts of traffic, especially for different estimates for weekdays versus weekends, hourly estimates, or micro analysis of issues such as boat ramp queues or boating traffic at busy docks or marinas. Vessel traffic simulations can provide detailed information about traffic in different management zones, at different times of the day, differences between weekdays and weekends or seasonal differences in traffic. This gives the public and managers a comprehensive understanding of boating traffic and allows consideration of both local and system-wide issues.

11.4.7 Establish LSA Classes

The Level of Sustainable Activity (LSA) framework is adapted from the US Federal Highway Administration's Level of Service framework (Transportation Research Board 2000). The basic idea is that different levels of traffic density

Table 11.4 Level of sustainable activity (LSA) traffic density levels and relationship to quality of service

Level of sustainable activity (LSA)	Traffic density	Quality of service
A	Lowest	Highest
B		
C		
D	Highest	Lowest

have implications for traffic safety and user experience in terms of stress and the ability to safely navigate traffic: The higher the traffic density, the lower the "Level of Service," and vice versa. Table 11.4 shows how the LSA level relates to traffic density and quality of service. Letters (A, B, C, etc.) are used to designate LSA rather than descriptive terms such as "High Density" in order not to affect user perceptions by use of labels that may be interpreted differently by different users.

The first step is to define four or more traffic density classes. These density classes not only serve to typify traffic volumes but also serve as management targets and a framework for users to conceptualize different traffic densities and respond to the impact of density levels on user experience or "Quality of Experience." Different density levels will have different impacts on quality of experience for different user groups. For example, independent motorized boaters withstand higher density traffic than rowers before their quality of experience declines. Not only is there a relationship between traffic density and user experience, but there is also differences in impact on quality of experience depending on the type of traffic. Rowers in Melbourne can tolerate higher densities with other rowers but quality of experience declines rapidly if passenger ferries are in the mix of traffic.

The LSA classes serve a number of purposes:

- They provide a way of translating traffic counts or simulation outputs to traffic density classes.
- When visualized in the form of graphic displays, stakeholders can readily relate to how densities would affect safety and quality of service.
- It provides management a framework for setting targets for managing vessel traffic.

11.4.8 Define Level of Sustainable Activity for Each Stakeholder Group and Each Management Zone

Once issues have been defined and future traffic volumes are forecast, each stakeholder group is briefed on the study findings to date with opportunities to ask questions and provide more input. The group is then introduced to the concept of LSA traffic density levels with illustrations of each level in the form of aerial photographs or maps. Upon viewing these illustrations, each focus group is asked the following questions:

1. What experience are you looking for as a water user in the study area?
2. What are the key factors that you look for to attain this experience?
3. What experience are you trying to: achieve while recreating or training? (clubs and recreational boaters); provide your clients? (commercial operators)
4. What is the current LSA during peak periods?
5. What is the ideal LSA?
6. What is the maximum tolerable LSA?

If there are known conflicts between different boating groups, questions 4, 5 and 6 are altered and repeated, but expressed in different contexts. The first time the questions are asked the stakeholders are asked to imagine that their group has exclusive use of the waterway (no competing uses). In subsequent repetitions, the stakeholders are asked the same three questions only in reference to sharing the waterway with another specific competing group. This determines if different traffic mixes impact the quality of service at varying levels of traffic density.

11.4.9 Define Management Options

Continuing on from defining the LSA for each focus group, the group is then asked to reflect on the impacts of increasing traffic on safety, quality of experience, and any other issue that arises in discussion. At this stage, the focus groups are fully engaged and are directed to come up with alternative management actions to mitigate impacts. These are compiled, and when all focus groups have completed their input, the management options suggested by the focus groups are summarized, organized and presented to management for feedback. The following are management actions that are typically recommended by stakeholders:

- Controlling use levels for one or more groups
- On-site traffic management
- Improve or relocate infrastructure
- Boater education
- New navigation rules
- Enforcement of existing rules

11.4.10 Develop a Vessel Traffic Management Plan

The Vessel Traffic Management Plan is a compilation of the data, the issues, and the management options. The LSA process gives both managers and stakeholders a complete picture of current conditions and expected impacts of future traffic volumes.

The draft vessel management plan is presented as a set of alternative management scenarios, each evaluated with a set of positive and negative outcomes. Management may then select one of the scenarios or may combine aspects of different management scenarios for a final decision. Over the longer term, conditions need to be monitored with ongoing consultation with stakeholders to flag new issues or alert management to emerging conditions that may require action.

The rest of this chapter will focus on two studies that have used the LSA framework. The first study is the most comprehensive implementation of the LSA framework for an urban waterway in Melbourne, Australia. The second is a partial implementation of the framework for a large wilderness waterway in Prince William Sound, Alaska. The two studies demonstrate the flexibility of the approach to very different environments (urban vs wilderness) with very different sets of issues and stakeholders.

11.5 Case Study A: Vessel Traffic Management in an Urban Waterway

The Melbourne Waterways Committee commissioned a study in 2006 to determine the traffic capacity of the Maribyrnong and Yarra Rivers (see Fig. 11.1) to develop a traffic management plan on the basis of the current level of river traffic and the projected traffic for the next 5- and 10-year periods. The project, called the "Two Rivers Vessel Traffic Management Plan" (GeoDimensions 2006, 2011; Itami 2008), specified a vessel traffic simulation model to quantify hourly traffic patterns for typical peak summer days in 2005 and to forecast traffic volumes to 2010 and 2015.

The study brief presumed that a traffic simulation in itself would determine appropriate traffic capacity levels. However, in reality, traffic simulation outputs in themselves *do not* "define" river traffic capacity. As indicated earlier, capacity has a large subjective component to it and may be different for different user groups sharing the waterway. In order to interpret the results of simulation it is necessary to use social science techniques to elicit from waterway users, their "quality of service" or "quality of experience" objectives and then determine the relationship between traffic density levels in respect to their quality of experience. Thus, the LSA framework described in the previous section was developed to provide a comprehensive method of integrating the quantitative outputs of vessel traffic simulation with the qualitative motivations, desired experiences, and concerns of waterway users.

Rather than discussing the implementation of the LSA approach to all seven river management zones indicated in Fig. 11.1, the framework will be discussed in the context of one of the most complex and dynamic river management zones at the heart of the study area: the "Marina/Transit Zone."

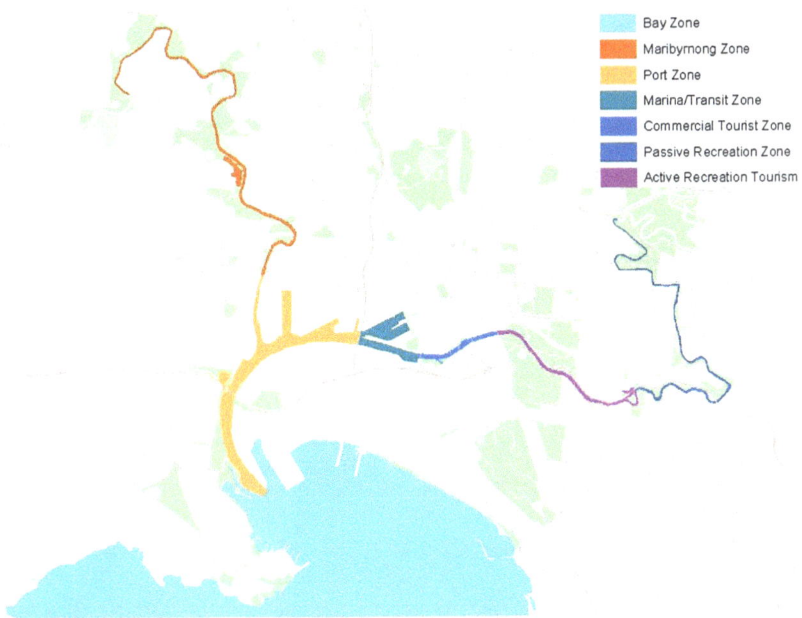

Fig. 11.1 Two Rivers Study area, Victoria, Australia showing seven management zones

11.5.1 Waterway Classification

The Two River's study area was subdivided into seven river management zones by an earlier study. The seven zones represented distinct differences in boating traffic, river characteristics, adjacent land uses, and boating infrastructure. The seven zones were maintained for the vessel traffic management study.

11.5.2 Marina/Transit Zone Inventory

The Marina/Transit Zone (see Fig. 11.2) comprises two water bodies: the Yarra River and Victoria Harbor. Both water bodies were historically used as shipping docks for large commercial shipping. However this area is under redevelopment for residential and commercial land uses and is seen as an extension to the City of Melbourne urban core. In 2005, the Marina/Transit Zone—now known as the "Docklands" was only partially developed with only a single marina developed at "Yarra's Edge" on the Yarra River. By 2010, plans were underway to complete the development of the Docklands along with a total of almost 1000 public and private marina berths taking up almost 37 % of the navigable water in the Docklands.

Description	*Value*
Typical Width	100-120m
Zone Length	1.4 km
Area of Navigable Water	59 ha
Bank Erosion Risk Rating	Low
Major Facilities 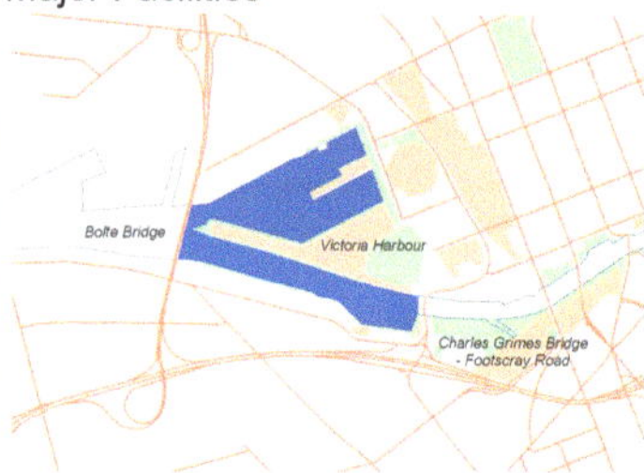	Bolte Bridge Telstra Dome Central Pier Commercial Berths MAB Docklands including private marina and public berths Victoria Harbour including commercial, private marina and public berths Yarra's Edge including private marina and public berths Spencer Street Station

Fig. 11.2 Marina/Transit Management Zone

Also during the 5-year period, new boating clubs including a sailing club, a dragon boat club, and an outrigger canoe club had established themselves at Victoria Harbor and were holding regular training and racing events there. These new recreational uses, along with proposed new marinas, presented some issues and conflicts unanticipated in 2005.

11.5.3 Selection of Stakeholder Groups

Stakeholders are separated into groups with common interests. In the two Rivers' study, previous consultants had defined the following stakeholder groups:

- Commercial shipping, including container ships, tug boats, and pilot boats
- 42 Commercial Operators, including ferries, water taxis, chartered cruises, restaurant boats and public and private marinas
- Recreational motorized boats
- 14 Rowing Clubs, 23 School Rowing Groups, and 5 Canoeing clubs including kayaks and dragon boats
- Sailing clubs including keel boat sailing, junior sailing, and motorized yachts
- 8 Local Government Areas
- 10 State Statutory Bodies

These groups were retained in for the LSA stakeholder groups. For the Marina/Transit management zone there was no commercial shipping or school rowing and only two Local Government Areas.

11.5.4 Define Issues

The project brief provided by the Melbourne Waterways Committee, provided a list of issues that highlighted the problems of lack of data on boating traffic, impacts on bank erosion, conflicts between user groups, and concerns about boating safety and the adequacy of boating infrastructure. However, each of these needed detailed analysis to determine the nature and significance of each issue. Face-to-face and phone interviews with representatives of each stakeholder group were conducted at the beginning of the study. Questions focused on the nature of the waterway use, the type and number of vessels, concerns relating to environmental impacts including bank erosion and noise pollution, waterway safety, and the nature of conflicts with other waterway users. Stakeholders were also given an opportunity to provide open-ended comments relating to any issues they felt were missing from the interview process. Detailed information was collected on patterns of use including user patterns during different seasons of the year, and specific user patterns during peak season weekdays and weekends.

Interviews showed a high degree of interest and concern by the boating community about the need for improved traffic management and there was generally a high degree of participation with only two refusals to participate. The major issues were:

- Safety issues for mixed use especially between larger motorized craft and rowers. School rowers are especially at risk.
- On water conflicts between waterway users resulting in verbal abuse creating unpleasant experiences for passengers on commercial boats and longer term animosity between groups
- Lack of empathy between different user groups on the impacts of boat wake on boating safety and enjoyment
- Impacts of boat wake on shoreline erosion and aquatic plants
- Ignorance of boating rules, especially by drivers of independent motorized vessels.
- Conflicts at shared commercial berths
- Conflicts between independent recreational motorized boats and club sailing and rowing training and events.
- Noise impacts on residential areas due to use of megaphones by rowing coaches.
- Perceived lack of enforcement of rules by agencies
- All the above aggravated during peak use periods.

11.5.5 Pattern of Use Analysis

To explore temporal and spatial distribution of boating use, the Recreation Behavior Simulator (RBSim) was employed. It is a software program fully integrated with GIS that has been specifically developed by researchers from GeoDimensions Pty Ltd and the University of Arizona for studying patterns of recreation use (Itami et al. 2003).

Data for the simulation was gathered using on-site observations, GPS tracking of vessels, and interviews with the boating community. Projected traffic was derived from interviews with the boating community and interviews and documents from planning authorities. Traffic projections were estimated separately for commercial shipping, commercial tour and ferry services, water taxies, rowing clubs, sailing clubs, and private motorized vessels.

The vessel simulation produced traffic volumes and densities for seven river management zones for three types of vessels: Motorized Vessels, Non-Motorized vessels, and Commercial Shipping.

The simulation outputs provide a good indication of current and future traffic patterns with 90 % confidence intervals within 4.75 % of total hourly traffic volumes during hours with the highest traffic variation and 95 % confidence intervals within 5.67 % of total hourly traffic volumes during hours with the lowest traffic variation (11 am to 8 pm).

11.5.6 Forecast Use Trends

Three time periods were simulated for 2005 (baseline conditions), 2010, and 2015. Simulation outputs included hourly traffic volumes and densities for each river management zone for a peak summer day for the three time periods as shown in Table 11.5.

Table 11.5 Hourly traffic volumes and densities for the Marina/Transit Zone with projections for 2010 and 2015

Hour	2005	2010	2015	2005 density	2010 density	2015 density
6:00	2	4	7	0.04	0.07	0.12
7:00	12	21	23	0.20	0.36	0.38
8:00	24	34	41	0.40	0.58	0.69
9:00	22	50	61	0.37	0.84	1.03
10:00	25	55	70	0.42	0.93	1.18
11:00	20	53	72	0.34	0.89	1.22
12:00	18	58	68	0.31	0.98	1.15
13:00	28	63	65	0.48	1.06	1.10
14:00	24	59	67	0.41	1.00	1.14
15:00	17	50	52	0.29	0.85	0.88
16:00	17	43	50	0.29	0.73	0.84
17:00	18	36	46	0.31	0.61	0.77
18:00	18	41	46	0.30	0.69	0.78
19:00	16	41	43	0.28	0.70	0.73
20:00	18	37	41	0.30	0.62	0.69
21:00	6	28	27	0.10	0.46	0.46
22:00	7	25	25	0.11	0.41	0.42
23:00	1	14	15	0.02	0.24	0.25

Densities are vessels per hectare

11.5.7 Establish LSA Classes

Table 11.6 shows four LSA classes used in the Docklands Vessel Traffic Management Plan. Note that the traffic density doubles with each level and the impact on Quality of Service decreases as traffic density increases. These classes were generated by analysing the output from the vessel simulation and discussion and review by managers.

11.5.8 Marina/Transit Zone Level of Sustainable Activity

A key issue in developing a traffic management plan for mixed-use environments like the Docklands is to gain an understanding of the motivations and quality of experience visitors are looking for and commercial operators are trying to provide for their clients. This includes objectives, expectations, the environmental setting, and interaction with other visitors. The Level of Sustainable Activity framework is designed to develop a user-based understanding of environmental and social capacity and ideas to optimally manage visitor experiences. To explore these relationships, stakeholders are first separated into three groups: commercial operators, boating clubs, and marina managers.

Each group is asked the following questions:

- What experience are you looking for as a water user?
- What are the key factors that you look for to attain this experience?
- What experiences are you trying to achieve while recreating or training (clubs and recreational boaters) and providing for your clients in the Docklands Study area?

Users are asked to view four levels of traffic density and make judgements about these densities for the following three circumstances:

- LSA at Peak-use period for non-event days;
- Preferred or Ideal LSA;
- Maximum tolerable LSA and LSA on the "busiest" event days such as Melbourne Cup or New Year.

LSA levels shown in Table 11.6 were represented in photo simulations showing proposed marinas and traffic densities for LSA levels A through D. Figure 11.3 shows the simulated image for LSA level D (6–8 boats per hectare).

Table 11.6 Level of sustainable activity (LSA) traffic density levels

Level of sustainable activity (LSA)	Area/Boat	Boats/Ha
A	10,000 sq. m	1
B	5000 sq. m	2
C	2500 sq. m	4
D	1250 sq. m	8

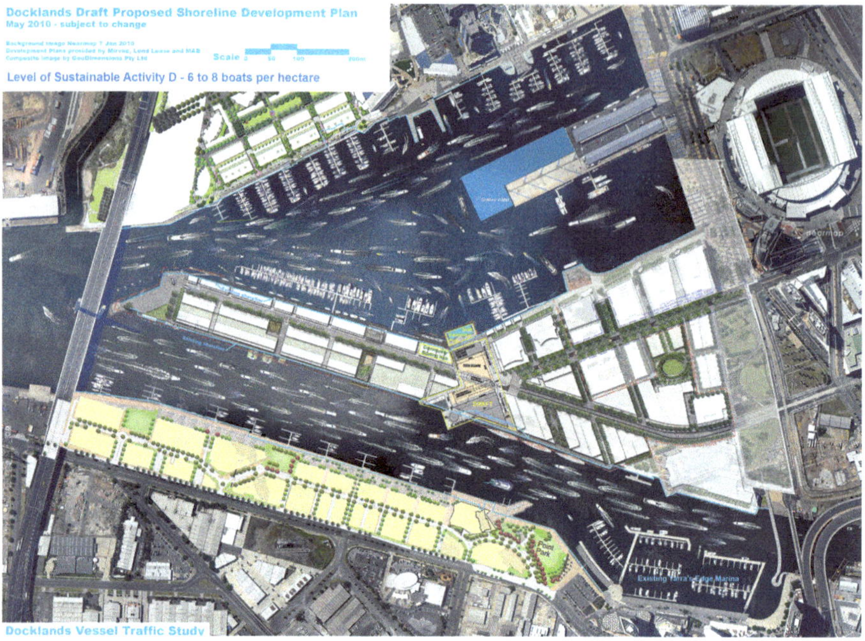

Fig. 11.3 LSA Level D for Marina/Transit Management Zone

11.5.9 Results of LSA Workshop

11.5.9.1 Quality of Experience

Table 11.8 shows the result of the LSA focus groups for the Quality of Experience questions. Quality of experience is influenced by factors such as speed of travel, competing traffic, traffic density, and other factors such as weather, availability of facilities, and condition and maintenance of vessels. When the three stakeholder groups were asked what type of experience they were looking for and what factors contributed to these experiences, most responses were fairly general with very few issues raised. This is probably due to the low traffic volumes in the Docklands area and the relatively safe conditions for existing boating activity in the area. However, requirements among the boating clubs vary because of specific requirements for good quality training and competition. For example, dragon boat users prefer light wind conditions for competition and dinghy sailing requires more wind. This has implications for scheduling training and competition in the Docklands, as wind conditions change in a fairly predictable way with the calmest time of the day in the early morning hours (Table 11.7).

Table 11.8 shows the results of the LSA ratings each group provided for: Ideal LSA, Maximum Tolerable LSA, LSA on busy summer non-event days, and LSA for busiest event days.

Table 11.7 Results of LSA workshops for boating clubs, commercial operators and marina managers

Group	Quality of experience
Boating clubs	• Safety a top priority for all clubs • Outrigger Canoe Club—tolerant to a wide range of conditions, most training occurs outside of Docklands • Dragon Boats—low traffic volumes, specific requirements for competitions • Docklands Yacht Club—need wind to sail, low traffic volumes, open water
Commercial operators	• Provide an enjoyable experience for passengers; provide education, entertainment, and unique views of Melbourne
Marina managers	• Most recreational boaters "out for pleasure cruising," enjoying boating, and to see views of Docklands. Visiting groups prefer to stay together in casual berths when staying overnight • Boats in permanent berths rarely move and owners often have variable boating skills, sometimes needing assistance to berth boat

Table 11.8 Results of LSA evaluations for three boating groups for the Marina/Transit Management Zone

Group	Ideal	Maximum tolerable	Non-event summer days	Event days	Conclusions
Clubs	LSA A	LSA A	LSA B	LSA C	Very sensitive to traffic
Commercial	LSA C	LSA C to <D	LSA A to B	LSA B	Most tolerant to traffic
Marina managers	LSA B	LSA B to C	LSA A to B	LSA B to C	Nearing capacity

Boat clubs are very sensitive to traffic during training and during competitive events. This is indicated in Table 11.8 showing that the Ideal LSA for this group is Level A and Maximum Tolerable LSA is also A. As a consequence this group also perceives the Docklands to be already at capacity during their training and event days. New marinas will result in increasing pressure on this group by reducing available water area and potentially increasing traffic during training and competition. The results also illustrate that dragon boats users may have to change their training times to the early morning. This is not a realistic option for sailing due to wind requirements. Commercial operators are extremely tolerant to increasing levels of traffic and there is currently good cooperation with clubs during events. It is suspected that this may change as space is lost to the development of new marinas. Marina managers are aware of safety issues relating to speed, boater competency, and traffic volumes. Recreational boaters create the greatest boating safety hazard. Victoria Harbor will reach capacity soon after all marinas are built.

11.5.10 Implications of LSA Results in Relation to Simulation Outputs

If the simulation projections are correct, traffic levels for motorized recreation boaters and commercial vessels will remain well within safe levels into 2020 as well as within capacity levels defined by these groups in the LSA workshops. However, for the sailing club and dragon boat club, even these low traffic densities may diminish quality of experience for training and events in Victoria Harbor. The Dragon Boat Club may have to reschedule training to the early morning hours to avoid busy afternoon traffic. The Sailing Club has few options for rescheduling events since they must sail when there is wind (which excludes early morning hours when traffic is lightest) this suggests that on-site vessel traffic management may be necessary during sailing events.

The LSA results show clear differences in boating conditions and crowding tolerances between the three stakeholder groups. This study shows that a single "Carrying Capacity" standard will not satisfy all user groups and, in fact, may create safety hazards if set too high.

11.5.11 Summary and Conclusions

The LSA framework provides a much more detailed understanding of the quality of experience objectives for each user group, and also a more detailed understanding of the boating conditions required by each group, resulting in more fine-tuned management decisions. In addition to the simulation and LSA workshops, boating schedules for each group were examined to determine if temporal separation was possible between groups. Also, a careful analysis of the pattern of use for each group provided guidance for marina design and layout to minimize conflicts between marina vessel traffic in Victoria Harbor.

LSA

- Disaggregates complex problems into simpler subsystems based on stakeholder groups and drives the process forward by constant feedback and engagement of stakeholders and management.
- Engagement of stakeholders through whole process including:
  - definition of issues
  - data collection
  - defining capacity
  - developing management recommendations
- Requires constant feedback to managers and stakeholders
  - the method is labor intensive, requiring good people skills;
  - however, it can provide early identification of potential pitfalls between the stakeholders and management

- The combination of objective traffic counts and computer simulation models with social science approaches lends credibility to the entire management process and gives stakeholders and management a clear contrast between objective data and the perceived needs and concerns of different stakeholders.
- Results in a strong buy-in by management and stakeholders into the process and the subsequent management recommendations
- Sets a baseline for future monitoring and planning studies
- Effectively integrates quantitative and qualitative criteria into the waterway-management process.

River capacity is a multidimensional concept for both river users and managers. Though it is tempting to define a single traffic density to manage across all user groups, the LSA method demonstrates that this strategy is doomed to fail in complex mixed-use environments. Capacity as it has been used in the recreation literature reviewed at the beginning of this chapter does not capture the desires, needs, and safety issues important to different users. This requires more sophisticated management strategies, and the LSA approach is a way to arrive at very specific management options for formulating a robust river traffic management strategy.

11.6 Prince William Sound: LSA in a Wilderness Waterway

Whereas the Two Rivers and Docklands projects are at the extreme urban end of the scale of recreational environments, Prince William Sound in Alaska is at the extreme wilderness end of the scale. The study areas could not be more different with Prince William Sound providing opportunities for vast areas of pristine waterways and uninhabited shorelines. Yet Prince William Sound is under increasing pressure from recreational boating as the area is promoted for unique experiences including whale watching and other wildlife experiences, viewing tidewater glaciers, fishing, hunting, kayaking, and pleasure cruising in the stunning fjords of Alaska. Even with these extreme differences in context, many of the management questions remain the same. Who is using Prince William Sound? How many users are there? What type of experience are they looking for? What level of use is appropriate for ensuring a high quality user experience in a wilderness waterway? Given the fundamentally similar set of questions, there was interest in seeing if the LSA framework would yield the same kinds of insights in Alaska as were discovered in the urban waterways in Australia. The University of Arizona, in association with the US Forest Service and GeoDimensions Pty Ltd, partnered to explore the nature of the dramatic increase in recreation in Prince William Sound since the Exxon Valdez oil spill. The study, funded by the Exxon Valdez reparation fund, was an opportunity to use LSA to explore with residents and visitors to Prince William Sound the type of experiences they were seeking, the activities they engaged in, and the nature of impacts between different users in the wilderness waterway.

The application of the LSA approach in Prince William Sound was not a full implementation as described for the Two Rivers study, so pattern of use and forecasting of use is not covered in this case study. The intent of the study did not include the development of a vessel management plan, so the case study ends with conclusions about the applicability of the LSA framework to wilderness waterways.

11.6.1 Waterway Classification and Inventory: Selecting Representative Bays

Prince William Sound, including the fjords and islands, extends approximately 160 km north to south and 165 km east to west. Because of the enormous extent of the geographic area and the limited time and budget, it was necessary to select three "Analysis Areas" that represented differences in size, use, and distance from Anchorage (the main origin of visitors). The location of the three analysis areas for the LSA workshops. Blackstone Bay is on the west side of Prince William Sound. Unakwik Inlet is north central, and Sheep and Simpson Bays are north of Cordova. These three locations were selected for their geographic distribution, differences in size and shoreline configuration, level of use, and relative distances to the access ports of Cordova, Valdez, and Anchorage. The rationale for selection is to determine how these factors affect local perceptions of quality of service. *Blackstone Bay* is on the west side of Prince William Sound and is the closest of the three Analysis Areas to Whittier and Anchorage. Therefore, it is closest to the largest urban population and consequently has heavier recreational use than the other two analysis areas in this study. Blackstone Bay is 64.1 sq. km. *Sheep and Simpson Bays* on the other hand are three bays north of Cordova and therefore receive a moderate level of use. Together they are 64 sq. km and contain small islands and small coves. Finally, *Unakwik Inlet* is 109.9 sq. km and is the largest of the three analysis areas with a distance of approximately 30 km from the north end of the Inlet to the mouth. It is the furthest from an access port and has the lowest relative amount of use. It also has the most diverse shoreline, with four smaller bays along its perimeter.

11.6.2 Selection of Stakeholder Groups

LSA workshops were conducted in three communities around Prince William Sound. Cordova is a small remote community with a population of 2242 (2008) accessible only by boat (including a ferry service) or plane. Valdez is a small remote community with a population of 3787 (2008) accessible by boat, road, and plane. Finally, Anchorage is Alaska's largest city with a population of 279,243 (2008); it is well serviced by all transportation modes and is the source of most boat traffic into Prince William Sound. Within these three communities three recreational user groups (Kayakers, Recreational motor and sail boaters, as well as Hunters) were selected with separate workshops for each group for a total of nine workshops.

11.6.3 Define Issues

In order to characterize visitors to Prince William Sound, a survey tool was constructed and on-site contacts were made by Forest Service and contract personnel. The survey comprised a carefully organized set of questions that were structured to elicit responses from visitors on a number of experience-based questions, situational responses to encounters with others, and characteristics of their trips. Included in the survey was a trip diary to document spatial and temporal trip patterns. The trip diary was a map of the Sound with instructions on how to document a visitor's own trip from the time of departure, where visitor camped, where visitor stopped, duration of these activities, routes taken, perceptions of place, and displacement (Poe and Gimblett 2015).

11.6.3.1 Results of the Survey

Primary recreation activities reported by respondents were: fishing (54 %), kayaking (13 %), and sightseeing (11 %). When asked about desired recreational experiences during these activities, respondents prioritized based on: (1) enjoying natural beauty; (2) spending time with family and friends; (3) fishing (primarily saltwater); and (4) being in a wild/undeveloped place. Specific choices about destinations were made based (1) good fishing (in saltwater); (2) glacier viewing; and (3) wildlife viewing. The ability to view wildlife was the only activity identified as "very important" to three categories of users (kayakers, small motorized boaters, yacht and sail boaters) (Poe and Gimblett 2015).

In terms of being able to achieve experiences associated with wildland experiences, respondents specifically reported they were able to experience solitude, with 78 % selecting "4" or "5" on a five-point scale indicating their ability to achieve solitude. Similarly, solitude was a strong motivator for survey respondents, but only 10 % identified it as a prominent reason for choosing their destination, suggesting that the ability to achieve solitude is not a limiting factor throughout the Sound as a whole. There was no correlation between longer trips and the desire for solitude. Similarly, when asked "if respondents were to feel crowded, what would they typically do," the overwhelming response (86 %) was to relocate to another location, suggesting that solitude opportunities are not limited (Poe and Gimblett 2015).

Our study illustrates that most visitors to the Sound can still get to their planned destinations and obtain the type of solitude and other opportunities that they desire.

Table 11.9 Levels of service for Prince William Sound

LSA levels for Prince William Sound	
LSA level	Vessels/sq. km
A	0.00
B	0.03
C	0.06
D	0.12
E	0.25

Contrary to what might be considered a popular sentiment, too many visitors or crowding was barely mentioned by questionnaire respondents. This conclusion could then be tested in LSA focus groups.

11.6.4 Establish LSA Classes

To determine realistic LSA levels for Prince William Sound, user survey data from 2005 was used (Wolfe 2007). Daily use levels for kayaks, small motorized boats (<30 ft) and large motorized yachts and sail boats were examined for peak summer days for Blackstone, Unakwik Inlet, and Sheep and Simpson Bays. Table 11.9 shows the density of vessels across all three analysis areas. Levels B, C, and D reflect actual empirical measures of vessel density from Unakwik, Sheep and Simpson Bays, and Blackstone respectively with E being a theoretical geometric progression of use to represent an extreme future possibility, and A being an opposite extreme of no use. Once these density levels were determined, images were prepared for each analysis area using survey locations where actual vessels had been observed over the summer season in 2005.

11.6.5 Define Level of Sustainable Activity for Each Stakeholder Group for Each of Three Bays

The basic premise of the Level of Sustainable Activity framework is that appropriate levels of recreational use density is specific to different user groups, their mode of travel, the geographic location of their destination, and the environmental context. All these factors have an impact on quality of experience. To gain an understanding of how these factors interact, groups of kayakers, recreational boaters, and hunters in Anchorage, Valdez, and Cordova were asked to evaluate five LSA density levels for three locations (Blackstone Bay, Unakwik Inlet, and Sheep and Simpson Bays). Participants in each user group were asked to make evaluations in three different contexts: Their preferred or *Ideal LSA*, expected LSA at busy times, and the *Maximum Tolerable LSA*. They were first asked to make the above evaluations for their own user group (kayakers, recreational boaters, or hunters) and then asked to repeat the evaluations for competing user groups for a total of nine evaluations. Workshop participants were instructed not to consider commercial traffic such as fishing fleets or commercial tour boats and ferries. Individual evaluations were recorded and summarized by taking the median score for each group.

11.6.5.1 Kayakers

Recreational kayakers are self-sufficient and generally paddle close to the shoreline, camp on shore, and are the most sensitive to noise, wake, and shoreline competition from other members of the boating community. They place high value on solitude, pristine, settings and contact with nature and wildlife.

The evaluations for *Ideal LSA* were consistent across the three kayak communities for all vessel types with high preference for low density use (LSA A—no other vessels; or B, 0.03 vessels/sq. km) across all three locations. These ratings are consistent with the high value kayakers expressed for solitude and wilderness experience in the Quality of Experience discussion.

The median values for *Maximum Tolerable LSA* ratings were fairly consistent between communities for each bay. Also the evaluations were fairly consistent by vessel type. Anchorage kayakers have the highest tolerance for other kayakers in Blackstone Bay (LSA D). This is largely due to the large size of the bay and the fact that this level of use is already expected for Blackstone Bay. Valdez kayakers also had high tolerance for other kayakers in Sheep and Simpson Bays. Anchorage kayakers had the lowest tolerance for other vessel types at Unakwik Inlet (LSA B for both small motorized boats and yachts and sailboats). Cordova kayakers also had low tolerance for yachts and sailboats at Unakwik Inlet. Otherwise, there is a high level of consistency for other vessel types by kayakers in all three communities with an LSA rating of C for small motorized boats and motorized yachts and sailboats for all three locations.

11.6.5.2 Recreational Motorized Boaters

Most recreational boaters in the LSA workshops have vessels that are self-sufficient in that they provide facilities for sleeping, eating, and entertaining on-board. The fact that this group does not tend to camp onshore, means that they do not compete with kayakers for campsites. They do, however, come onshore for other land-based activities like berry picking, hiking, and picnicking.

Generally, recreational boaters have a more social orientation than either kayakers or hunters; this is reflected in their generally higher LSA ratings for *Ideal LSA* (LSA B and C) than the other user groups. The reasons expressed for the higher ratings include the concept that other boats in the area are good (in case of emergencies) and that kayaks cause little or no conflict. Valdez boaters had higher *Ideal LSA* ratings for kayaks and small motorized boats at Blackstone Bay (LSA C and D respectively). Their reasoning was that Blackstone is the busiest bay and they expect it to be busy, and it may be appropriate to have heavy traffic in this setting.

The perception that kayaks have little impact on the feeling of wilderness is also reflected in the high LSA ratings for kayaks by all three recreational boating communities during peak season (LSA C, D, and E for Blackstone Bay; B, C, and D for Unakwik Inlet; and B and C for Sheep and Simpson Bays).

Motor yachts and sailboats are seen to have more impact than motor boats of the smaller size; however, recreational boaters as a group are much more tolerant of other boats than kayakers and hunters across all vessel types and all locations.

Maximum Tolerable LSA ratings (C, D, and E) are generally high for Blackstone Bay for kayaks as recreational boaters do not see them as having a big impact either on traffic or on sense of solitude. Also, there is an expectation that Blackstone Bay will be busy because of its proximity to Whittier and the opportunity to view glaciers. Respondents tend to be more tolerant of higher use levels in areas where higher use is expected and are known destinations for commercial tour boats and recreational boats. One of the key factors determining LSA levels for large boats is the availability of safe anchorages. Local knowledge and maps and guidebooks that show anchorages are valuable resources for recreational boaters in Prince William Sound (PWS).

11.6.5.3 Hunters

Hunters generally hunt from small motorized boats. Because hunting season is in the spring and fall, they are usually not in competition with the heavy summer kayaking and recreational boating season. Valdez hunters failed to attend the workshop. For Anchorage and Cordova hunters there is a high level of agreement on the *Ideal LSA* with level A and B ratings across all locations for all vessel types. This is consistent with the expressed value hunters place on solitude and their preference for low competition when hunting. Anchorage hunters expressed concern that PWS was getting too much pressure from hunters and that they were already getting displaced by heavy hunting use or depletion of game stocks.

Hunters see little problem with kayakers camping onshore as they are quiet and do not interrupt game hunting. Therefore, the LSA ratings for small motorized boats are of most relevance when analyzing hunters' perceptions. Cordova hunters had no opinion about conditions in Blackstone Bay for kayaks and motor yachts and sailboats; they did not express the same level of hunting pressure as Anchorage hunters. Their scores show only Blackstone Bay under unacceptable traffic levels for small motorized boats. Interestingly, LSA scores for *Maximum Tolerable LSA* were higher for Cordova hunters for Blackstone and Unakwik Inlet (Anchorage LSA = B, Cordova LSA = C). This may be because Anchorage hunters have more experience with high levels of hunting traffic and are better able to judge the impact of vessel density on hunting quality of experience.

11.6.5.4 Summary

The LSA workshops generally support the intercept survey results relating to lack of concern by boaters in PWS about crowding. However, much more detail about the motives and experiences sought by different users were discovered along with the nature of potential conflicts and compatibilities between different groups into the future. This gives wilderness managers a much clearer insight into the nature of recreational use.

11.6.6 The Applicability of the LSA Framework to Wilderness Waterways

The applicability of the LSA framework to wilderness waterways provided a way to:

- Reveal differences in wilderness qualities and experiences among different recreational groups (e.g., *natural beauty, plentiful wildlife for viewing, fish and game for harvesting,* and *access to solitude and viewing glaciers*).
- Recognize that the successful recovery of recreation in the Sound is likely dependent on facilitating key recreation opportunities sought by users in the region while maintaining a spectrum of available wilderness experiences.
- Offer a means for users to clearly articulate the type and nature of impacts with other user groups.
- Reveal and ease tensions between management and the boating community when issues are clearly defined by users rather than management.
- Access information from focus groups, which allows more in-depth exploration of issues, causes, and identification of potential management solutions.

11.7 Conclusions

This chapter has provided an alternative approach to examining, understanding, and deriving capacity from a user point of view in two polar extremes: small urban rivers at one end and vast wilderness waterways at the other. The chapter has reviewed the literature on recreational boating capacity studies and found them to focus on a single capacity number for a water body, usually measured in area/boat (acres/boat). Where there are mixed-boating uses, a weighted average has generally been used. We argue that these methods do not properly explore the implications of setting a single capacity for a water body from a user's perspective. By coupling waterway simulations with a strategic user-based approach such as LSA, it is possible to move away from a single "normative metric" for defining social carrying capacity to one that explores the acceptability of a range of social and biophysical norms related to visitors. By using surveys followed by focus groups comprised of representatives of each boating type, it is possible to define a more comprehensive set of issues and explore the motivations of each group, the type of experiences they value, and the nature of their conflicts with other users. Using the LSA approach, it is not only possible to determine and simulate the capacity limits for each group, but also to understand the factors that create conflicts between these groups whether they have to do with social, aesthetic, safety, functional, or environmental factors. This provides resource managers with deeper insight into the experience of each group and also provides an opportunity to develop and test much more sophisticated management strategies that can be implemented in the future. The LSA framework demonstrated here shows promise in both urban and wilderness settings.

These studies have found that users are not only fully engaged in the process and they also provide a transparent method for developing management strategies. This more collaborative nature of stakeholder involvement results in a more holistic approach to recreation management. The LSA framework provides a clear way to interpret the outputs of visitor simulations making simulation outputs much easier to interpret both from a management viewpoint and for stakeholders. As demonstrated in the Docklands Project, the method is also useful for exploring new recreational uses in an area and providing detailed guidance to the development of new facilities infrastructure.

The LSA method is a social science approach to waterway management and thus requires good skills in design of surveys and focus group workshops. Excellent skills in presentation, communication, and interviewing are required as well as a team with the technical skills to collect traffic count data and generate valid traffic volumes using computer simulation techniques. The integration of these disciplines provides a much more robust, defensible and transparent basis for waterway management.

References

Ashton P (1971) Recreational boating capacity: a preliminary study of three heavily used lakes in southeastern Michigan. Unpublished Doctoral dissertation, Michigan State University

Bosley HE (2005) Techniques for estimating boating carrying capacity: a literature review. Prepared for Catawba-Water Relicensing Coalition

Clark R, Stankey G (1979) The recreation opportunity spectrum: a framework for planning, management, and research. U.S. Department of Agriculture Forest Service. Pacific Northwest Forest and Range Experiment Station. General Technical Report. PNW-98 December

Environmental Resources Management Inc (2004) Deep Creek Lake boating and commercial use carrying capacity study. Prepared for Maryland Department of Natural Resources

GeoDimensions Pty Ltd (2006) Two rivers traffic management plan: a strategy for sharing Melbourne's rivers and bays. Prepared for Parks Victoria, December

GeoDimensions Pty Ltd (2008) Williamstown vessel traffic management plan. Prepared for Parks Victoria, October

GeoDimensions Pty Ltd (2011) Docklands vessel traffic management plan. Prepared for Melbourne Waterways Working Group, March

Gimblett H, Durnota B, Itami RM (1996) Spatially-explicit autonomous agents for modeling recreation use in complex wilderness landscapes. Complex Int J 3

Gimblett HR, Richards M, Itami RM (1998) A complex systems approach to simulating human behavior using synthetic landscapes. Complex Int J 6
Haas G, Aukerman R, Lovejoy V, Welch D (2004) Water recreation opportunity spectrum (WROS) users' guidebook. United States Department of the Interior/Bureau of Reclamation/Office of Program and Policy Services/Denver Federal Center, Lakewood
Itami RM (2002) Mobile agents with spatial intelligence. In: Gimblett HR (ed) Integrating GIS and agent based modeling techniques for understanding social and ecological processes. Oxford University Press, Oxford
Itami RM, Raulings R, MacLaren G, Hirst K, Gimblett HR, Zanon D, Chladek P (2003) RBSim 2: simulating the complex interactions between human movement and the outdoor recreation environment. J Nat Conserv 11(4):278–286, Urban & Fischer
Itami RM (2005) Port Campbell National Park: Australia: predicting the effects of changes in park infrastructure and increase use. In: Computer simulation modeling of recreation use: current status, case studies, and future directions. Compiled by David N. Cole. USDA Forest Service, Rocky Mountain Research Station, General Technical Report RBRS-GTR-143, September 2005. http://www.fs.fed.us/rm/pubs/rmrs_gtr143.pdf
Itami RM (2008) Level of sustainable activity: moving visitor simulation from description to management for an urban waterway in Australia. In: Gimblett HR, Skov-Petersen H (eds) Monitoring, simulation and management of visitor landscapes. University of Arizona Press, Tucson
Jaakson R, Buszynski MD, Botting D (1989) Carrying capacity and lake recreation planning (part I). The Michigan Riparian, pp 11–12
Kusler JA (1972) Carrying capacity controls for recreation water uses. Upper Great Lakes Regional Commission
Kiser B, Lawson S, Itami RM (2008) Assessing the reliability of computer simulation for modeling visitor use in low use areas. In: Gimblett HR, Skov-Petersen H (eds) Monitoring, simulation and management of visitor landscapes. University of Arizona Press, Tucson
Manning RE, Itami RM, Cole DN, Gimblett HR (2005) Overview of computer simulation modeling approaches and methods. In: Computer simulation modeling of recreation use: current status, case studies, and future directions. Compiled by David N. Cole. USDA Forest Service, Rocky Mountain Research Station, General Technical Report RBRS-GTR-143, September 2005. http://www.fs.fed.us/rm/pubs/rmrs_gtr143.pdf
Poe A, Gimblett HR (2015) Characterizing the spatial and temporal patterns of recreation use and experience opportunities in the sound: perceptions of the stakeholders. In: Poe A, Gimblett HR (eds) Integrating science and community perspectives in Prince William Sound: a case study on managing for sustainable use of wildlands. University of Arizona Press (in review)
Shelby B, Heberlien TA (1986) Carrying capacity in recreation settings. Oregon State University Press, Corvallis
Transportation Research Board (2000) Highway capacity manual 2000. National Research Council/National Academy of Sciences, Washington
Wagner KJ (1991) Assessing impacts of motorized watercraft on lakes: issues and perceptions. In: Proceedings of a national conference on enhancing the states' lake management programs, pp 77–93
Warbach JD, Wyckoff MA, Fisher GE, Johnson P, Gruenwald G (1994) Regulating keyhole development: carrying capacity analysis and ordinances providing lake access regulations. Planning and Zoning Center, Inc.
Warren R, Rea P (1989) Management of aquatic recreation resources. Publishing Horizons, Inc., Columbus
Whittaker D, Shelby B, Manning R, Cole D, Haas G (2010) Capacity reconsidered: finding consensus and clarifying differences. National Association of Recreation Resource Planners, Marienville
Wolfe PE (2007) Boating in Alaska's Prince William Sounds: modeling and assessment of recreation use. Unpublished Master's in Forest Resources Thesis, Oregon State University
Xia J, Arrowsmith C (2008) Techniques for counting and tracking the spatial and temporal movement of visitors. In: Gimblett HR, Skov-Petersen H (eds) Monitoring, simulation and management of visitor landscapes. University of Arizona Press, Tucson

Chapter 12
Engaging Stakeholders in Environmental and Sustainability Decisions with Participatory System Dynamics Modeling

Nuno Videira, Paula Antunes, and Rui Santos

Chapter Highlights

Approach: Participatory system dynamics modeling is presented as an approach to engage broad stakeholder groups in the development of scoping models that foster learning about complex problems in socio-ecological systems.

Participant Engagement: A case study in the Ria Formosa Natural Park in Portugal illustrates the active engagement of over 70 participants in four workshops aimed at building a simulation model that supports analysis of management policies for the protected area.

Models/Outcomes: The Ria Formosa participatory model allows testing of several scenarios for sustainable development of the Natural Park. Process evaluation showed a favorable reaction from participants and positive outcomes with respect to the usefulness of the method in promoting group communication and social learning among stakeholder groups in the protected area.

Challenges: To promote higher impact at the institutional level, a combination of modeling with other tools in a participatory decision-making context is suggested. This challenge is illustrated with the combination of participatory modeling with visioning workshops and multi-criteria analysis.

N. Videira (✉) • P. Antunes • R. Santos
CENSE – Center for Environmental and Sustainability Research, Faculdade de Ciências e Tecnologia, Universidade Nova de Lisboa, Campus de Caparica, 2829-516 Caparica, Portugal
e-mail: nmvc@fct.unl.pt

S. Gray et al. (eds.), *Environmental Modeling with Stakeholders*,
DOI 10.1007/978-3-319-25053-3_12

12.1 Introduction

In the context of environmental decision-making, *participation* refers to any organized process adopted by elected officials, government agencies, or other public or private sector organizations to engage the public and stakeholders in environmental assessment, planning, management, monitoring, and evaluation (NRC 2008).

Active stakeholder participation in decision-making processes is advocated for three fundamental reasons (Blackstock et al. 2007): (1) *normative*–both society and individual citizens are enriched through the encouragement of social and individual learning, (2) *substantive*–accommodating multiple views improves understanding of the issues and subsequently the selection of more appropriate solutions; (3) *instrumental*–success of policy implementation is promoted through the encouragement of collaborative relationships.

As stakeholder engagement in modeling becomes increasingly popular in current decision-making modes (Voinov and Bousquet 2010; Kelly et al. 2013), participatory modeling with system dynamics arises as a promising platform to promote learning in teams dealing with complex environmental and sustainability issues.

This chapter highlights the distinctive elements of the system dynamics modeling approach and how the method is applied in participatory contexts to support learning and implementation of high-leverage policies. With a core shell of flexible and iterative modeling tasks, successive integrative layers unfold to accommodate stakeholder participation and explore the combination with other decision-making methods and tools.

12.2 System Dynamics Modeling Process and Tools

System dynamics is a computer-aided methodology developed in the 1960s for understanding complex systems that change over time (Ford 2010). Interdisciplinary by design, system dynamics draws on theories of nonlinear dynamics and information-feedback control to support an experimental approach to system analysis whereby simulation models are build to support a given decision-making process (Forrester 1961; Richardson and Pugh 1981; Sterman 2000).

Following this perspective, system dynamics models are understood as tools to foster learning and derive new insights about the real systems they represent. Unlike forecasting models, system dynamics models are not designed to be point predictive (Ford 2010). The intended purpose of a system dynamics model is to improve understanding on the patterns of change in a system. This is usually achieved by comparing and interpreting the results obtained through multiple model simulations over a selected time horizon and under different scenario conditions (Ford 2010).

Judging the usefulness of system dynamics models is a process of confidence building and iterative model testing. Since all models are simplified descriptions or abstractions of reality, they are inevitably incomplete or *wrong* (Sterman 2000). The goal is then not to prove that a given model is *right*, but rather to improve the confidence in the model's ability to support understanding of dynamic behavior patterns and the identification of high leverage policies for intervening in the system (Sterman 2000). Useful system dynamics computer simulation models bring value by tackling the *bounded rationality* (Simon 1964) and information-processing limitations of the mental models upon which all decisions are made (Sterman 1988).

12.2.1 Fundamentals of the System Dynamics Approach

To understand how systems behave dynamically, the system dynamics approach is based on three key tenets:

- **Information feedback**: whenever the (decision) environment leads to a decision that results in action that, in turn, affects the environment and thus influences future decisions (Forrester 1961).
- **Time delays:** the time between making a decision and its effects on the state of the system (Sterman 2000).
- **Nonlinearities**: characterizing relationships among system components where the effect is not proportional to the cause and/or separate components have impacts on decisions that are interdependent (Forrester 1961; Sterman 2000).

The interaction of these concepts, together with other complementary sources of dynamic complexity is responsible for the behavior of natural and human systems. Since many scientific, engineering, and societal issues can only be adequately described in terms of complexity and complex systems (Meyers 2009), system dynamics recognizes that embracing complexity is essential to support a holistic understanding of dynamic problems. Systems exhibit dynamics due to changes occurring at many different and sometimes interacting time scales. Together with information feedback, time delays and nonlinearities, other sources of dynamic complexity include: tight couplings between actors and the natural environment, path dependence, self-organization, and adaptability of systems (Sterman 2000).

System dynamics computer simulations are developed to learn how to manage a dynamic problem by allowing *virtual experimentation* (i.e., simulation) with alternative policies for solving that problem (Forrester 1961). One of the main advantages of the system dynamics approach is to enhance understanding of complex dynamic behavior in systems to support the design and testing of strategies that are able to create high leverage, and often counterintuitive, intervention policies. By looking at the causal relationships describing the feedback structure underlying a certain system behavior, decision-makers are better equipped to face recurrent situations of policy resistance and unanticipated side effects. It should be emphasized that the relationships in a system dynamics model must be defined to capture (what the modeler perceives as) *causality* and not *correlation* among variables. Under this approach, only causal relationships should be used to represent the referred underlying structure of the system. On the other hand, correlations may emerge eventually from the behavior simulated by a system dynamics model (Sterman 2000).

12.2.2 Overview of the System Dynamics Modeling Process

The fundamental steps in a system dynamics modeling process are described in detail in several textbooks, including Richardson and Pugh (1981), Sterman (2000), and Morecroft (2007). Publications addressing system dynamics applications to environmental issues have also been presented by Hannon and Ruth (1994), Deaton and Winebreak (1999), and Ford (2010). Building on these reference works, the purpose of this section is then to provide a brief overview of a typical system dynamics modeling process to subsequently *explain* and *illustrate*, in Sects. 12.3 and 12.4 respectively, how a participatory modeling process accommodates model-building activities within a group setting.

The typical question driving a system dynamics model-building process is "*Why does a system behave dynamically in a given way?*" The fundamental postulate is that this question is answered by studying the internal structure underlying the problematic behavior observed in the real world (Richardson and Pugh 1981). Structure hereby designates the "feedback loops, stocks and flows, nonlinearities and delays created by the interaction of the physical and institutional structure of the system with the decision-making processes of the agents acting within it" (Sterman 2000, p. 107).

Since a system dynamics model is developed for a specific purpose, embedded in a given organizational and societal context, another relevant issue is to identify *whose* purpose and *whose* needs are addressed with the modeling effort. This is what Sterman (2000, p. 84) defines as the *clients* of the study (i.e., those people who experience the dynamic problem in a real world context and who can act upon it.) Hence, "clients" for a system dynamics modeling study may be individuals, public and private organizations, communities, or even the public at large (Sterman 2000).

The construction of a system dynamics model follows a continuous progression of iterative and interconnected steps depicted in Fig. 12.1.

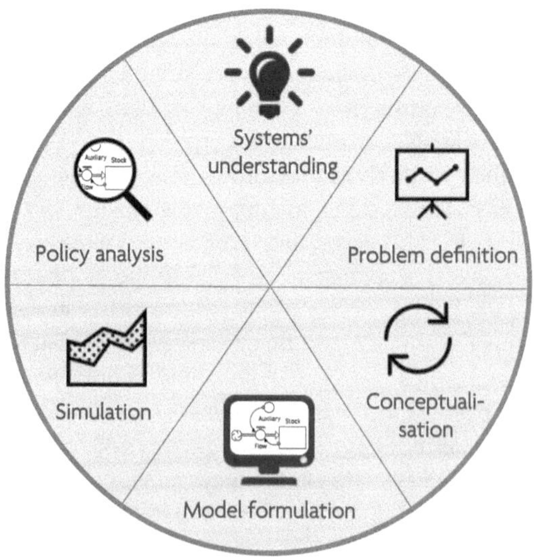

Fig. 12.1 System dynamics modeling process—modeling as a means to improve systems' understanding (Adapted from Richardson and Pugh 1981)

Systems' Understanding The process starts by observing a problematic behavior in the real world, recognizing relevant patterns of behavior and developing a preliminary understanding of the system. The subsequent modeling steps will iteratively improve and build systems' understanding, up to a point where, after a simulation model has been built and policies analyzed, decision-makers (hereby broadly understood as any of the possible model "clients" mentioned above) may implement recommendations back in the real world. As emphasized by Richardson and Pugh (1981), "the model is a means to an end, and that end is understanding". The insights fostered by system dynamics models may be used by individuals and organizations (i.e. model "clients") to support the formulation of policies, plans, programs and inform decisions made in the real world (Sterman 2000). The extent to which the modeling process and model results are embedded in formal planning, assessment, or management processes may vary depending, among several factors, on the kind of agreement/contract established between modelers and "client" organizations or individuals.

Problem Definition The issues to be modeled should be adequately described and framed from a system dynamics perspective. Hence, understanding the domain of study and defining the purpose of the modeling exercise takes shape at this stage. The modeler, often in consultation with the "clients" and using primary and/or secondary data sources (Sterman 2000), starts by building the so-called *reference modes* (i.e., behavior-over-time graphs of important variables characterizing the issues at hand.) This initial sketch provides focus on behavior patterns of the variables that are considered relevant for analyzing the problem and defining policies to address it (Richardson and Pugh 1981; Sterman 2000). Typically, these dynamic patterns translate into simple or combined representations of six main shapes: exponential growth, exponential decay, exponential approach to equilibrium, s-shaped growth, overshoot, and oscillations (Ford 2010). While defining the reference modes, modelers need to select a relevant *time horizon* for analysis, by asking "How far back in the past should relevant behavior patterns be considered?" and "How far in the future will relevant policies be analyzed?". A *problem statement* may be produced at this stage, capturing the purpose of the modeling process, as well as a preliminary hypothesis, which brings forward a cursory interpretation of the causes of behavior patterns depicted in the reference mode.

Conceptualization This step involves abstraction of a simplified structure translating the dynamic problems identified in the real world system. This requires keeping focus and perspective, defining the appropriate *level of aggregation* for the modeling purpose and the *system boundaries*. Modelers identify candidate model components (e.g., variables, causal links) and employ conceptualization tools that will yield a *dynamic hypothesis* (i.e., a cause-effect representation of the structural components that appear to explain the observed problem behavior) (Richardson and Pugh 1981). There are two main conceptualization tools used in system dynamics modeling: causal loop diagrams (CLDs) and stock-and-flow diagrams (SFDs) (see Box 12.1).

Model Formulation This element is the centerpiece of the system dynamics modeling process, whereby a formal, quantitative computer model is built as a laboratory tool to allow experimentation and policy analysis (Richardson and Pugh 1981). This step usually entails the selection of a system dynamics software package, which will support the construction of the SFD and formulation of the model's equations and parameters. As accumulations, stocks are defined in a standard way–they integrate the difference between inflows and outflows over the simulation interval and add that result to the stock's value at a previous time. Inflows and outflows describe the rate of change of stocks and are often formulated with the aid of auxiliary variables and constants. Validation and verification tests are needed to build confidence in the usefulness of the model.

Box 12.1: Notation of Causal Loop (CLDs) and Stock-and-Flow Diagrams (SFDs)
Source: Adapted from Richardson and Pugh (1981); Sterman (2000).

CLDs are conceptual tools to represent the causal chain of effects between a set of variables characterizing a dynamic issue. Arrows are used in CLDs to represent causal links between variables. A positive link (+) indicates that, all else equal, variables change in the same direction (e.g., if "Variable A" is causally linked to "Variable B," when A increases then B also increases). A negative link (–) expresses that the connected variables change in opposite directions. When time delays are considered relevant they are represented in CLDs using a double line mark crossing a causal link arrow connecting two variables.

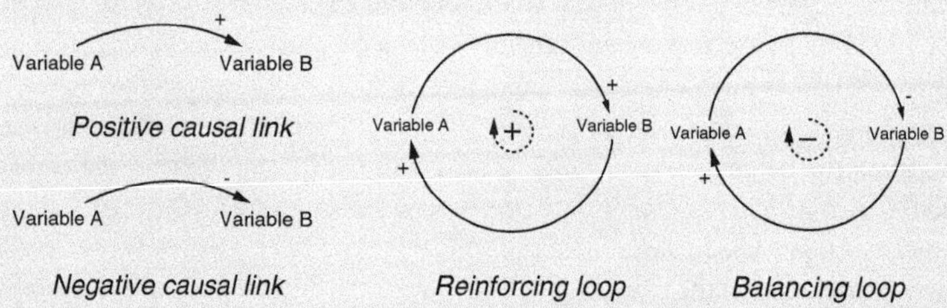

Two types of feedback loops may be created around two or more variables. These are classified as: reinforcing loops (also called positive feedback loops), when tracing the effect of a change around the loop reinforces the initial change; or balancing loops (also called negative feedback loops) when the effect of a change in any of the variables around the loop opposes the initial change. Balancing loops characterize self-correcting or stabilizing behavior, while reinforcing loops are sources of growth, erosion and collapse in systems.

The building blocks of SFDs include stocks (which represent accumulations), inflow and outflow rates (into and out of a stock) and auxiliary vari-

(continued)

Box 12.1 (continued)

ables or converters (which help define the inflow and outflow rates). Converters and flows may be linked to each other and to stocks through arrows depicting cause-effect relationships (also called "action connectors" in STELLA™).

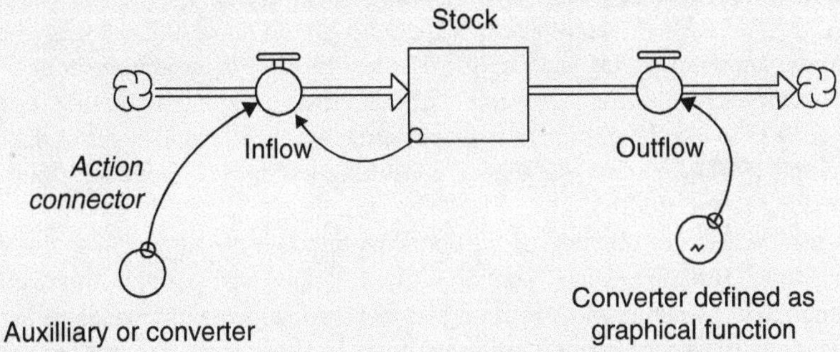

At the conceptualization stage, SFDs may be sketched manually or by using one of the several system dynamics user-friendly software packages, such as STELLA™ and iTHINK™ (from ISEE Systems), VENSIM™ (from Ventana Systems Inc.) or POWERSIM STUDIO™ (from Powersim Software AS). With simple drag-and-drop actions, the different types of variables included in SFDs are selected from the software menus onto a working sheet. At the model formulation stage, dialogue boxes for setting up the equations and parameter values are prompted by clicking on the model elements. Converters may be formalized as constants, equations or defined as graphical functions. Finally, the model may be run to observe behavior of variables over a specified time horizon.

Simulation When all elements are formalized, the model may be simulated to show behavior over a specified time horizon. The modeler may perform several tasks at this stage, such as comparing simulated behavior with reference modes, analyzing which feedback loops dominate behavior, determining the causes underlying possible shifts in loop dominance, and identifying high leverage points of intervention (Sterman 2000). From the analysis of simulated results, insights are derived on how the structure drives behavior of the system.

Policy Analysis Once a validated system dynamics model is obtained, the modeler may experiment with it and test policies for achieving improved system's performance. The model is then used as a virtual environment where "what-if?" questions are simulated and policy recommendations are produced. The modeling cycle is closed when model-based recommendations are implemented in the real world.

12.3 Participatory System Dynamics Modeling

System dynamics is one of the methodologies that recognized the value of stakeholder-based modeling processes at the earliest stages in its history. System dynamicists often emphasize that "modeling with problem-owners" is a defining feature of the method (Lane 2010; Rouwette and Vennix 2006).

In retrospect, several examples of experiences with stakeholder engagement in system dynamics modeling may be found since the 1970s (Andersen et al. 2007). These examples include the involvement of executives in setting marketing strategies (Morecroft 1984), collaboration of management teams in strategic business decisions (Richmond 1997), and development of consulting methods to facilitate group decision-making (Lane 1992).

As the theory and practice of participatory modeling built up over the years, Group Model Building gained support as one of the most popular approaches to develop system dynamics models through direct involvement with managers in client organizations (Richardson and Andersen 1995; Vennix 1996). Similar approaches, such as Mediated Modeling (van den Belt, 2000; van den Belt, 2004), were also developed focusing predominantly on environmental applications and inter-organizational cooperation of broad stakeholder groups.

Recently, participatory system dynamics modeling approaches have been used as a platform for structured deliberation, involving stakeholders in the design, assessment, and implementation of public and private decisions (Stave 2002; van den Belt et al. 2010). Ford (2010, p. 313) summarized the reasons why system dynamics is a suitable method for collaborative modeling of environmental systems:

> The clarity of the stock and flows and the emphasis on the feedback control provide a common language that can be understood by scientists from many disciplines. And system dynamics software aids in the formulation and testing of models in an iterative fashion. The approach stresses clarity and transparency, and is ideally suited for cooperative modeling involving participation by experts from multiple disciplines and by stakeholders.

Developing a participatory system dynamics model implies that the modeling steps previously presented in Fig. 12.1 are embedded in broader processes of collaboration among stakeholder groups. In such contexts, development of system's understanding within the modeling process may also contribute to satisfying other purposes, such as social learning and consensus building (Kallis et al. 2006; van den Belt 2004; Videira et al. 2009). Modeling for learning among stakeholder groups often puts emphasis on the plausibility of outcomes and understanding of dynamic behavior patterns, rather than the predictive accuracy of the model (Kelly et al. 2013; van den Belt 2004).

Participants usually engage in the construction of models at a so-called *scoping level* (i.e., high generality and low resolution models), which may be subsequently iterated and developed to obtain more detailed *research* or *management* models (van den Belt 2004). Whereas scoping models may be constructed in as few as two or three workshop-days (van den Belt 2004; Videira et al. 2011), involving participants in the progressive iteration of such models usually requires a relatively higher time effort from modelers and participants, as described below with the Ria Formosa Natural Park case study.

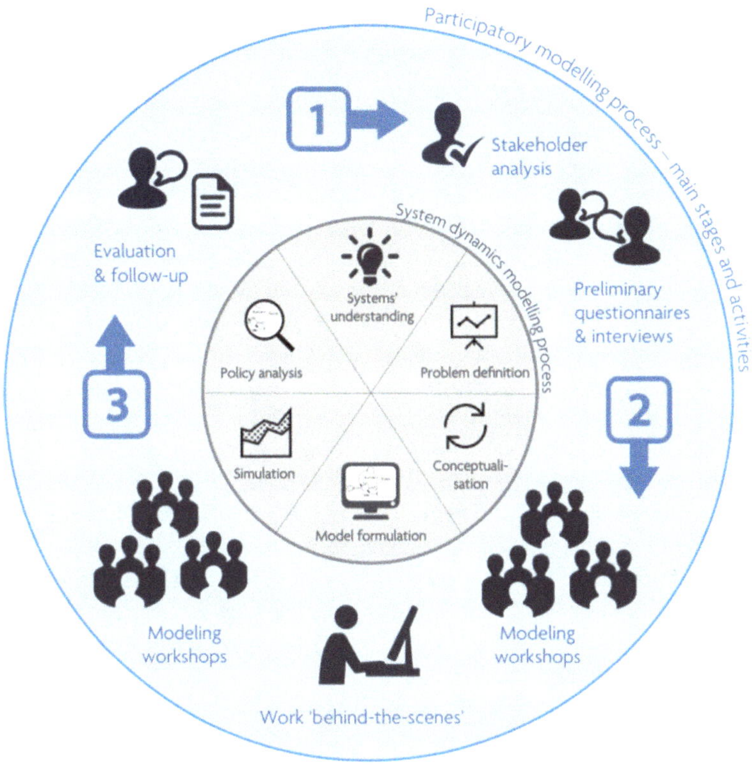

Fig. 12.2 A generic participatory system dynamics modeling process–main stages and activities

A participatory modeling process may have different initiators and sponsors. Some authors have described case studies initiated for research purposes and experimentation with the approach (Antunes et al. 2006; van den Belt 2004; Videira et al. 2009), wherein researchers are the process initiators, often in collaboration with local sponsors or partner organizations. In other situations, participatory modeling projects have been promoted by stakeholder advisory groups convened by governmental agencies (Stave 2002; van den Belt 2004), and "bottom-up" community-based groups (Tidwell et al. 2004). Kallis et al. (2007) recommend that for any deliberative process a *steering group* be established, including core representatives from the sponsoring/initiating organizations, the modeling and/or facilitating professional team, and other relevant stakeholder forums. The division of responsibilities among members of the steering group will depend on the local context in which the participatory modeling process takes place (Kallis et al. 2007). Richardson and Andersen (1995) proposed a set of five roles for group modeling projects, which may be defined (non-exclusively) among members of the steering group: gatekeeper (helps locally to initiate the process), facilitator (conducts group discussions), modeler (develops the computer model), process coach (focuses on group dynamics), and recorder (records workshops' proceedings).

Typically, a generic participatory modeling framework involves three main stages (Fig. 12.2) (Antunes et al. 2006; van den Belt et al. 2010; Videira et al. 2011).

Stage 1: Preparatory Activities The steering committee starts the participatory modeling process with the analysis of stakeholders who should be invited. Stakeholders are hereby understood as societal actors who are or will be affected by, or who have a strong interest in, the outcome of a decision (Beierle and Cayford 2002; NRC 2008). Participatory modeling projects usually follow a *stakeholder analysis* procedure whereby participants are recruited following an identification of the interest groups and networks associated with the selected modeling issues (Videira et al. 2011). A "snowball" referencing method, according to which confirmed participants are asked to suggest other interested parties, complements the invitation process (van den Belt 2004). Several criteria may be employed to decide whom to involve in participatory modeling workshops, such as selecting those participants who bring diverse perspectives on the problem and are able to implement solutions (Vennix 1996; van den Belt 2004).

The *desired level of participation* also needs to be set by the steering group at this stage. A participatory system dynamics modeling process typically aims at active collaboration of stakeholders in the modeling process (i.e., stakeholders working with modelers from the problem definition to the policy analysis step). However, participatory system dynamics modeling approaches may be adapted to other levels of participation (Videira et al. 2011), following the participatory impact spectrum proposed by Arnstein (1969).

Before the modeling workshops occur, it is extremely useful to draw a preliminary assessment of stakeholders' perceptions regarding the issues to be modeled. Hence, *preliminary questionnaires and interviews* with invited stakeholders provide a rich entry point to explain the modeling process and build rapport with participants (Vennix 1996). These results may also be used to build a preliminary system dynamics model or collect the diversity of opinions on driving questions to be addressed at the inception workshop.

Stage 2: Modeling Workshops A typical participatory modeling process develops in a series of workshops, usually intertwining small group (5–12 participants) with large plenary sessions (more than 20 participants) (van den Belt 2004; Kallis et al. 2006). In the Mediated Modeling approach, each of the main stages of the system dynamics modeling approach presented in Fig. 12.1 (from problem definition to policy analysis) is usually developed in a distinct workshop (Antunes et al. 2006; van den Belt 2004). Nevertheless, other configurations have been tested and participatory system dynamics approaches have revealed flexibility in adapting to different contexts, goals, available resources, and time constraints of participants (Videira et al 2011). For example, in several Group Model Building experiences, the perspective of simulating the collaborative model as soon as possible in the workshop series is emphasized (Beall and Ford 2010). Thus, a "running model" may be achieved early in the process and successive iterations improving the model's structure can be based on insights gained from the simulation.

With respect to time requirements for stakeholder participation in the modeling process, the Mediated Modeling cases described by van den Belt (2004) show a range from 8 to 48 workshop-hours. The length of stakeholder involvement depends on several aspects, such as the type of models (e.g., scoping versus research models; simulation models versus qualitative causal loop diagrams), history of the stakeholder group in working together, available resources, and complexity of the modeled problems (Beall and Ford 2010; van den Belt 2004).

Between workshops, modelers perform several supporting tasks *behind-the-scenes* of collaborative modeling activities (Stave 2002). For example, Vennix (1996) describes the elaboration of workbooks, which report on intermediate progress, as a useful way to document the process. Developing detailed scripts for workshop activities, collecting historical data on variables represented in the model, refinement of reference modes, model validation, and verification tests are also examples of behind-the-scenes activities that are essential to a participatory modeling process.

Stage 3: Follow-Up Activities After the workshops conclusion of workshops, an evaluation of results should be carried out by the steering group to bring insights into the impact produced by the participatory modeling process. A systematic evaluation based on four outcome levels is usually suggested (van den Belt 2004; Rouwette et al. 2002; Videira et al. 2009): (1) methodological lessons (e.g., was the method efficient compared to others? Are the models useful?); (2) individual impacts (e.g., did the process foster learning? Did participants react positively?); (3) group effects (e.g., did the process improve communication and promote a shared view?); and (4) organizational changes (e.g., are modeling results implementable in the real world? Were institutional and system changes promoted?). An illustration of possible evaluation measures is provided with the Ria Formosa Natural Park case study presented in the following section. Evaluating participatory modeling outcomes may be based on several instruments, such as, observations, questionnaires or interviews conducted with participants. The follow-up stage may also entail dissemination activities (e.g., reporting final results back to stakeholders) and training stakeholders to use the models autonomously.

12.4 Participatory System Dynamics Modeling in Action: Engaging Stakeholders in Nature Conservation Planning and Management

The Ria Formosa Natural Park is a protected area created in 1987. This coastal park is located in the south of Portugal, spreading along 55 km of low-lying coastline in the Algarve region. The interconnected system of saltmarshes, channels, and sandy barrier islands is a refuge for bird species and an attractive tourist destination area. The Ria Formosa is extensively used for recreational purposes (e.g., beach, golf, yachting, water sports, game fishing) (Videira et al. 2011). With a total area of 18,400 ha, the natural park was established to regulate human activities and protect the unique ecosystem features, in line with the promotion of the cultural, social, and economic development of local populations (Decree-Law No. 373/87 of 9 December 1987). In a complex and multiple-use area such as the Ria Formosa, increasing pressures on natural resources arise from various sources and activities such as tourism, urbanization, aquaculture, bivalve cultivation, fishing, wastewater discharges, and navigation.

After the first Mediated Modeling scoping model was developed in 1998 in the Ria Formosa area (Videira et al. 2004), a follow-up research project was initiated three years later, sponsored by the Portuguese Foundation for Science and Technology (PNAT/1999/GES/15010). The project aimed to contribute to the study of sustainable development of the Ria Formosa Natural Park and was led by researchers from the New University of Lisbon and the University of Algarve in collaboration with park managers. An overview of the process is depicted in Fig. 12.3. This figure highlights the correspondence between each of the four workshops and the system dynamics modeling steps, the number of participants involved at each stage, the duration of each workshop and the activities developed "behind-the-scenes."

12.4.1 Preparatory Stage: Stakeholder Analysis and Interviews

Stakeholder analysis was performed in a series of preparatory meetings with the staff of Ria Formosa Natural Park and the modeling team members. This steering group identified relevant driving forces and pressures affecting management of the protected area, including factors related with (Videira et al. 2011):

- **Institutional context**, namely the definition of property regimes and land-use rules. Jurisdiction overlap was identified as an important barrier to integrated environmental management of the area since several authorities are responsible

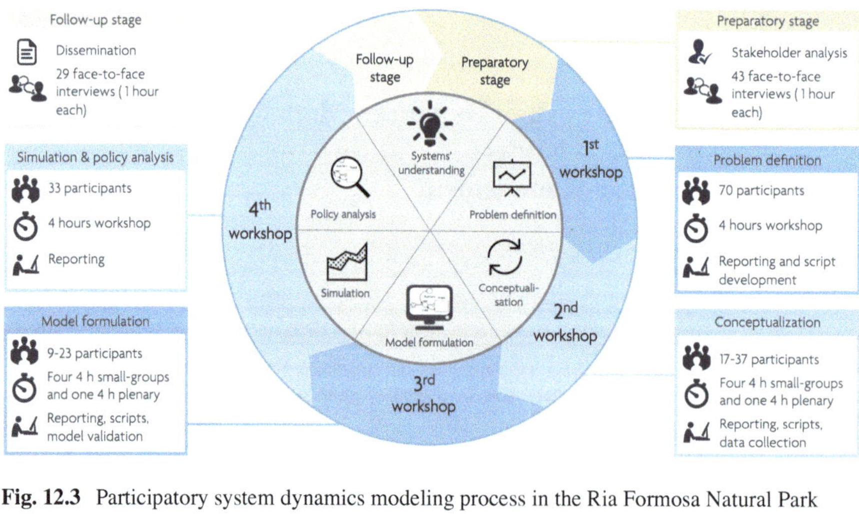

Fig. 12.3 Participatory system dynamics modeling process in the Ria Formosa Natural Park

for seemingly uncoordinated decisions affecting the ecosystem (e.g., licensing of activities using natural resources) (Videira et al. 2003).

- **Economic activities**, such as urban sprawl, fishing, cultivation of fish and bivalves, dredging, and tourism.
- **Natural processes**, including hydrodynamics, migratory and reproductive cycles of species, climate variations, dynamics of sediment transportation.

Subsequently, the steering committee identified stakeholder groups associated with these driving forces and the Park's nature conservation decisions. That procedure resulted in a preliminary list of stakeholders to invite for the participatory modeling process. The diverse interest groups interacting in the area included environmental administration (at national, regional, and local levels), tourism organizations, industrial firms, trade associations, fishing companies, maritime authorities, municipalities, and several NGOs (Videira et al. 2011).

Approximately 100 invitations were sent together with a request to schedule a preliminary interview. Forty stakeholders were available for face-to-face interviews that were conducted prior to the first workshop. A questionnaire was mailed to the remaining stakeholders who had responded positively to the workshop invitation. The interviews and questionnaires provided valuable information for framing the problems to be addressed and capturing "mental models" of stakeholders before the participatory process events. The interview script included questions related to the pressures affecting the Ria Formosa Natural Park, identification of projects to implement in the Park, and perceptions of the adequacy of the protected area nature conservation objectives.

12.4.2 Workshop 1: Problem Definition

Seventy participants from sixty different organizations attended the kick-off workshop. Presentations focused on clarifying the goals of the modeling process, methods deployed, expected outcomes, and collaborative role of stakeholders (see Box 12.2). In the subsequent discussion period, participants debated the Ria Formosa Natural Park management objectives, main drivers of ecosystem change, and current problems in the area. This debate led to an agreement on the problem statement to be addressed: *Develop a system dynamics model to support analysis of management policies contributing to achievement of nature conservation objectives in the Ria Formosa*. The defined time horizon for analysis was set to 1980–2002 (reference modes) and 2002–2015 (scenario runs). It was also agreed that lessons gained from the participatory modeling process would support the development of a management plan for the protected area. At the end of the session, the modeling team proposed creating workgroups (WG) to structure the group modeling tasks to be performed in subsequent meetings. These groups were organized around coherent themes, which roughly corresponded to the sectors of the previously developed Ria Formosa scoping model (Videira et al. 2004).

Box 12.2: Script for the Problem Definition Workshop
Source: Adapted from Videira et al. (2003).

1. Present the participatory modeling process goals and expected outcomes.
2. Discuss the rationale for stakeholder engagement in environmental decisions and clarify the role of participants in the modeling process.
3. Present the fundamentals of the system dynamics approach.
4. Discuss the results from the preliminary interviews.
5. Draw the reference behavior modes for the main problems observed in the protected area.
6. Elaborate a problem statement to be addressed in the modeling process, thus setting system boundaries, the model's time horizon, and target level of aggregation.
7. Establish thematic working groups for the subsequent modeling workshops.

12.4.3 Workshop 2: Conceptualization

The participatory events for the model-conceptualization task were organized in four thematic sessions of approximately four hours each: demography, accessibilities, and urban development (WG1); fisheries and aquaculture (WG2); tourism and recreation (WG3); nature conservation and environmental management (WG4). Using the model developed in the previous Mediated Modeling scoping exercise as a point of departure (Videira et al. 2004), each session aimed at representing key stocks and flows describing feedback processes governing each theme. The preliminary model structure built with the POWERSIM™ software was projected onto a screen and explained to participants who then proposed new variables and causal relationships, thus performing the necessary structural changes to build a model that supported simulation of management scenarios for the Park. A fifth session was subsequently organized to bring together results from each WG. In this plenary session, participants discussed the interrelationships among model sectors, which significantly expanded and detailed the structure of the preliminary scoping model, as summarized in Fig. 12.4.

12.4.4 Workshop 3: Model Formulation

Between workshops, the modeling team collected data needed for the quantification of approximately 300 variables (including 40 stocks and 65 flows) obtained in the conceptual SFD model developed in the previous workshop. Data were collected through interviews with stakeholders, access to databases, and official statistics. With the collected data, some of the model parameters and equations were set.

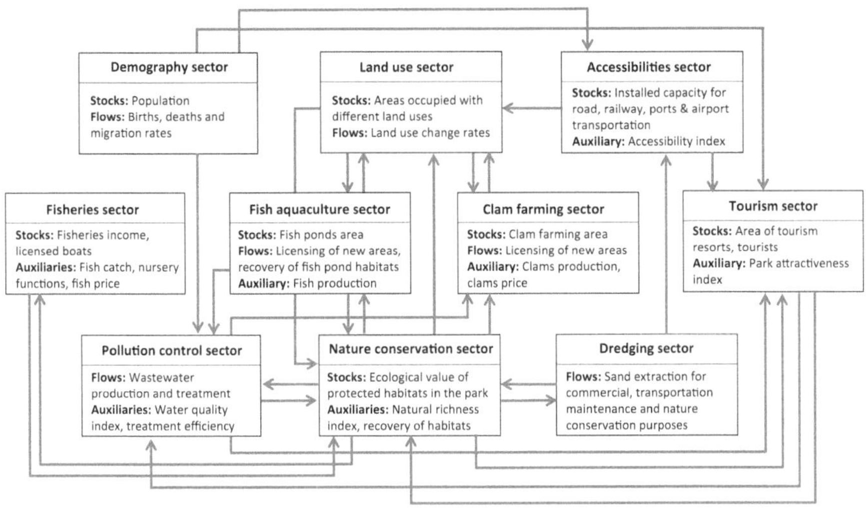

Fig. 12.4 High-level overview of key variables, model sectors, and interrelationships in the Ria Formosa Natural Park participatory model (Adapted from Videira et al. 2003, 2011)

Consistency tests were performed, as well as a qualitative assessment of modelers' confidence in the data substantiating the formulation of each variable, which was later communicated to participants (Videira et al. 2003).

Similar to the second workshop, the third event followed the consecutive thematic working-group structure over a total period of two-and-a-half days (i.e., four WG sessions followed by a final plenary meeting). In each WG, participants discussed proposals for the formulation of mathematical equations and quantification of model parameters. For example, WG1 examined the formulation of an index describing the overall protected area attractiveness. Participants considered that this index should aggregate information on the accessibility to the area, the conservation status of natural habitats, and the water quality in the lagoon system. Following group deliberation, stakeholders proposed to formulate the "Park attractiveness" auxiliary variable as a linear weighted sum of the variables "Accessibility index," "Water quality index," and "Natural richness index." The weights agreed by the group for each sub-index were 0.3, 0.4, and 0.3, respectively (Videira et al. 2003).

In the plenary session, participants gathered to provide feedback on work developed in each of the thematic groups. The third workshop concluded with the development of an outline of the scenarios to be simulated in the final event (Table 12.1).

12.4.5 *Workshop 4: Simulation and Policy Analysis*

A final workbook was prepared "behind-the-scenes" and distributed to participants, documenting the final model and providing guidelines for experimentation of alternative management scenarios for the Ria Formosa Natural Park (Videira et al. 2003).

Table 12.1 Policy analysis experiments developed with the Ria Formosa Natural Park participatory model (Adapted from Videira et al. 2011)

Scenarios	Rationale	Policy analysis
Scenario 1	Maintaining existing management policies for the protected area	Effects on natural values, development of economic activities, and population dynamics
Scenario 2	Increasing investment on nature conservation and water treatment policies	Effects on water quality, conservation value of natural habitats, and overall area attractiveness
Scenario 3	Creating a nature conservation fund using a share of tourism revenues	Effects on conservation of natural habitats and development of tourism activities
Scenario 4	Approving new licenses for clam farming and fish aquaculture	Effects on ecosystem production functions, land use changes and conservation value of intertidal saltmarsh habitats
Scenario 5	Increasing mass tourism in the natural park area	Effects on water quality, quality of life, and conservation value of protected habitats
Scenario 6	Joint implementation of policies in scenarios 2–5	Combination of all effects indicated above

During the first part of the meeting, the goal was to show the model's structure and how it drove the behavior of the system. Hence, each sector of the SFD was projected onto a screen next to the visualization of simulated *behavior over time* of corresponding key variables (from 1980 to 2002).

Subsequently, participants discussed the results from each policy scenario compared against the *business-as-usual* run (Scenario 1). For example, Fig. 12.5 depicts the effects of reinforcing public investments in nature conservation and efficiency of wastewater treatment plants (Scenario 2). As compared to the base run, these environmental policies would improve water quality (i.e., the simulated water quality index increases). They would also lead to an increase in the simulated natural richness index, due to conservation of saltmarshes, intertidal areas, dunes, and other valuable habitats. By reducing pollution and conserving natural values, the overall park attractiveness index, as simulated by the model, showed a significant gain over the policy analysis period (from 2002 to 2015).

During the final workshop, several insights were gained from simulation and policy analysis, particularly since the project allowed for uncovering often unperceived or distant interrelationships between nature conservation and economic sectors in the Ria Formosa. By creating a knowledge *level playing field* (Stave 2002) for discussion among stakeholders and a platform for virtual experimentation with different "what-if" scenarios, the model facilitated an open exploration, over a time scale and across model sectors, of possible consequences emerging from opposing perspectives. For example, by comparing simulation results from scenario 2 ("nature conservation perspective") with those obtained in scenarios 4 and 5 ("economic development perspective"), participants were able to openly discuss implications and trade-offs of each perspective. This example illustrates the argument presented by van den Belt (2004) that a participatory modeling process helps to accommodate disagreeing

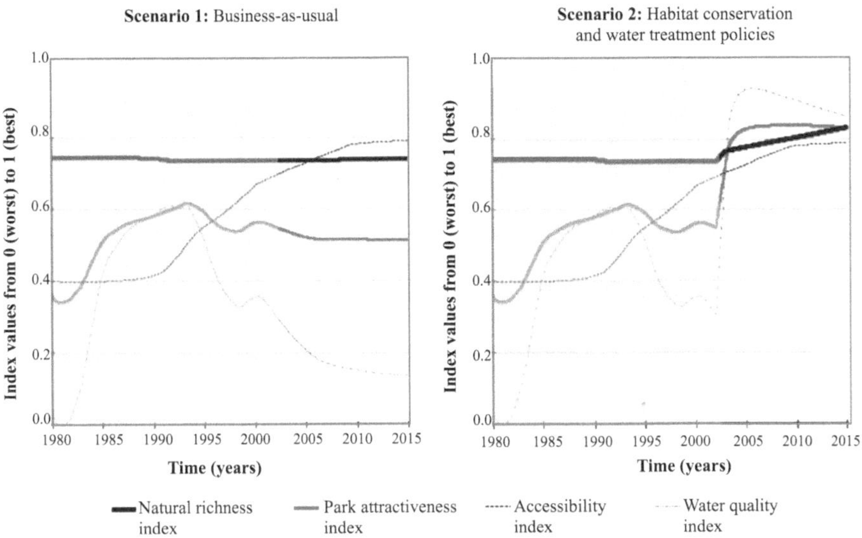

Fig. 12.5 Example of two scenarios simulated with the Ria Formosa Natural Park participatory model

views and resolve conflicts, essentially by establishing a shared holistic view of the relevant system's structure and by promoting the quantification and simulation of stakeholders' assumptions during the modeling process. This does not mean that a full consensus on problems and solutions is to be achieved (or that it is even realistic or desirable in all cases), nor that improved systems' understanding will automatically translate into concerted action. However, through direct engagement in the model-building process, increased ownership towards problem scoping creates favorable conditions for raising commitment towards implementation of shared and knowledge-based solutions (van den Belt 2004).

12.4.6 Follow-Up Stage

The participatory modeling process was concluded with a series of dissemination and evaluation activities. First, project deliverables were distributed to participants, including a synthesis report and the simulation model, for which a user-friendly interface was produced (Fig. 12.6).

To evaluate project outcomes on the previously mentioned four evaluation levels (i.e., methodological lessons, individual impacts, group effects, and organizational changes), a questionnaire was prepared. The goal was to collect perceptions of participants on several dimensions of the participatory modeling process, such as group communication, consensus building, learning effects, and stakeholder commitment towards implementation of model results. In the period following the fourth work-

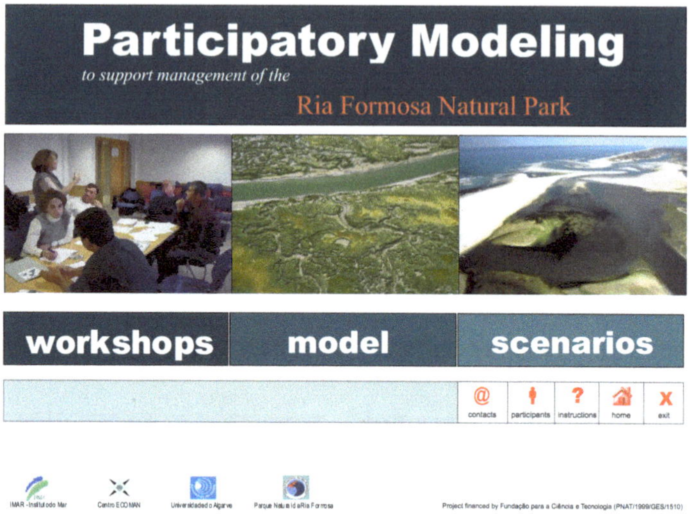

Fig. 12.6 User interface for the Ria Formosa Natural Park participatory model

shop, this questionnaire was used in semi-structured interviews conducted with 29 participants, or sent by e-mail to stakeholders who were not available for a face-to-face meeting.

Evaluation results showed a positive reception by the majority of participants who declared satisfaction with the "openess of discussions," "diversity of representation and involvement of so many stakeholders," and a "better understanding of feedback processes" (Videira et al. 2003). When asked to rate their level of agreement towards a series of statements on a five-point Likert scale (from "1-strongly disagree" to "5-strongly agree"), the majority "agreed" or "strongly agreed" that workshop discussions were constructive, fair, and well structured and that the model is a useful tool for discussing management options for the protected area. Nevertheless, a lukewarm result was observed in relation to a few criteria. For example, participants declared a mild agreement with respect to the achievement of consensus on future actions and regretted that time was short in some of the workshops.

Overall, the Ria Formosa Natural Park case study illustrates well the strengths and limitations of the participatory modeling approach. On one hand, it accommodated a broad stakeholder group in a collaborative learning environment where a shared and holistic understanding of the protected area issues was achieved (Videira et al. 2003). Over approximately 50 h of workshops and several months of "behind-the-scenes" work, the process yielded strong positive results in terms of individual reactions and group communication. The fact that the project aimed at iterating a previous scoping model towards a more detailed model for supporting discussion of park management scenarios, justified the relatively high effort of stakeholder involvement, as compared to other experiences (Beall and Ford 2010; van den Belt 2004). On the downside, there were difficulties in sustaining involvement of some stakeholders over a long time period, particularly since participation in the research project was voluntary and there was no official mandate for implementation of the results. Possible strategies for promoting a stable participation rate in long deliberative processes have been recommended by Kallis et al. (2007). These include paid compensation to participants (e.g., compensation for time or travel costs) and embedding the process in real world decision-making processes to foster institutional commitment and engagement of participants.

Finally, it should be emphasized that while the Ria Formosa participatory modeling process promoted social learning, further resources would be needed to monitor outcomes at the institutional level and assess whether improved systems' understanding was translated into effective decision rules governing the protected area (e.g., assess which organizational changes occurred as a direct or indirect consequence of the participatory modeling process) (Videira et al. 2003).

12.5 Combination of Tools in Participatory Integrated Assessments

Modeling approaches are one of the methodological options for engaging stakeholders in facilitated dialogues and debates (Reed 2008). This section aims to illustrate further possibilities for integrating participatory system dynamics with other platforms promoting stakeholder participation in environmental decision-making.

A first relevant example of integrative potential is the combination of model conceptualization activities and causal loop diagramming with visioning workshops. Visioning is concerned with eliciting desirable futures to assist in strategy development and providing decision-making guidance (O'Brien and Meadows 2001). Olabisi et al. (2010) explored the complementarity of visioning and participatory system dynamics modeling to assist regional leaders in Minnesota in the development of sustainability strategies for local communities. They found that although rarely used together, the two methodologies complement one another since they both acknowledge the complexity within systems and promote a holistic understanding of causal relationships. In another application, Videira et al. (2012) developed an integrated sustainability assessment framework where CLDs were combined with visioning activities. In this case study, a visioning exercise was required to develop common ground among a group of stakeholders interested in maritime issues in Portugal. CLDs obtained in a preceding conceptualization workshop were used as input for the visioning event, wherein desired goals and sustainability criteria for evaluating alternative maritime policies were defined (see Box 12.3).

A second integrative option for enhancing the application of system dynamics modeling in a participatory decision-making context is the combination with multicriteria assessment techniques. As argued by Antunes et al. (2006), to be more effectively used for decision-making, participatory modeling approaches may be combined with some kind of formal appraisal of alternatives. This would enable expanding the

Box 12.3: Participatory Modeling and Visioningfor Sustainability Assessment of Integrated Maritime Policies
Source: Videira et al. (2012).

In 2007, the European Commission presented its plan for Europe's Integrated Maritime Policy. In a case study developed in Portugal within the scope of the SUSTAINAMICS research project (PTDC/AMB/66909), participatory modeling and visioning workshops were conducted to test new approaches supporting active collaboration of stakeholders in the development of national maritime policies. Following the generic methodology presented in Fig. 12.2, a group of relevant stakeholders were first selected and invited for a conceptualization workshop.

A mix of 22 representatives from central public administration, ocean-related businesses, academia, and NGOs were convened during four hours to map causal relationships underlying key maritime issues, such as overexploitation of marine resources, spatial planning conflicts, and impacts on coastal areas. Working in small groups, participants developed causal loop diagrams and identified variables representing the main leverage points to steer sustainable maritime policies. Unlike the model described in the Ria Formosa Natural Park case study, these conceptual diagrams do not allow computer simulation. Using the notation presented in Box 12.1, these system maps depict cause-effect relationships between variables as perceived by participants of each small group. The fact that stakeholders were able to draw feedback loops and develop a systems view of the structure underlying each issue was an important result from this experience. Nevertheless, some participants rated the diagrams as incomplete during the evaluation questionnaire, and a full agreement on the variables and completeness of relationships was not achieved for all CLDs. Still, the majority of participants agreed that the systems-thinking tools deployed helped to structure discussions and analysis of maritime problems. The process created a common language and provided useful qualitative models for problem conceptualization, which may be further iterated.

Subsequently, stakeholders were invited again for a visioning workshop, which gathered 16 participants. The goal was to develop a shared vision and broad sustainability criteria and indicators for the assessment of maritime policies. Instead of using scenarios as a typical point of departure for discussion, like in scenario workshops (Andersen and Jaeger 1999), the methodology considered the use of CLDs from the previous workshop. This option fostered an understanding of the feedback processes governing current trends in key maritime issues. Next, participants deliberated on desired objectives for 2030 for each theme addressed in the CLDs. This deliberation then led to the discovery of common ground leading to a shared vision statement for maritime and costal environments. Wrapping up the meeting, participants worked on the definition of broad criteria and indicators against which alternative maritime policies may be assessed, with the subsequent development of simulation models.

above mentioned objective of promoting understanding towards the use of this improved knowledge for effective support to decision-making processes.

Multi-criteria analysis (MCA) is a decision-making method used to evaluate problems when one is faced with a number of different alternatives and expectations and wants to find the "preferred" solution with regard to different—and often conflicting—objectives. The ability of MCA to deal with complex and unstructured sustainability related decision problems—which involve a number of conflicting environmental, societal, and economic objectives and multiple interest groups—has been widely acknowledged (Paruccini 1994; Beinat and Nijkamp 1998; Kiker et al. 2005).

Participatory modeling may complement multi-criteria methods by accommodating the dynamic nature of socio-ecological systems in the decision process (Antunes et al. 2006), which most of the MCA techniques can not do. Hence, an integrative modeling approach combining participatory model building with multi-criteria evaluation creates an operational framework for decision-making, adding further analytical capabilities to the problem scoping advantages of collaborative system dynamics modeling.

A structured approach to decision making regarding environmental and sustainability issues needs to address the following fundamental tasks (Gregory 2000; Antunes et al. 2006):

- Framing the decision: defining the problem to make sure the "right" problem is addressed. *Right* here means a problem formulation that reflects the main issue at stake in a decision-making process in the perspective of those involved (either decision makers or a more extended stakeholder group);
- Defining key objectives and criteria: what values matter most to stakeholders;
- Establishing alternatives and considering the relevant constraints;
- Identifying consequences: the most important impacts that can affect the stated objectives and associated uncertainties;
- Evaluating the desirability of the consequences according to the proposed criteria;
- Clarifying trade-offs: identifying important conflicts across the desired objectives to use this knowledge to inform decision-making.

Such a process can be undertaken by one single decision-maker, by a small group of decision-makers, or by an extended stakeholder group. In cases of participation of more than one individual there is a need for some form of mediation and consensus building in order to arrive at an agreed upon problem structuring and to assess the desirability of the different options. In situations of controversial issues, where there are different perspectives and conflicting values at stake, approaches such as social multi-criteria evaluation (Munda 2004; Scolobig et al. 2008; Antunes et al. 2011) may be useful. These approaches do not aim to arrive at a consensus, but to shed light on the issues at stake and the points of divergence among different actors.

The combination of participatory system dynamics modeling with MCA can contribute effectively to the development of the proposed tasks (Antunes et al. 2006), as depicted in Fig. 12.7.

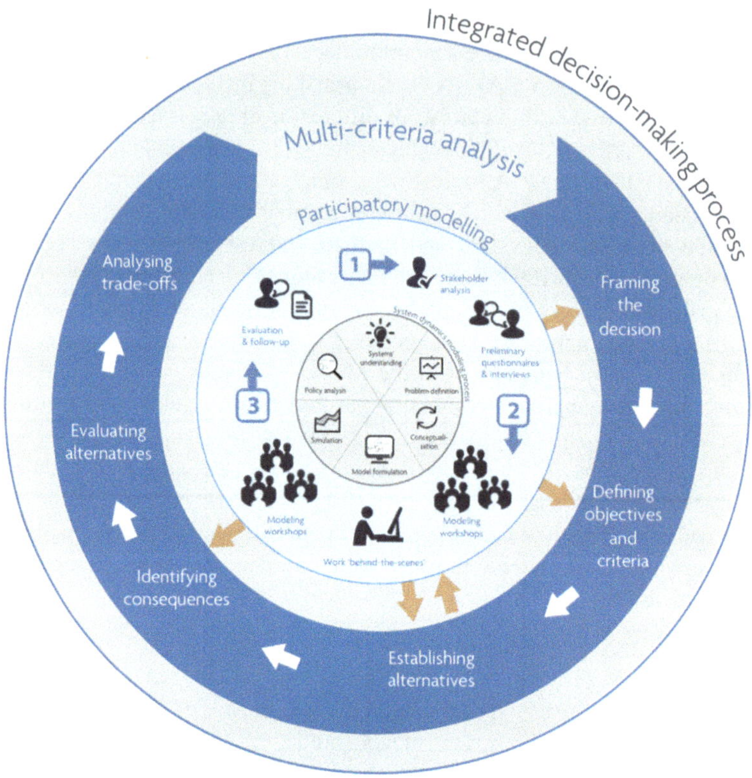

Fig. 12.7 Combining participatory system dynamics modeling with multi-criteria analysis in a decision-making process (Adapted from Antunes et al. 2006)

The group model-building process can be seen as a structuring exercise during which stakeholders develop new insights that are used to frame the decision problem. The mapping of the key variables and relationships in the model-building workshops, and the learning about the system's structure and behavior gained throughout the whole process, can be very helpful for the identification of points of view (represented as decision criteria) and also for the formulation of alternatives. The resulting system dynamics model, developed by the group of stakeholders, can then be used to help them identify and test alternative policy scenarios and discuss their possible consequences for the selected decision criteria, in a dynamic (short- and long-term) perspective.

A participatory MCA methodology may be subsequently used to evaluate the different alternatives in relation to the adopted criteria and also to evaluate trade-offs among conflicting objectives and points of view. Both during the system dynamics modeling and the MCA processes, new (and hopefully better) alternatives can be formulated, which will be considered in the whole process in an iterative way.

The examples presented above bring new avenues for extending the practice of participatory system dynamics modeling towards integrative approaches that create synergies with other deliberative platforms in different stages of planning, assessment and management processes.

Despite the flexibility of the presented participatory modeling approach with respect to the length of workshops, size of stakeholder groups, and size/type of models built, there are some important limitations to consider when selecting this method. For example, the approach may be overly dependent on the presence of an experienced system-dynamics modeler, particularly if simulation models are envisaged. Previous experiences also show that when a participatory modeling process requires a significant time effort for model development, and the process is not embedded in institutional frameworks conducive to implementation of results, additional strategies are needed to sustain participation of voluntary stakeholders throughout the series of workshops.

On the other hand, the strengths of participatory system dynamics modeling–such as fostering group learning processes, accommodating participants with dynamic complexity, and deploying a systems approach to complex issues–indicate that the approach is particularly suited for modeling dynamic issues with stakeholders when co-production of knowledge and learning purposes are envisaged. If combined with other tools that are particularly fit for goal setting and evaluation of alternatives, an integrative toolkit may be developed to support decision-making processes. Hence, by unraveling the underlying feedback structures and improving stakeholders' understanding of complex socio-ecological systems, participatory system dynamics modeling has a lot to offer in the context of integrated environmental and sustainability decision-making.

References

Andersen I, Jaeger B (1999) Scenario workshops and consensus conferences: towards more democratic decision-making. Sci Public Policy 26:331–340

Andersen D, Vennix J, Richardson G, Rouwette E (2007) Group model building: problem structuring, policy simulation and decision support. J Oper Res Soc 58(5):691–694

Antunes P, Santos R, Videira N (2006) Participatory decision making for sustainable development – the use of mediated modeling techniques. Land Use Policy 23:44–52

Antunes P, Karadzic V, Santos R, Beça P, Osann A (2011) Participatory multicriteria analysis for the evaluation of irrigation management alternatives. The case of Caia irrigation area. Portugal Int J Agric Sustain 9(2):334–349

Arnstein SR (1969) A ladder of citizen participation. J Am Inst Plann 30(4):216–224

Beall A, Ford A (2010) Reports from the Field: Assessing the art and science of participatory environmental modeling. Int J Inf Syst Soc Change 1(2):72–89

Beierle T, Cayford J (2002) Democracy in practice. Public Participation in Environmental Decisions. Resources for the Future, Washington

Beinat E, Nijkamp P (eds) (1998) Multicriteria evaluation in land-use management: methodologies and case studies. Kluwer Academic, Dordrecht

Blackstock K, Kelly G, Horsey B (2007) Developing and applying a framework to evaluate participatory research for sustainability. Ecol Econ 60:726–742

Deaton M, Winebrake J (1999) Dynamic modeling of environmental systems. Springer, New York

Ford A (2010) Modeling the environment. Island Press, Washington

Forrester JW (1961) Industrial dynamics. Productivity Press, Portland
Gregory R (2000) Using stakeholder values to make smarter environmental decisions. Environment 42(5):34–44
Hannon B, Ruth M (1994) Dynamic modeling. Springer, New York
Kallis G, Videira N, Antunes P, Guimarães-Pereira Â, Spash C, Coccossis H, Corral-Quintana S, del Moral L, Hatzilacou D, Lobo G, Mexa A, Paneque P, Pedregal B, Santos R (2006) Participatory methods for water resources planning. Environ Plan C Gov Policy 24(2):215–234
Kallis G, Videira N, Antunes P, Santos R (2007) Integrated deliberative decision processes for water resources planning and evaluation. IWA Publishing, London
Kiker GA, Bridges TS, Varghese A, Seager TP, Linkov I (2005) Application of multicriteria decision analysis in environmental decision making. Integr Environ Assess Manag 1(2):95–108
Lane DC (1992) Modelling as learning: a consultancy methodology for enhancing learning in management teams. Eur J Oper Res 59(1):64–84
Lane DC (2010) Participative modelling and big issues: defining features of system dynamics? Syst Res Behav Sci 27(4):461–465
Meyers R (2009) Encyclopedia of complexity and systems science. Springer, New York
Morecroft JDW (1984) Strategy support models. Strateg Manage J 5:215–229
Morecroft J (2007) Strategic modelling and business dynamics, a feedback systems approach. Wiley, Chichester
Munda G (2004) Social multi-criteria evaluation: methodological foundations and operational consequences. Eur J Oper Res 158(3):662–677
NRC (2008) Public participation in environmental assessment and decision making. In: Dietz T, Stern PC (eds) Committee on the human dimensions of global change. National Research Council of the National Academies, The National Academies Press, Washington
O'Brien F, Meadows M (2001) How to develop visions: a literature review, and a revised CHOICES approach for an uncertain world. J Syst Pract Action Res 14(4):495–515
Olabisi L, Kapuscinski A, Johnson K, Reich P, Stenquist B, Draeger K (2010) Using scenario visioning and participatory system dynamics modeling to investigate the future: lessons from Minnesota 2050. Sustainability 2:2686–2706
Paruccini M (1994) Applying multiple criteria aid for decision to environmental management. Kluwer Academic, Dordrecht
Kelly (Letcher) RA, Jakeman A, Barreteau O, Borsuk M, ElSawah S, Hamilton S, Henriksen H, Kuikka S, Maier H, Rizzoli A, van Delden H, Voinov A (2013) Selecting among five common modelling approaches for integrated environmental assessment and management. Environ Model Softw 47:159–181
Reed M (2008) Stakeholder participation for environmental management: a literature review. Biol Conserv 141:2417–2431
Richardson G, Andersen D (1995) Teamwork in group model-building. Syst Dyn Rev 11(2):113–137
Richardson GP, Pugh AL (1981) Introduction to system dynamics modeling with dynamo. Productivity Press, Portland
Richmond B (1997) The strategic forum: aligning objectives, strategy and process. Syst Dyn Rev 13:131–148
Rouwette E, Vennix J (2006) System dynamics and organizational interventions. Syst Res Behav Sci 23:451–466
Rouwette E, Vennix J, Mullekom T (2002) Group model building effectiveness: a review of assessment studies. Syst Dyn Rev 18(1):5–45
Scolobig A, Castán Broto V, Zabala A (2008) Integrating multiple perspectives in social multicriteria evaluation of flood-mitigation alternatives: the case of Malborghetto-Valbruna. Environ Plan C Gov Policy 26(6):1143–1161
Simon H (1964) Theories of Bounded Rationality, CIP Working Paper 66, March 12
Stave K (2002) Using system dynamics to improve public participation in environmental decisions. Syst Dyn Rev 18(2):139–167

Sterman JD (1988) A Skeptic's guide to computer models. In: Grant L (ed) Foresight and national decisions: the horseman and the bureaucrat. University Press of America, Lanham, pp 133–167

Sterman JD (2000) Business dynamics - systems thinking and modeling for a complex world. McGraw-Hill, Boston

Tidwell VD, Passell HD, Conrad SH, Thomas RP (2004) System dynamics modeling for community-based water planning: application to the Middle Rio Grande. Aquat Sci 66(4):357–372

van den Belt M (2000) Mediated modeling: a collaborative approach for the development of shared understanding and evaluation of environmental policy scenarios. With case studies in the Fox River Basin, Wisconsin and the Ria Formosa, Portugal. Ph.D. Dissertation, University of Maryland

van den Belt M (2004) Mediated modeling: a systems dynamics approach to environmental consensus building. Island Press, Washington

van den Belt M, Kenyan J, Krueger E, Maynards A, Galen Roy M, Raphael I (2010) Public Sector administration of ecological economics systems using mediated modeling. Ann N Y Acad Sci 1185:196–210

Vennix J (1996) Group model building: facilitating team learning using system dynamics. Wiley, Chichester

Videira N, Antunes P, Santos R, Gamito S (2003) Participatory modelling in environmental decision-making: the Ria Formosa Natural Park Case Study. J Environ Assess Policy Manage 5(3):421–447

Videira N, van den Belt M, Antunes P, Santos R, Gamito S (2004) Decision support in coastal zone management in the Ria Formosa, Portugal. In: van den Belt M. Mediated modeling: a systems dynamics approach to environmental consensus building. Island Press, Washington, pp 205–229

Videira N, Antunes P, Santos R (2009) Scoping river basin management issues with participatory modelling: the Baixo Guadiana experience. Ecol Econ 68(4):965–978

Videira N, van den Belt M, Antunes P, Santos R, Boumans R (2011) Integrated modeling of coastal and estuarine ecosystem services. In: van den Belt M, Costanza R (eds) Volume 12, Ecological economics of estuaries and coasts. In: Wolanski E, McLusky DS (eds) Treatise on estuarine and coastal science. Academic Press, Waltham

Videira N, Lopes R, Antunes P, Santos R (2012) Mapping maritime sustainability issues with participatory modelling. Syst Res Behav Sci 29(6):596–619

Voinov A, Bousquet F (2010) Modelling with stakeholders. Environ Model Software 25:1268–1281

Chapter 13
Participatory Modeling and Community Dialog About Vulnerability of Lobster Fishing to Climate Change

Thomas Webler, Esperanza Stancioff, Rob Goble, and Jessica Whitehead

Chapter Highlights

Approach: We used dialog-based group concept mapping called VCAPS (Vulnerability, Consequences, and Adaptation Planning Scenarios) and system dynamics modeling to elicit and organize local knowledge and expert knowledge about vulnerability to climate change.

Participant Engagement: Lobstermen, community members, and scientists engaged in two concept mapping workshops and three system dynamics modeling workshops to build a model of how fishing effort and climate change relate to new income for lobstermen.

Models/Outcomes: VCAPS models of climate vulnerability were made for several climate stressors including warming oceans, heat waves, increased storms, heavy precipitation, extended droughts, and sea level rise. A system

T. Webler (✉)
Department of Environmental Studies, Keene State College,
229 Main Street, Keene, NH 03435, USA
e-mail: thomas.webler@keene.edu

E. Stancioff
University of Maine Cooperative Extension and Maine Sea Grant,
377 Manktown Rd, Waldoboro, ME 04572, USA

R. Goble
Environmental Science and Policy Program, Clark University,
950 Main Street, Worcester, MA 01610, USA

J. Whitehead
North Carolina Sea Grant, North Carolina State University,
1575 Varsity Drive, Varsity Research Building, Module 1, Raleigh, NC 27695, USA

S. Gray et al. (eds.), *Environmental Modeling with Stakeholders*,
DOI 10.1007/978-3-319-25053-3_13

dynamics model with six modules was made to summarize how climate change and fishing effort might impact the income of lobstermen.

Challenges: Over the 24-month period of the project, the issues of concern to lobstermen and the community evolved; although they remained interested in ocean warming, they wanted to see the system dynamics model expanded to include ocean acidification. We tried, but were not able to engage younger lobstermen in the meetings.

13.1 Introduction

Environmental decisions are, by their nature, interventions in human and natural systems. To be successful, these decisions must be informed with appropriate and competent knowledge. As decision makers weigh their options and consider what actions to take, understandings of how these systems will likely react to a proposed action are vital. Decisions informed with appropriate and competent knowledge are more effective, more accepted, and more sustainable because they lead to fewer surprises (Susskind et al. 2001; NRC 1996).

There are many types of environmental decision makers: governmental officials, private companies, non-governmental organizations, households, and individuals. Each has different knowledge needs, different abilities to use available knowledge, and different knowledge to contribute. An effective decision-making process elicits and organizes knowledge in a manner that is useful to the decision maker.

Knowledge about the problem is needed to enable decision making and the generation of decision options that could address the problem. Since human actions to address environmental problems bridge social and natural worlds, decision makers must understand the social system and the natural system. Increasingly, it is common to recognize this as a coupled human-natural system (or social-environmental system) (Liu et al. 2007).

Readers of the academic and practitioner literature on participatory policy making will be familiar with the notion that there are different ways of knowing. Scientific knowledge is gained through a process of reducing the system to constituents and using systematic and repeated observations to study each element independently. Local or experiential knowledge is acquired by non-scientists or lay people who develop sophisticated and competent understandings of a local place through extended experiences and careful observations. The word "expert" can apply equally to scientists or local people whose knowledge is highly regarded by their peers. A significant and growing academic literature continues to confirm the correctness of local knowledge and its relevance and importance for policy-making. To support accurate understandings of the system, both local and scientific knowledge need to be integrated. Appropriate generalization is also needed to allow translation of knowledge gained in one context to another. That is, highly specific knowledge needs to be linked up in such a way that, together, it leads to a reasonable representation of how the system functions. This chapter illustrates how participatory modeling can be used to structure and organize local and scientific knowledge in a manner useful to environmental decision makers.

13.1.1 The Problem of Local Climate Vulnerability Assessment

Although climate change is threatening communities, already—with some experiencing significant impacts—adaptation has not been occurring at the rate or scale deemed necessary by experts (NRC 2010). The projected costs of inaction in terms of money and lives are enormous. Failure to adapt to climate change can also exacerbate drivers of other social, economic, and political challenges, such as political conflict, food security, and poverty reduction (Intergovernmental Panel on Climate Change 2014).

There are many reasons for this failure to prepare. One of particular interest to our work is the insufficient connection between decision making (by public and private bodies) and knowledge of vulnerability and adaptation actions (Mastrandrea et al. 2010). Many decision makers do not understand the threats associated with climatic change or what to do about them (Tribbia and Moser 2008). This is particularly challenging at local levels because local governments lack the professional staff who can interpret climate information, climate projections are not readily available at a scale of interest to a community, and because groups in communities interpret climate change through different knowledge and value frames.

13.1.2 The Idea of an Analytic-Deliberative Process

Decision making about climate change adaptation has much in common with managing risks of natural and technological hazards, a topic that has been the focus of many committees at the US National Research Council (NRC). A committee in 1996 generated a roadmap for how to link together knowledge from the public, governmental officials, and scientists to support decision making (NRC 1996). They called their model "analytic-deliberation." Several NRC committees have advocated this model since it was originally proposed (NRC 1999a, b, 2002, 2008) and the academic literature about the model is growing (North et al. 2014; Karjalainen et al. 2013; Perry 2012; Rodriguez-Piñeros and Lewis 2013; Burgess et al. 2007; Renn 1999; Tuler and Webler 1999; Webler and Tuler 1999; Bradbury 1998).

Analytic deliberation (AD) consists of coordinating and integrating two ways of making sense of the world: analysis and deliberation. Lay people and scientists both engage in analysis and in deliberation. Analysis refers to information gathering and the construction of knowledge through observation and making sense of that information through some form of processing. Scientists do this through systematic data collection and analysis. Local knowledge is accrued through observations and the everyday practice of recognizing patterns and connections. Both local and scientific knowledge are valid and relevant to decision making. Deliberation refers to knowledge sharing and synthesis that happens through communication. Local people—such as fishermen—talk among themselves, sharing and confirming their observations. Scientists also talk among themselves at professional conferences or in research teams as they gather and analyze data. Deliberation by local people and scientists are both needed for good decision making. The AD model pays attention to how all modes of learning by all kinds of parties are coordinated and used to inform environmental decision making.

Analytic deliberation is a promising idea because it can clarify how different kinds of people—the public, interest groups, governmental officials, and experts—each play a mutually beneficial role in generating sound knowledge to support effective decisions. However, there is no single accepted recipe for how to realize an AD process in a given context. Our research illustrates how these ideas can be put into practice to support effective decision making.

13.1.3 Applying the Ideas of Analytic-Deliberation to Climate Vulnerability Analysis

Our participatory modeling approach is an attempt to help local communities threatened by climate change to use their local knowledge assets and the best available scientific knowledge to characterize the threats they face and to strategize possible solutions. Our focus is on decisions made by local government, small private economic operators, local organizations, and individuals. We report here on work with the small coastal community of South Thomaston, Maine. Our work contributes to accumulating efforts at dialog-based approaches to local climate change adaptation (Frazier et al. 2010; Kirshen et al. 2008; Sheppard 2012).

13.2 Participatory Modeling Using Group Concept Mapping and System Dynamics

13.2.1 The Idea of Participatory Modeling

Participatory modeling (Mendoza and Martins 2006; Mendoza and Prabhu 2005), mediated modeling (Van den Belt 2004), cooperative modeling(Tidwell and Van Den Brink 2008), group model building (Andersen et al. 2007; Rouwette et al. 2002), or computer-mediated collaborative decision-making (Cockerill et al. 2009) are all ways of bringing stakeholders together to summarize how a system works. The summary is designed to be useful for decision-makers. Humans make sense of complex systems by representing them as simplified mental models (Morgan et al.

2001), and mediated modeling exploits this tendency by bringing scientists and stakeholders together to enhance system understanding by organizing group interactions around building models (Van den Belt 2004).

As these models proliferate, a discussion about the proper way they should be used in policy making has emerged (Paolisso et al. 2013; Hare 2011; Webler et al. 2011; Voinov and Gaddis 2008; Korfmacher 1998). Debate about citizen and stakeholder involvement in the modeling and decision making processes, and wider discussion of the use of models in policy making is common. While it is widely acknowledged that models are useful tools for advancing science and for framing policy, there is also agreement that significant obstacles to the effective use of models in the policy process exist, a conclusion also illustrated by Heather Squires and Ortwin Renn in their study of fisheries management in Europe (Squires and Renn 2011). They noted that even a robust collaborative modeling approach can be buffeted by outside political forces, which can upset pre-existing agreements or alter attitudes toward cooperation and learning (p. 414). On this same point Thomas Dietz perspicaciously noted that, to effectively unite science and policy decisions, questions about how to properly address uncertainty and the limits of knowledge must first be answered (Dietz 2013). According to van den Belt (Van den Belt 2004), mediated modeling has two key advantages. First, it is a powerful tool in encouraging group learning and increasing shared understanding. Second, it assists participants in developing a consensus on the structure and function of the system in question, resulting in more productive group thinking and discussion. Together, these advantages promote a sense of ownership over the resulting recommendations or decisions. Professional knowledge experts, such as academics, consultants, and industry or government scientists, are integral participants in mediated modeling. Social scientists may serve both as facilitators for bringing together multiple stakeholder perspectives and as consultants on relevant scientific information (Van den Belt 2004). As a result, the collaborative bottom-up process of creating a model through a carefully designed and well-managed mediated modeling exercise can promote group learning and consensus building, ground the model in stakeholders' reality, and enhance the likelihood that the model will remain useful in decision-making (Tidwell and Van Den Brink 2008; Cockerill et al. 2009; Costanza and Ruth 1998; Größler 2007). Unsuccessful mediated modeling has been documented in the literature (Größler 2007) and, in addition to the impact of outside political pressures noted by Squires and Renn, there are other dangers that dominant group members will intentionally or unintentionally coerce or intimidate other participants. Scholars have cataloged several potential problems with group-based learning (Isenberg 1986; Mutz 2006; Thompson 2008) and they apply to mediated modeling. These should not be ignored, but neither does their existence justify abandoning group-based modeling or learning enterprises. There are very many examples of useful mediated modeling and group system dynamics modeling exercises (Otto and Struben 2004; Metcalf et al. 2010) and these suggest that shortcomings, especially when foreseen and accounted for, can be outweighed by the benefits.

13.2.2 Dialog-Based Vulnerability Mapping and the VCAPS Process

The VCAPS (Vulnerability Consequences Adaptation Planning Scenarios) process combines structured discussion with interactive concept mapping to create visual summaries of local knowledge about vulnerability and resilience (Webler et al. 2014; Kettle et al. 2014). It helps stakeholders depict how the community is impacted by hazards and how mitigation actions could help reduce those impacts.

VCAPS elicits and organizes local knowledge together with expert knowledge and assessments to depict the nature, magnitude, and likelihood of threats associated with climatic change. Dialog allows the group to assert and confirm knowledge claims through an immediate process of challenge and justification. Concept mapping is used by the group to track the progress of this deliberation and to summarize group consensus, (Flanagan and Christakis 2010).

The process involves a research team working together with a group of local participants to construct a consensual conceptual model of how a specific stressor (e.g., ocean warming) might alter the coupled human-natural system in which the community is embedded. VCAPS differs from mental modeling because it is group-based. It differs from common notions of concept mapping because it uses a very limited "vocabulary" of concepts from hazard management and vulnerability analysis to summarize the threats. However, like mental modeling and concept mapping, VCAPS produces qualitative diagrams of how beliefs interconnect (Webler et al. 2014).

VCAPS happens in meetings with local decision makers and local experts. Scientific experts such as climate scientists, economists, or ecologists may be present to provide help as needed. The group performs the following:

- reviews the recent science on the issue,
- defines a policy scenario to work from,
- elicits and summarizes the local knowledge and local perspectives on the issue,
- characterizes the threats and vulnerabilities,
- generates decision options, and
- characterizes the resources needed to carry out decision actions.

The consensus understanding of the community's vulnerability is summarized in flow-chart type figures that document how the hazard creates harm in the human-natural system. Groups are able to quickly learn the language of this modeling structure and to characterize their system in considerable detail. In addition to mapping the causal chains that link stressors with consequences, the participants also identify context-specific features that modify vulnerability (e.g., depth of water at a specific dock), and identify decision options or "management actions" that might protect against harm and enhance resilience.

13.2.3 System Dynamics Modeling

Climate change will impact human-natural systems in non-linear and dynamic ways. Human actions further complicate projections. System dynamics (SD) programming makes it possible to represent understandings of these relationships. In SD, the system is characterized by stock-flow diagrams that include feedback loops (Lane 2008). We use a program called STELLA (ISEE Systems 2014).

SD modeling is ideal when a complex set of factors and feedbacks from multiple stressors impact a system's function (Winz et al. 2009). Modeling begins with initial assumptions and a conceptual model of the dynamic system. A combination of stocks (state variables) and flows (controls) are used to represent variables and relations among variables in the system. This structure permits system dynamics models to accommodate empirical data about known relationships, as well as qualitative data, assumptions, or expert judgments about system variables (Ford and Flynn 2005). This ability to integrate physical principles with expert models makes system dynamics an intuitive methodology well-suited to bringing stakeholders into modeling processes through cooperative modeling groups (Rouwette et al. 2002) that are aimed at addressing many environmental and resource management issues (Cavana and Ford 2004).

13.3 Participatory Modeling of Climate Change Vulnerability in South Thomaston

13.3.1 Background Interviews with Decision Makers

Our intention was to focus on how climate change impacts the coupled human-nature system of South Thomaston, Maine USA. One of the major commercial fishing ports in Knox County and in Maine, South Thomaston, lies on the western shores of Penobscot Bay. The municipality of 1558 inhabitants is the regional urban center and county seat and houses one of the largest lobster cooperatives in the state. For much of the town's 400 years of recorded history, fish landings included a diversity of commercial species such as herring, cod, shrimp, sea urchins, and sea scallops. A statewide decline in natural populations, of these species has forced commercial fishermen to rely primarily on the lobster fishery. Lobster landings in Maine have steadily increased during the last 30 years, reflecting both a growing population of lobsters and an increase in offshore fishing effort. Over the last 5 years for which data are available (2009–2013), South Thomaston has been one of the top ten ports for landings (all species). In 2013, it landed a total catch worth $11.37 million, most of which was lobster (Maine Department of Marine Fisheries 2013).

Fishing provides a median income of $43,846 in South Thomaston, a number whose significance is underrated inasmuch as fishing is a part-time occupation for many (U.S. Census Bureau 2012). Lobstermen supplement their income in a wide variety of ways. In South Thomaston, more people are employed in management occupations, and that survey category has a higher median income, but fishing is an extremely profitable business for individuals without professions that require higher education degrees. An estimated 100 boats regularly use the commercial fishing facilities in South Thomaston/Spruce Head. The captains of these boats employ an additional 100 to −150 crew members. Shore-side wharves, transport and processing facilities employ 40–50 seasonal workers. This sector has also shaped and preserved the unique cultural and social fabric of coastal communities along the Maine coast and continues to define an important segment of the maritime living heritage.

In the spring of 2012, historically warm water in the Gulf of Maine during the winter contributed to lobsters shedding their shells in March and April rather than in July or August as fishermen expected based on previous seasons. High volume catches of soft-shell lobsters began to be landed in South Thomaston/Spruce Head and throughout Maine in the spring when Canada was also catching large numbers of lobsters. Soft-shell lobsters (or shedders) are mostly sold to processing plants in Canada where they are canned. However, the Canadian plants operate on a very restricted and planned timetable. They could not adjust to the sudden appearance of Maine shedders. This, together with the fact that the lobsters preceded the summer tourists, produced a glut of shedders. Prices plummeted in Maine and dealers could not sell all the lobsters being landed. Lobstermen continued to fish despite the low price, due to fear that this might be their only opportunity to make an income. Many lobstermen suffered significant losses. This episode created uncertainty and a high sense of vulnerability among the fishermen and the community.

In early 2013, we conducted individual in-person interviews of 18 lobstermen and 5 community members. The individuals were recruited by a former lobsterman from the community who was employed by Maine Sea Grant. The lobstermen and community members were eager to talk with us and most had a great deal to say about their observations of the ocean, the lobster fishery, and the local economy. Most asked several questions about climate change.

Participants were very concerned about the impact of climate change on the lobster fishery and community. Many mentioned warming water temperatures as a likely cause of the 2012 episode. Respondents also noted changes in algae or plankton and expressed concern for increases in shell disease, a disfiguration of the shell due to stress. Lobsters with this condition can still be sold, but at a lower price. However, increases in shell disease preceded the collapse of the lobster fishery in Long Island Sound. Only two fishermen acknowledged a concern for ocean acidification, although in latter meetings this became an increasingly important topic of interest and discussion.

We asked our 23 interviewees if they could think of some actions that are being taken or could be taken to cope with or lessen these impacts, including immediate or long-term adaptive changes. Reducing fishing effort by limiting the fishing season was the most common answer (45 % of respondents). Twenty-seven percent recommended increasing the minimum gauge size of traps (so that more small lobsters escape). Interestingly enough, two respondents (18 %) suggested increasing fishing effort (through higher trap limits or more licenses) as a solution.

When asked what could be done to increase the capacities and overcome the barriers to these actions, there was no consensus. Fishermen tended to focus on reducing regulations and better science and management. Community members tended to have more specific responses, including increasing public pressure on government, holding more local meetings, developing a comprehensive plan, and assembling an individual stakeholder action list. Increasing the money available to support actions was the only recommendation listed by both the fishermen and the community members.

13.3.2 The VCAPS Process for Group Concept Mapping

The research team used the results of the community interviews to prepare for the VCAPS process. Over the course of two meetings held a month apart, we facilitated a conversation among lobstermen, local government (volunteer town officials), and other community members about climate vulnerability. We invited experts from the University of Maine and the Maine Department of Marine Resources to share the latest science about climate change and how lobsters and the community may be impacted by warmer water and air temperatures, ocean acidification, sea level rise, increased storms, and longer periods without rain.

Workshop participants discussed impacts of a changing climate on the community and on the lobster fishery. During the discussions, the research team diagrammed the causal pathways and added contextual factors and management actions to the diagrams as participants mentioned them. These diagrams were projected onto a wall in front of the participants.

Stakeholders identified several stressors but focused intensely on increasing water temperatures as a climate-related stressor that they suspected would precipitate a number of impacts for lobster fishing. Figure 13.1 is a small segment of the VCAPS diagram for this stressor. The overriding concerns were for the unpredictable timing of the first shed, the vulnerability of the shedders, and the inability of Canadian processors to purchase the catch. The arrow-shaped boxes describe states of affairs. The octagons at the right end are consequences of concern. The ovals beneath the arrow-shaped boxes are factors that shape sensitivity. The rectangular nodes describe possible actions that could reduce vulnerability. Some of these actions give birth to a new chain of outcomes. For example, if lobstermen do not fish and instead wait for lobster shells to harden (at which point they can survive transport to distant markets), then there is also a danger that the lobsters could move off-shore and outside of some people's fishing territories.

VCAPS diagrams are made in real-time during the discussion and serve as a record of the group discussion. We also produced lists and narratives, summarizing the content. During the VCAPS process, lobstermen and community members, generated lists of impacts and potential adaptive actions. These were often unrelated. Examples are presented in Table 13.1.

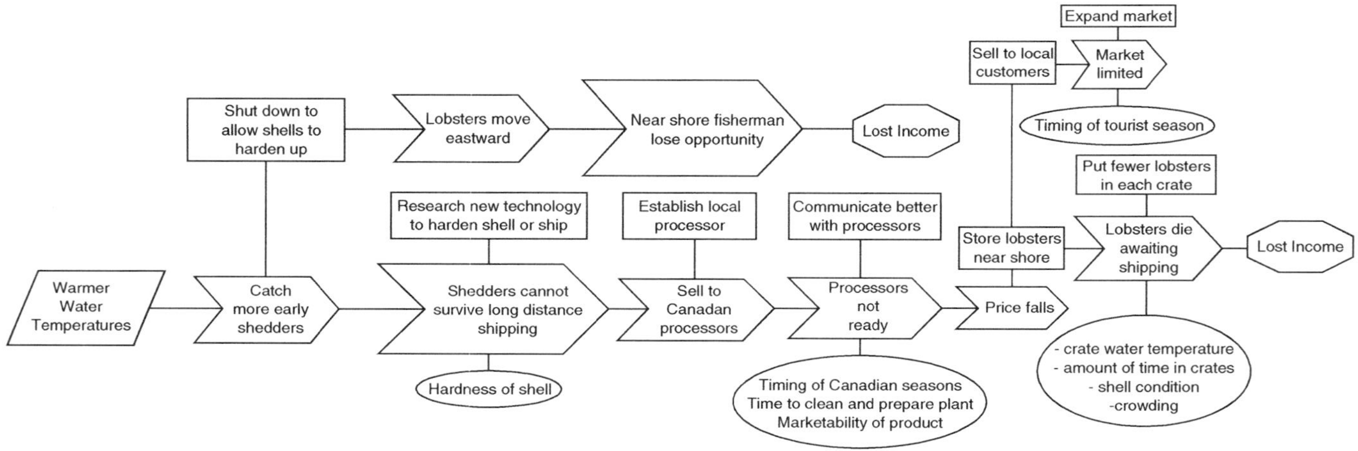

Fig. 13.1 Portion of VCAPS diagram having to do with lobster shedding time

Table 13.1 Impacts and adaptive actions identified by the community

Potential impacts of warming water included:	Potential actions to address impacts included:
Increased frequency of molting in lobsters	Develop an on-board grading system
Higher number of shedders	Work with Co-op to establish a local processing plant
Increased lobster landings—especially early shedders	Temporary stop to fishing to allow shells to harden
Lobsters moving "Down East"	Research new technologies for hardening shells
Increased algae growth, and increased bio-fouling of lobster gear	Establish a lobster holding system that does not result in overcrowding and mortality
Introduction of new species of fish	Establish experimental fisheries and promoting recreational fishing
Introduction of new diseases	Improve communication with Canadian processors regarding local needs
	Expand the market by more aggressively promoting local lobster consumption

13.3.3 Supplementing Dialog Mapping with System Dynamics Models

After reviewing the VCAPS report, South Thomaston lobstermen met again with the research team to learn about system dynamics modeling and to begin collaborative work on building an SD model. We suggested that SD could be used to better understand a specific part of the coupled human-natural system summarized in the VCAPS diagrams. Unsurprisingly, lobstermen wanted to understand how the timing of the first shed affects net income. During a day-long meeting, we facilitated a participatory modeling effort in which we began to build an SD model on the screen while lobstermen discussed their understandings of the system. A lobster biologist and an oceanographer were on-hand to answer questions and contribute their knowledge. After the meeting, the research team continued to revise the model and scheduled a second meeting a month later to share the revised model. At this meeting, a smaller group of lobstermen carefully reviewed the model and made suggestions for how it could be improved. A few individuals contributed sensitive personal financial and catch data under the condition that their identities not be associated with the data. The research team eventually settled on a model comprising six modules, as shown in Fig. 13.2. Lobstermen fish different numbers of traps and haul them at different rates. Our model placed considerable effort on characterizing individual fishermen's effort. However, the core of the model is the Ecology and Catch Module, a simplified version of which is shown in Fig. 13.3. The model contains three compartments for shedders, as they harden, and captures the fact that the shedders are more likely to wander into traps as they age. The size of these sub-populations of lobsters varies each week. We built a generic model and designed the interface panel so that fishermen could input data about their fishing effort and landings.

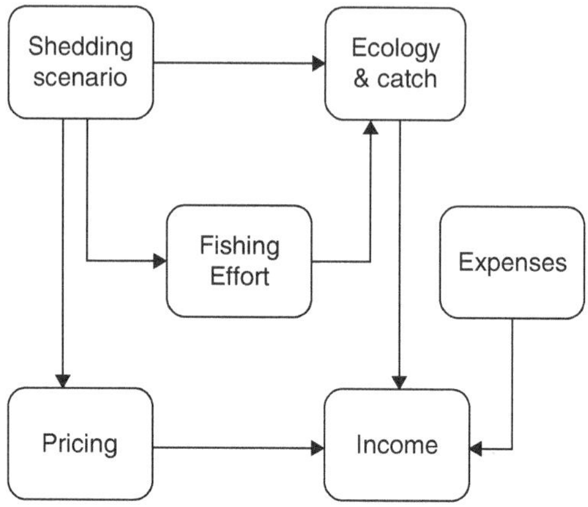

Fig. 13.2 Architecture of the system dynamics model

In Maine, lobstermen have individual territories that they alone can fish. We simulated this by modeling a unit volume of ocean, which has an initial age-stratified population of existing and immigrating lobsters. Lobsters of various ages and molting status are removed through fishing, natural death (via predation and disease), and by aging (lobsters exceeding a specified size are illegal to harvest). Over several weeks, a molted lobster's shell hardens up (right side of Fig. 13.3). Hard-shell lobsters usually command a higher price. Since each lobsterman's area is unique, we adjusted initial conditions in the model to produce the landing rates that were reported by individual fishermen. Ten weeks later and with two individualized models in hand, we scheduled a third meeting with the lobstermen.

At this third meeting, a core group of seven lobstermen and community members heard a brief presentation about the model. The model is based around various "shed scenarios." Each scenario is summarized as a graph of the percentage of lobsters shedding at any given date of the year (see line 2 in Fig. 13.4, it corresponds to the y-axis labels from 0 to 1). STELLA allows users to re-draw this scenario using a cursor or pen. Our participants explored various shed scenarios, and examined how changes in fishing effort (i.e., number of traps fished) could help protect lobstermen from dire economic consequences. The dependent variable in all these discussions was net income (line 1). The top two diagrams in Fig. 13.4 shows how two different shed scenarios (a customary or pre-2012 scenario, and the unusual 2012 scenario) affect net income with fishing effort held constant. The lower two diagrams show the same two scenarios but with a fishing effort shifted to fall and winter. The take-home message was that income could be resilient to changes in shedding, if lobstermen changed *when* they fish. The sticky policy point was that all Maine lobstermen would need to change their behavior in a coordinated fashion. This led to some very spirited discussions. Enthusiasm was high because lobstermen felt that these results represented their experiences. Several other individuals

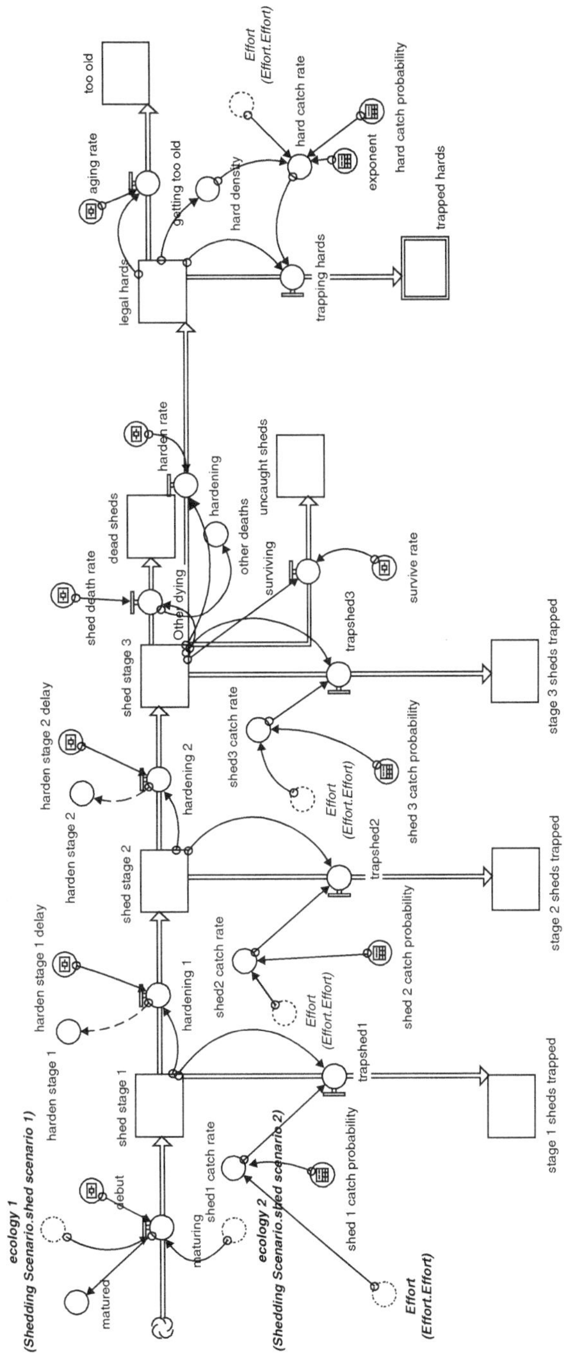

Fig. 13.3 Ecology and catch module in the model

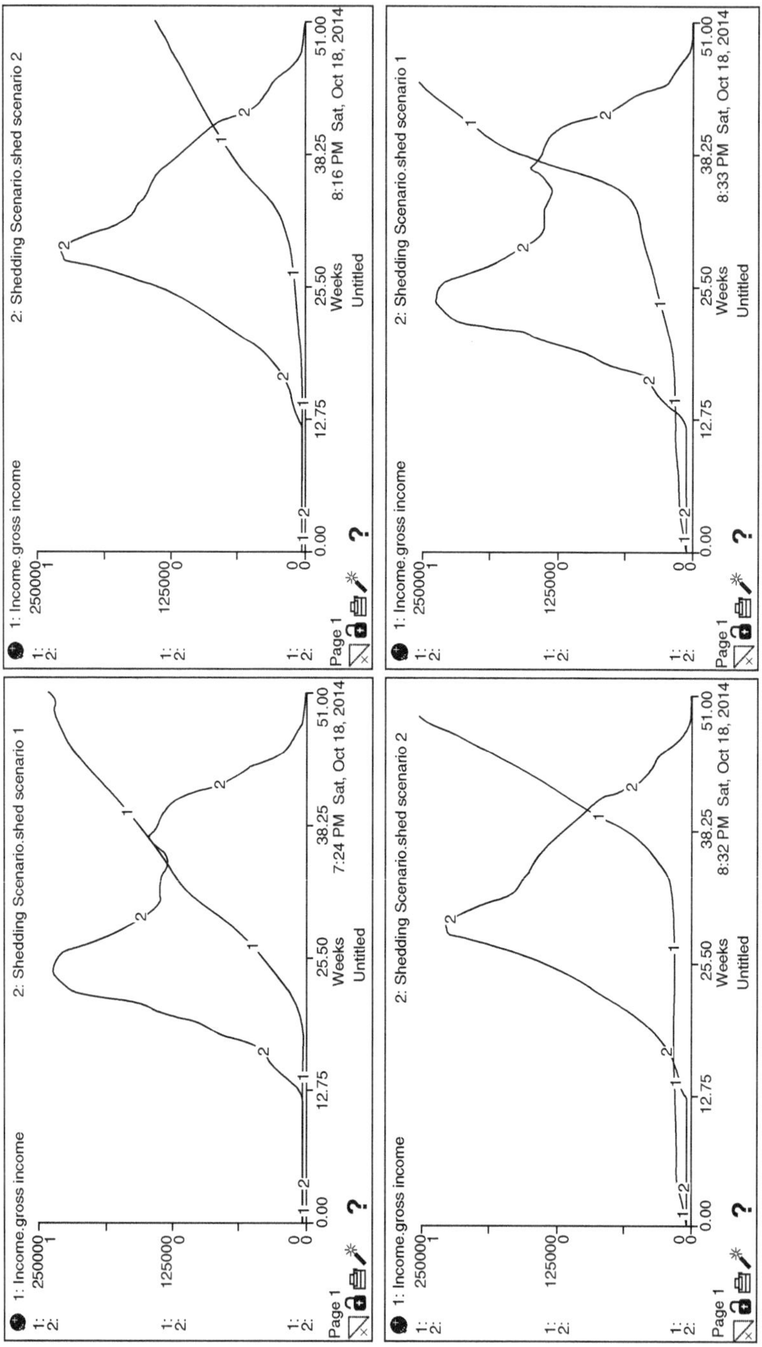

Fig. 13.4 Relationship between shed scenario and net income under different fishing efforts. Line 1 represents income, line 2 represents the percent of lobsters that have shed. *Upper left*: customary (pre-2012) shed scenario and summer fishing effort. *Upper right*: 2012 shed scenario with customary summer fishing effort. *Lower left*: 2012 shed scenario with fishing effort shifted to autumn. *Lower right*: Pre-2012 shed scenario with fishing effort shifted to autumn

agreed to share their personal data with the research team to contribute to the model's development. They also requested a new module that would factor in the incidence of shell disease. A fourth meeting was scheduled for late November 2014 (after the completion of this manuscript). Lobstermen are unavailable to meet during the fishing season, so it is not unusual for 6 months to pass between meetings.

13.4 Outcomes of the Participatory Modeling Dialogs

Here we share a number of observations of the participatory modeling effort and the subsequent dialog that followed during the use of the SD model. Participants in the VCAPS process discussed and documented how climate change was affecting the fishery and the community. While doing this, they made sense of their personal experiences and their collective local knowledge. The process was unequivocally egalitarian in that not only did each individual have the opportunity to speak, we went out of our way to elicit input from everyone present. The process was also democratic in that it allowed—indeed encouraged—each individual to contribute his or her observations and have them checked, and perhaps confirmed, by others. While some people remarked on the discouraging nature of the climate projections, others applauded the value of putting together one "big picture" that showed how things were connected. One excited individual exclaimed, "I'd like to see you do this with every town on the Maine coast!"

Community members and lobstermen used the VCAPS diagrams to emphasize their concerns, but concerns also evolved. In our initial interviews, only one person mentioned ocean acidification as a strong concern. Even after listening to a presentation about ocean acidification from a university scholar during the first VCAPS meeting, the community seemed to have reached the opinion that it could be a problem in the future, but it was not one they felt they had any control over, and they focused their attention on warming oceans instead. However, a year later, at the third SD modeling meeting, ocean acidification had become a major concern. Lobstermen had noticed a large increase in the percentage of lobsters with shell disease and were concerned that acidification was becoming a bigger issue. One noted with alarm the disappearance of mussels growing on floating gear and the unprecedented appearance of clams growing on ropes and floats—places where they had never seen a clam grow before. Another noted the wholesale disappearance of starfish along the coast. The research and regulatory community should take note of their interest in ocean acidification as a possible cause or correlate of species change and also shell disease, which they see as a harbinger of collapse.

Lobstermen used the SD model to discuss and explore policy options to mitigate their vulnerability to economic losses from an early shed. In our April 2014 meeting, the conversation was electrified by observations of an unusual absence of harvestable sized lobsters. Although it was still early in the season, concern was present. Several noted that, for the past 40 years, "you could set your watch" by the appearance of the first molt. This was no longer the case. The research team made it clear

Table 13.2 Variables used to compute fishermen's expenses

Cost item	Period of variable
Bait	Weekly
Fuel	Weekly
Crew	Weekly
Boat payments	Monthly
Insurance	Monthly
Licenses and registration	Annual
Gear	Annual
Repairs	Annual
Trap tag fees	Annual
Depreciation	Annual
Truck loan	Annual
Other	Annual

that the model should not be expected to mimic reality, rather it is an approximation of some of the relationships among variables in the system, and the lobstermen clearly understood this. They did not over-interpret the model results. They used the model to run possible scenarios and then shifted to a discussion about policy options. Those discussions frequently cited the model results, indicating that the model did serve as a dialog aid, but their discussions deviated well beyond what the model could convey. As we noted above, one of the policy discussions was about whether or not to limit fishing during summer when lobsters have low value and to fish more in the late autumn. The model clearly helped lobstermen make their points. It helped the group take the proposal seriously simply because we could run that policy scenario in the model and people could see net income rise dramatically as lobstermen delayed fishing until the lobsters hardened up and increased in value (lower two diagrams in Fig. 13.4). In other words, the model served as a catalyst for a conversation about what is happening and what could be done about it.

All of the participants at the final meeting, were older lobstermen, many with children who have entered the industry. They lamented the poor financial decisions some younger lobstermen seemed to be making. They used this opportunity to ask about each other's kids. They used the SD model to examine the financial impacts of 1 year of poor "settlement"—the process that determines the number of lobsters that will shed and mature into legal catch in the coming years. We ran a scenario where settlement rates dropped, but expenses stayed the same and watched income drop to pitiful levels. "That should wake up some of these younger guys," said one fisherman. In this context, lobstermen saw value in the model as a teaching aid for financial resilience planning for the younger lobstermen. The model clearly showed how lobstermen's income is earned irregularly during the year, necessitating the need for financial planning, even in a good year. Table 13.2 itemizes the variables included in the model's module for expenses. Some were computed weekly, others annually. Older lobstermen suggested the model could help inspire youth to do a business plan and avoid taking on too much debt. Sea Grant personnel made the connection between this need and a Sea Grant resource to help fishermen do business planning. It was suggested that the model be amended to include a module for business planning and household economic management. In summary, the model demonstrated potential as a decision analytic aid in individual financial planning.

NOAA funded this project for 3 years, but the inclusion of Maine Sea Grant as a partner strengthens the likelihood that this work will continue. Maine Sea Grant has an obligation to work with fishermen and lobstermen all along the coast and this project established close ties among individuals. A key resource to collaborative research with communities is establishing a relationship of trust among the research team and the community. The existence of this trust represents an existing asset to the research community and should facilitate the birth of additional projects. Our team has already applied for additional NOAA funding to continue and expand this work in Maine.

Our participatory modeling process was democratic in that it was inclusive and fair. We were receptive to anyone in the community participating and we made sure all voices were heard. An appropriate environment for constructive dialog was ensured by outlining norms for dialog at the onset and including experienced facilitators of group dialog. There were no problems, possibly because this is a small New England town that relies on small group consensus building for most of its governance. The interaction between experts and stakeholders was non-problematic. The two experts we brought into the discussion from outside the project team were already highly regarded by the lobstermen. Both had many years of experience working directly with lobstermen. Their presence influenced the group dynamics in a very positive way. Speakers often made reference to the expert's presentations.

One might reasonably suspect that there is a high degree of subject-dependency in a project such as this. Only by running parallel, independent processes could we gain insight into how the models produced would vary due to group context. That would be a costly and demanding experiment to do with community participants. Moreover, to be useful for community-level decision making, the independently developed models would need to be reconciled, integrated, or combined into a consensus model or recommendation. This would have to happen through a group process, which would, again, introduce group context and subjectivity into the picture. The best way to proceed is simply to build one diverse group that functions well and have experienced facilitators run the meetings. Our group involved a multiplicity of viewpoints and experience from different people in the community.

Throughout all our meetings with the lobstermen and the community members, the research team learned a great deal about what was needed for the VCAPS diagram and the SD model to be more accurate and more relevant. For instance, we learned about the long-term perspectives this generation has on fishing and the deep concerns they have for their children's economic future. We also learned about regional trends in lobster behavior and abundance, something that was not included in the highly localized model we built. There were numerous facts that we incorporated into the model after conversations with the fisherman. The SD model was truly a collaborative effort.

13.5 Conclusions: Opportunities for Further Investigation and Application

This intriguing project began to address a significant gap in our understanding of how climate change will affect coupled human-natural systems. While the amount of data on climate change impacts to ecological systems is exploding, much less is being learned about the human dimensions. We know relatively little about how human behavior intervenes in ecological system performance. For example, one lobsterman noted that early fishing traps many sub-legal lobsters that must be thrown overboard. He suspected that this stresses the lobster, and he wondered if this stress, combined with increased stress from climate change, leads to more cases of shell disease. Another question arose about whether fishing offshore would affect settlement rates of lobsters near shore. There are numerous possibilities that need to be identified and the most important ones explored. Structured discussion around models such as the VCAPS and SD models of our project will facilitate this exploration.

We also know relatively little about how changes in ecological systems will translate into cultural, social, and economic impacts. For centuries, communities have celebrated the onset of productivity in natural resource systems with festivals. Does the cultural significance of a lobster festival change when there are no locally harvested lobsters? If income from lobster fishing drops precipitously and young people exit the industry, what will that do to community identity? What will it mean for the care of the elderly in a community where families still play a key role in supporting the retired generation?

The results from this case study in Maine are, of course, shaped by contextual specifics of that place. One cannot arrive at generalizable theory from a single case study. However, the Maine case study was only one of three cases in this larger research project. When case studies in Wellfleet, MA and Beaufort, SC are completed, we intend to identify generalizable outcomes from a cross-case comparison.

This project allowed us to investigate and experience how participatory modeling could be integrated with dialog-based policy and decision making. The notion of analysis and deliberation advanced in the risk policy literature suggests value in tightly linking multiple forms of knowledge making with deliberative processing. Participatory model making is one form that an analytic-deliberative process can take. Dialog that accompanied the model making was viewed as constructive by the participants in South Thomaston. The models also played an important role, but not because of their accuracy or precision. All models are only approximations of reality. “The map is not the territory,” is how Alfred Korzybski famously put it (Korzybski 1933). Like a map, participatory models are inspirations and guides for organizing, sharing, and deliberating about information and, ultimately, making decisions.

A key advantage of participatory modeling is that it provides a means to bring local knowledge into consideration along with expert knowledge. Local people have intimate experiential knowledge of their coupled human-natural system while experts have generalizable knowledge about nature and society. In the context of climate change, the Intergovernmental Panel for Climate Change noted that, "Integration of local knowledge with additional scientific and technical knowledge can improve disaster risk reduction and climate change adaptation" (IPCC 2014; Burgess et al. 2007). They noted that there is high agreement and robust evidence for this conclusion. The VCAPS approach, combined with system dynamics modeling, which we illustrated in this chapter, shows promise as an effective approach to use participatory modeling in gathering and assessing key information that can support environmental decision-making.

Acknowledgements This work was funded by the NOAA Climate Program under grant number NA12OAR4310106. We would like to thank all the lobstermen and community members in South Thomaston who took time out of their busy lives to patiently work with us. We would also like to recognize the important contributions to the project by Sherman Hoyt, formerly of Maine Sea Grant and now a consultant and to Sam Belknap, a University of Maine NSF IGERT Fellow. Both provided their local knowledge with the midcoast Maine lobster industry and the South Thomaston community of lobstermen. They conducted outreach and participated in meetings. Professor Robert Steneck of the University of Maine and Dr. Carl Wilson with Maine Department of Marine Resources generously donated their time in meetings, presenting current research and information as a framework for discussion. Kristen Grant of Maine Sea Grant assisted with interviews and meetings. Brian Seitzman of Clark University provided his expertise to help complete the SD Model. Many thanks to Anne Baker and Yue Sun, graduate students from Clark University who worked to make GIS data relevant and accessible to the working group. Seth Tuler was involved with general project coordination. Special thanks go out to Penny Alley, South Thomaston Librarian for her hospitality and her assistance with our meetings in the South Thomaston Historical Society. A project advisory committee comprising Joe Costa, Kirstin Dow, Troy Hartley, Michael Johnston, Patricia Pinto da Silva, and Dave Whitaker provided frequent insights and valuable recommendations.

References

Andersen DF, Vennix JA, Richardson GP, Rouwette EA (2007) Group model building: problem structuring, policy simulation and decision support. J Oper Res Soc 58:691–694

Bradbury JA (1998) Expanding the rationale for analysis and deliberation: looking beyond understanding risk. Hum Ecol Rev 5:42–44

Burgess J, Stirling A, Clark J, Davies G, Eames M, Staley K, Williamson S (2007) Deliberative mapping: a novel analytic-deliberative methodology to support contested science-policy decisions. Public Underst Sci 16(3):299–322

Cavana RY, Ford A (2004) Environmental and resource systems: editors' introduction. Syst Dyn Rev 20(2):89–98

Cockerill K, Daniel L, Malczynski L, Tidwell V (2009) A fresh look at a policy sciences methodology: collaborative modeling for more effective policy. Policy Sci 42(3):211–225

Costanza R, Ruth M (1998) Using dynamic modeling to scope environment problems and build consensus. Environ Manag 22(2):183–195

Dietz T (2013) Bringing values and deliberation to science communication. Proc Natl Acad Sci 110(3):14081–14087

Flanagan TR, Christakis AN (2010) The talking point: creating an environment for exploring complex meaning. Information Age Publishing, Charlotte

Ford A, Flynn H (2005) Statistical screening of system dynamics models. Syst Dyn Rev 21(4):273–303

Frazier TG, Wood N, Yarnal B (2010) Stakeholder perspectives on land-use strategies for adapting to climate-change-enhanced coastal hazards: Sarasota, Florida. Appl Geogr 30:506–517
Größler A (2007) System dynamics projects that failed to make an impact. Syst Dyn Rev 23(4):437–452
Hare M (2011) Forms of participatory modelling and its potential for widespread adoption in the water sector. Environ Policy Gov 21(6):386–402
IPCC (2014) Climate Change 2014: impacts, adaptation, and vulnerability. Working Group II. http://www.ipcc.ch/report/ar5/wg2/
ISEE Systems (2014) STELLA: systems thinking for education and research. Retrieved from: http://www.iseesystems.com/softwares/Education/StellaSoftware.aspx. Accessed 14 Jun 2014
Isenberg DJ (1986) Group polarization: a critical review and meta-analysis. J Pers Soc Psychol 50(6):1141–1151
Karjalainen TP, Marttunen M, Sarkki S, Rytkönen AM (2013) Integrating ecosystem services into environmental impact assessment: an analytic–deliberative approach. Environ Impact Assess Rev 40:54–64
Kettle N, Dow K, Tuler S, Webler T, Whitehead J, Miller K (2014) Integrating scientific and local knowledge to inform risk-based management approaches for climate adaptation. Public date (Web): August 14, 2014
Kirshen P, Ruth M, Anderson W (2008) Interdependencies of urban climate change impacts and adaptation strategies: a case study of Metropolitan Boston USA. Clim Chang 86(1–2):105–122
Korfmacher KS (1998) Water quality modeling for environmental management: lessons from the policy sciences. Policy Sci 31(1):35–54
Korzybski AN (1933) Science and sanity. An introduction to Non-Aristotelian Systems. Science Press Printing Co, New York
Lane DC (2008) The emergence and use of diagramming in system dynamics: a critical account. Syst Res Behav Sci 25(1):3–23
Liu J, Dietz T, Carpenter SR, Alberti M, Folke C, Moran E (2007) Complexity of coupled human and natural systems. Science 317(5844):1513–1516
Maine Department of Marine Fisheries (2013) Most recent Maine commercial landings. Retrieved from: http://www.maine.gov/dmr/commercialfishing/documents/09-13Top10PortsByValue.Table.pdf
Mastrandrea MD, Heller NE, Root TL, Schneider SH (2010) Bridging the gap: linking climate-impacts research with adaptation planning and management. Clim Chang 100:87–101
Mendoza GA, Martins H (2006) Multi-criteria decision analysis in natural resource management: a critical review of methods and new modelling paradigms. For Ecol Manag 230(1):1–22
Mendoza GA, Prabhu R (2005) Combining participatory modeling and multi-criteria analysis for community-based forest management. For Ecol Manag 207(1–2):145–156
Metcalf SS, Wheeler E, BenDor TK, Lubinski KS, Hannon BM (2010) Sharing the floodplain: mediated modeling for environmental management. Environ Model Softw 25(11):1282–1290
Morgan MG, Fischhoff B, Bostrom A, Atman C (2001) Risk communication: the mental models approach. Cambridge University Press, New York
Mutz DC (2006) Hearing the other side: deliberative versus participatory democracy. Cambridge University Press, Cambridge
North W, Stern PC, Webler T, Field P (2014) Public and stakeholder participation for managing and reducing the risks of shale gas development. Environ Sci Technol. Publication Date (Web): April 29, 2014. doi:10.1021/es405170k
NRC (1996) Understanding risk: informing decisions in a democratic society. National Academy Press, Washington
NRC (1999a) Making climate forecasts matter. National Academy Press, Washington
NRC (1999b) New strategies for America's watersheds. National Academy Press, Washington
NRC (2002) Drama of the commons. National Academy Press, Washington
NRC (2008) Public participation in environmental decision making. National Academy Press, Washington

NRC (2010) Advancing the science of climate change. National Academy Press, Washington
Otto P, Struben J (2004) Gloucester Fishery: insights from a group modeling intervention. Syst Dyn Rev 20(4):287–312
Paolisso M, Trombley J, Hood RR, Sellner KG (2013) Environmental models and public stakeholders in the Chesapeake Bay Watershed. Estuar Coasts 38(1):97–113. doi:10.1007/s12237-013-9650-z
Perry SL (2012) Environmental reviews and case studies: addressing the societal costs of unconventional oil and gas exploration and production: a framework for evaluating short-term, future, and cumulative risks and uncertainties of hydrofracking. Environ Pract 14(04):352–365
Renn O (1999) A model for an analytic-deliberative process in risk management. Environ Sci Technol 33(18):3049–3055
Rodriguez-Piñeros S, Lewis DK (2013) Analysis and deliberation as a mechanism to assess changes in preferences for indicators of sustainable forest management: a case study in Puebla, Mexico. J Environ Manag 128:52–61
Rouwette EA, Vennix JA, Mullekom TV (2002) Group model building effectiveness: a review of assessment studies. Syst Dyn Rev 18(1):5–45
Sheppard S (2012) Visualizing climate change: a guide to visual communication of climate change and developing local solutions. Routledge, Abingdon
Squires H, Renn O (2011) Can participatory modelling support social learning in marine fisheries? Reflections from the invest in Fish South West Project. Environ Policy Gov 21(6):403–416
Susskind L, Jain R, Martynuik A (2001) Better environmental policy studies: how to design and conduct more effective analyses. Island, Washington
Thompson DF (2008) Deliberative democratic theory and empirical political science. Annu Rev Polit Sci 11:497–520
Tidwell VC, Van Den Brink C (2008) Cooperative modeling: linking science, communication, and ground water planning. Groundwater 46(2):174–182
Tribbia J, Moser SC (2008) More than information: what coastal managers need to plan for climate change. Environ Sci Pol 11(4):315–328
Tuler S, Webler T (1999) Designing an analytic deliberative process for environmental health policy making in the US nuclear weapons complex. RISK Health Saf Environ 10(1):65–87
U.S. Census Bureau (2012) American Community Survey. Table S2401. Occupation by sex and median earnings in the past 12 months (in 2012 inflation-adjusted dollars) for the civilian employed population 16 years and over. Retrieved from: http://factfinder2.census.gov/faces/tableservices/jsf/pages/productview.xhtml?src=CF. Assessed 17 Oct 2014
Van den Belt M (2004) Mediated modeling: a system dynamics approach to environmental consensus building. Island, Washington
Voinov A, Gaddis B (2008) Lessons for successful participatory watershed modeling: a perspective from modeling practitioners. Ecol Model 216(2):197–207
Webler T, Tuler S (1999) Integrating technical analysis with deliberation in regional watershed management planning: applying the National Research Council approach. Policy Stud J 27(3):530–543
Webler T, Tuler S, Dietz T (2011) Modellers' and outreach professionals' views on the role of models in watershed management. Environ Policy Gov 21(6):472–486
Webler T, Tuler S, Dow K, Whitehead J, Kettle N (2014) Design and evaluation of a local analytic-deliberative process for climate adaptation. Local Environ. Public date (Web): July 17, 2014. doi:10.1080/13549839.2014.930425
Winz I, Brierley G, Trowsdale S (2009) The use of system dynamics simulation in water resources management. Water Resour Manag 23(7):1301–1323

Chapter 14
Participatory Modeling to Assess Climate Impacts on Water Resources in the Big Wood Basin, Idaho

Allison M. Inouye, Denise H. Lach, John R. Stevenson, John P. Bolte, and Jennifer Koch

Chapter Highlights

Approach: We used a co-production of knowledge approach to develop three different models with stakeholders to increase understanding of local-scale dynamics of water usage in a watershed in relation to climate and population changes.

Participant Engagement: Stakeholders and scientists created a Knowledge to Action Network (KTAN) by engaging in modeling efforts, a series of informational webinars, and small group meetings to build models of a river basin focused on research topics identified by potential users.

Models/Outcomes: Three models were developed and assessed by stakeholders for potential usability. The conceptual modeling process helped stakeholders characterize the complexity of the basin while the system dynamics model helped to more formally understand the components of water supply and demand in the basin. The integrated model helped the KTAN visualize and compare the impacts of water scarcity under alternative scenarios of climate and population change as well as learn about uncertainties in model projections. An interactive website allowed participants to explore the myriad results.

A.M. Inouye (✉) • D.H. Lach • J.R. Stevenson • J.P. Bolte
Climate Impacts Research Consortium, Oregon State University, Corvallis, OR, USA
e-mail: allisonminouye@gmail.com

J. Koch
Department of Geography and Environmental Sustainability, The University of Oklahoma, Norman, OK, USA

S. Gray et al. (eds.), *Environmental Modeling with Stakeholders*,
DOI 10.1007/978-3-319-25053-3_14

Challenges: Both engagement and technical challenges arose over the course of this project including a lack of urgency or champion for the project, a geographically dispersed team, long delays due to modeling challenges that included a small but highly variable/complex landscape, an inability to characterize and model all important variables, and visualizing model outcomes.

14.1 Introduction

Changing regulatory requirements at both the state and local levels calls for involving a group of what is commonly referred to as "stakeholders" in decision-making about natural resource management, environmental clean-up, and other wicked problems that emerge from the complexity of linked human and natural systems (e.g., Duram and Brown 1999; Korfmacher 2001). While regulations like the National Environmental Policy Act and the Clean Water Act require involvement of those affected, participation is also driven by increasing skepticism that scientists and technocrats have easy answers to complicated issues, especially those issues where the decisions are urgent; values are contested; stakes are high; and the scientific or technical knowledge is limited, uncertain, or contradictory.

Funtowicz and Ravetz (1991) proposed an approach to thinking about these kinds of problems called post-normal science. They argued that inclusion of an "extended peer community" can be critical to managing complex systems where both decision stakes and uncertainties are high; these peer communities may possess local expertise on land use or ecosystem processes that are not captured by scientific assessments. Multiple approaches have been introduced to engage this extended peer community to participate in both the production and evaluation of knowledge (e.g., Lach and Sanford 2010; Menon and Stafinski 2008; Petersen et al. 2011; Ravetz 2011). Cash et al. (2003) describe how an evolving group of participants can stimulate information exchange among technical and local experts to generate new knowledge that facilitates local planning and decision-making efforts. Acknowledging Cash et al. (2003), we call these people working together a Knowledge to Action Network or KTAN.

We developed a KTAN in this case study to (1) pose relevant research questions collaboratively and iteratively, (2) create a useable model to explore these questions, and (3) increase knowledge about how scientific information is generated within models in order to interpret the results. Due to the nature of the KTAN's research questions, modeling was identified as an appropriate methodology. Models have long been used by scientists and decision makers to gain understanding of the world by organizing data, synthesizing information from multiple sources, and making projections (e.g., Bredehoeft 2005; Canham et al. 2003). In the context of this paper, we defined the term "model" to describe any simplified representation of a system that involves relationships among different entities within a boundary. A model can be qualitative/conceptual, quantitative, or a combination of both. We used several different modeling approaches to describe local and regional conditions in ways that allowed us to think together about potential impacts of external drivers and management decisions.

As many have noted, there are multiple problems with scientific models arising from the general approach of simplifying complex systems including unsupportable assumptions, parameters that are incorrectly operationalized or missing, model uncertainties, and lack of supporting or relevant data (e.g., Oreskes et al. 1994; Shackley et al. 1998; Morgan 1999; Sarewitz et al. 2000). Yet, when communicating with decision-makers about the kinds of decisions they are contemplating in the face of climate change, population growth, and a globalizing economy, some models have been successful in helping them visualize, discuss, and weigh the outcomes of policy decisions and choices (e.g., Canham et al. 2003).

In this case study of participatory modeling, we considered both scientists and local decision makers as the producers of knowledge, with the expectation that all would participate in framing research questions, designing model implementation, and collecting and interpreting results. We used three different approaches to modeling, beginning with a conceptual model, or mind map, of the major issues facing the study system; followed by a system dynamics model to help quantify portions of the conceptual model; and finally a spatially-explicit, integrated model that examined several alternative scenarios. The goal was to co-produce knowledge on the most salient issues—as identified by the potential users—in a way that both scientists and decision makers found credible, legitimate, and useful.

14.2 Participatory Modeling Process

Our approach to participatory modeling emphasized both the process of knowledge generation and technical model outputs. Figure 14.1 outlines the major steps we followed from summer 2012 to fall 2014.

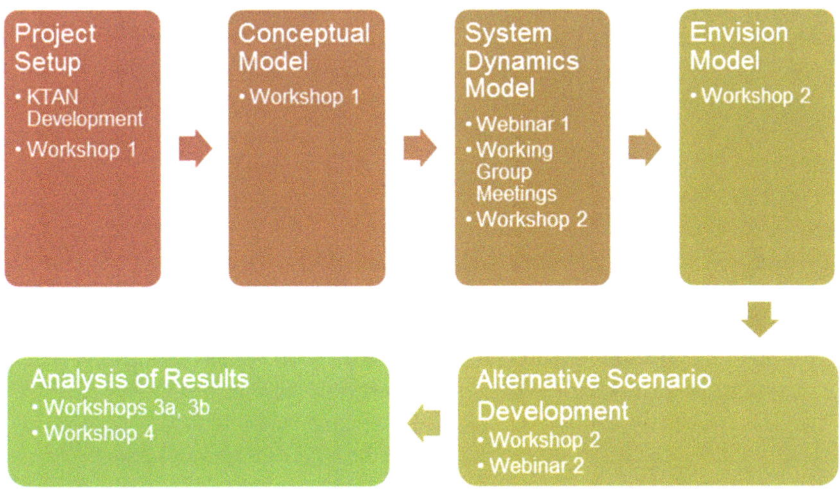

Fig. 14.1 Summary of participatory modeling approach

14.2.1 Project Setup

14.2.1.1 Development of a Knowledge to Action Network

In 2011, the Climate Impacts Research Consortium (CIRC) worked with decision makers across the Pacific Northwest to identify priority areas where climate research could help inform long-term planning. Building on these discussions, CIRC engaged water managers in Idaho in early 2012 to discuss priorities for increasing understanding of climate impacts on water resources. This event was attended by federal reclamation managers, state water rights administrators, hydroelectric utility providers, irrigation district managers, and university extension agents. The meeting highlighted the need for a comprehensive assessment of future changes to water resources that included climate change as well as socioeconomic factors such as population growth and residential development. CIRC proposed initiating a participatory modeling effort and conducting a case study for this type of research.

Continued outreach to these and other local water managers led to the identification of the Big Wood River Basin, a tributary to the Snake River in Central Idaho, as a pilot watershed. According to participants, the Big Wood Basin exemplifies the challenge faced by many areas of the Western United States to sustain traditional water uses, such as irrigated agriculture, while accommodating new demands, such as growing municipal and in-stream uses. Just as importantly, and unlike some basins, competing water uses had not yet reached a point that precluded users from sitting in the same room.

Over the summer of 2012, CIRC and these local partners contacted approximately 30 stakeholders who represented a broad spectrum of water users in the Big Wood Basin and agreed to collaborate on an integrated study. This group included university extension researchers; agricultural producers; irrigation district managers; conservation groups; landowners; recreational user groups; consultants; elected officials; and representatives from city, county, state and federal government in addition to CIRC researchers specializing in environmental modeling, public policy, hydrology, and outreach. We refer to this group as the KTAN. Network membership remained dynamic throughout the project—at times members chose to expand the network by inviting additional individuals or interest groups while other members elected not to participate for the entire project. A total of 59 individuals participated in the project in some manner (Fig. 14.2).

Sub-groups were formed throughout the project as necessary, including a technical team that was responsible for quantitative model development and three working groups that reviewed and critiqued the technical team's work. Ideally the technical team would represent the entire KTAN but due to the large geographic distance between the CIRC researchers and the local stakeholders, we were limited in the number of meetings we could host. Thus, in this project, the core technical team consisted of CIRC researchers and graduate students based in Oregon. The KTAN accepted this arrangement and in order to maintain transparency and foster co-production throughout the research, three working groups in Idaho were also formed to represent local expertise in hydrology, agricultural water demand, and land-use planning. The technical team collaborated with the working groups throughout the model development.

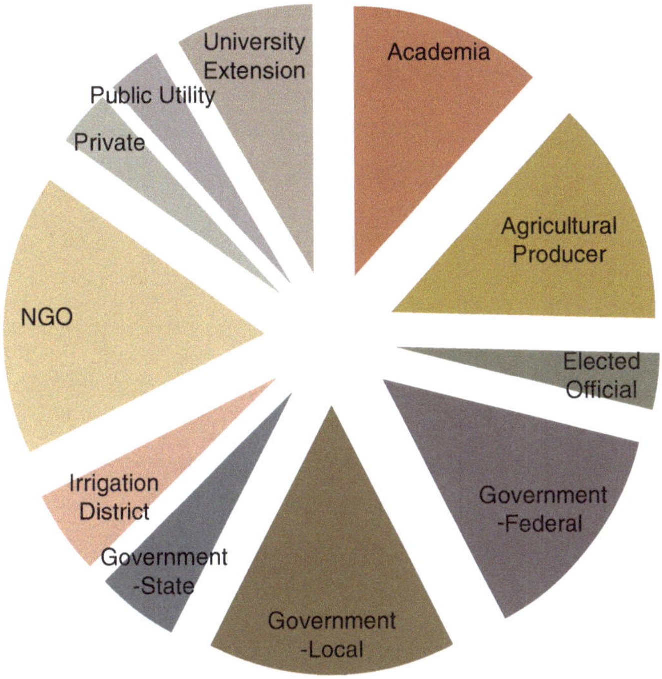

Fig. 14.2 Big Wood Basin KTAN by participant affiliation

14.2.1.2 Study Area

The study area centers on the approximately 8000 km^2 Big Wood River, Little Wood River, and Camas Creek drainages in Central Idaho (Fig. 14.3). The landscape varies from arid lava flow fields and shrubland interspersed with irrigated agriculture in the lower elevations to forested mountains at the higher elevations (Buhidar 2001). Winter snowfall serves as the dominant source of water, providing nearly 60 % of the annual supply. Surface and groundwater resources support many uses, including irrigation, livestock, municipal and domestic requirements, habitat, industry, recreation, and commercial use. Irrigation is the single largest water use, and three of the five counties in the study area are considered to be farming dependent (University of Idaho Extension 2006). The less agricultural portion of the basin is primarily supported by outdoor recreational tourism, including skiing and fishing.

The 2010 population was approximately 34,000 (US Census Bureau 2010). Development is generally concentrated in a small area, primarily due to land ownership and suitability. The Wood River Valley, home to most of the study area's population, experienced a growth rate of over 350 % from 1970 to 2005, leading to concerns over the sustainability of water resources, especially since newly developed areas utilize largely unregulated groundwater resources (Skinner et al 2007). Senior surface water rights holders worry supplies are being negatively impacted by the additional growth. These concerns are underscored by uncertainty over how climate change will affect snowpack.

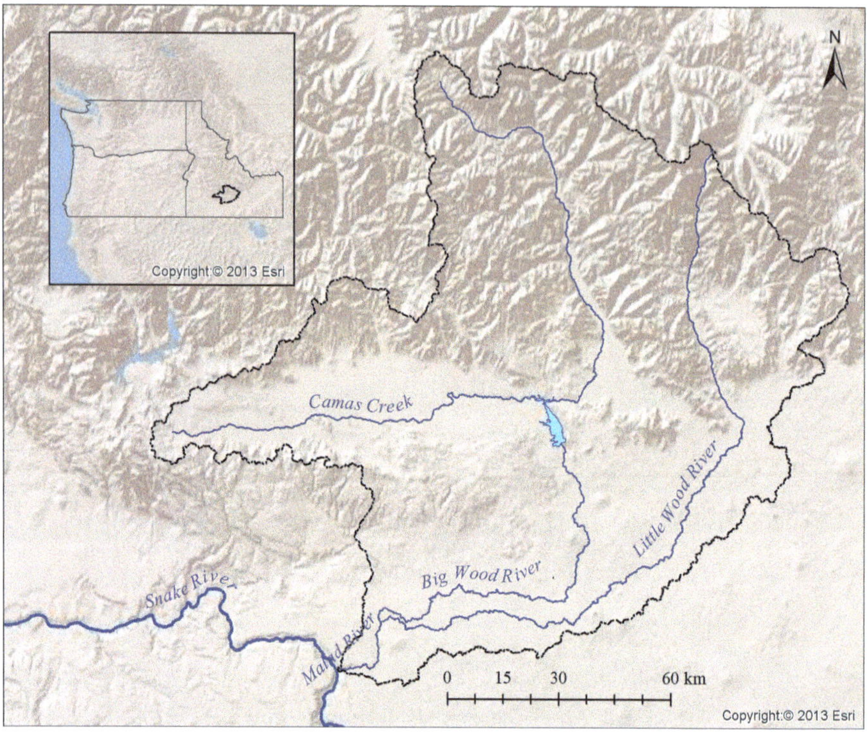

Fig. 14.3 Map of study area (Basemap used by permission. Copyright © 2013 Esri. All rights reserved.)

14.2.2 A Conceptual Model of the System

In August 2012, we began our first meeting (Workshop 1) with the KTAN by describing the project and seeking participation. We explained to the participants that the undertaking was designed as an experiment in co-producing knowledge that they would find usable in day-to-day decision making. This would require their participation throughout the process, willingness to critique and question both the science and each other, and an ability to work in a collaborative group with multiple and conflicting goals.

Next, we asked each participant to share a best-case and worst-case headline from the local newspaper 40 years in the future. This exercise was designed to have participants envision a desired future from which steps can be taken to attain those conditions (Holmberg and Robert 2000). Six main themes emerged: aquatic habitat protection, responsible development, economically resilient communities, community collaboration, sufficient water for agriculture and municipalities, and maintenance of an agricultural-based economy. We referred to these themes as endpoints for the basin—things people care about and would like to have more information about. After the headline

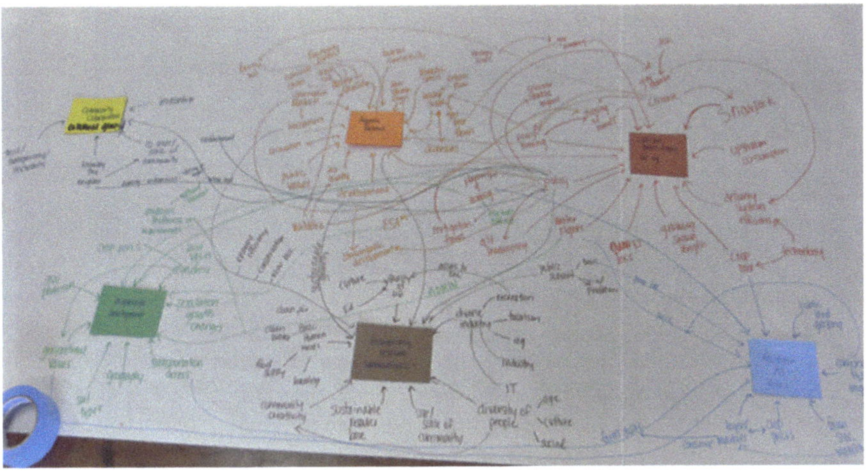

Fig. 14.4 Conceptual model of the Big Wood Basin study system

exercise we formed small groups, each tasked with hand-drawing a conceptual model of one endpoint. These conceptual models identified components, influences, and relationships associated with that endpoint. A brief demonstration was provided before group work started to help ensure some level of consistency across the groups, although no strict guidelines were imposed. Each small group then presented their completed figure, and the larger group worked to merge the figures, removing many duplicate components and adding missing components as well as linkages between the endpoints (Fig. 14.4).

This activity served two purposes. It allowed participants to quickly learn about the basin from each other and create a mind map highlighting connections and feedbacks among the diverse end points. Many of these stakeholders had never been in the same room before; this gave them the opportunity to view the basin through others' perspectives. The exercise also allowed the KTAN to move forward in framing the scope and priorities of the climate research efforts. Once the draft conceptual model had been completed by the large group, we discussed technical, logistical, and other limitations that could preclude certain sections of the conceptual model from being included in a more quantitative model. By explicitly discussing these limitations with the KTAN early in the process, we were able to collaboratively identify which endpoints would be critical to include in modeling efforts in order to ensure that the results would be useable, and which endpoints could be secondary. For example, while community collaboration emerged as a high-priority endpoint, the group discussed ways to characterize this through outputs of the model rather than attempt to model it directly. Creating the conceptual model and the resulting discussion helped open the black box that sophisticated modeling can appear to be and helped the KTAN develop a feasible project scope and research questions.

14.2.3 Translating the Conceptual Model into a System Dynamics Model

Given the interest in understanding future conditions in the study area, the KTAN members agreed that quantitative models would be appropriate for addressing the research questions. In the absence of a preferred modeling platform by the larger KTAN, the technical team proposed a two-pronged approach—an initial system dynamics model followed by a more comprehensive, spatially-explicit systems modeling platform called *Envision*.

A system dynamics model based on the conceptual model would be developed first. System dynamics models are gaining popularity for these types of collaborative projects (e.g., Stave 2003; Tidwell et al. 2004; Langsdale et al. 2009; Beall et al. 2011) because they are generally approachable even for non-modelers and are relatively quick to construct. Our goal in developing a system dynamics model was to allow participants to interact with a model early in the process, observe dynamics and feedback processes, and provide input that added to the saliency and credibility of the project output. As discussed below, the system dynamics approach was helpful in facilitating co-learning about what information is needed by managers and what can be provided by researchers, but ultimately was not sufficient to answer the most pertinent stakeholder questions and thus was not completed.

The technical team generated an initial framework for a system dynamics model using the Vensim PLE software (Ventana Systems, Inc.) during fall of 2012. The framework components—water supply, land development, and aquatic habitat—were based on endpoints from the concept map that the team concluded could be defensibly represented within the available timeframe and resources. These endpoints were represented as loosely coupled sub-models of hydrology, land-use transitions, and population growth (Fig. 14.5). Anticipating the eventual transition to the *Envision* model, the team utilized the structure and algorithms from that model whenever possible. The technical team presented the framework to the larger KTAN through Webinar 1, asking participants to consider if the reduced list of endpoints could adequately support planning for future climate conditions within the basin. While some concerns were voiced, particularly around the limited representation of groundwater within the modeling framework, participants agreed these endpoints were among the most important components of the conceptual model.

In March 2013, representatives from the technical team held three in-person meetings with the working groups. The goal of these meetings was to use local knowledge to refine the modeling approach and identify missing elements critical to decision making. The meetings started with presentations of the draft approach followed by facilitated discussions about the validity of the approach, missing factors, and trade-offs between additional or needed efforts. For example, the agricultural water demand group identified that soil type and evapotranspiration (ET) estimates for vegetation types were important factors to represent water demand in the basin. Following a facilitated discussion, the group concluded that improved ET estimates could be incorporated but that fine-scale variability of soil types could not be addressed without reducing efforts in other parts of the project.

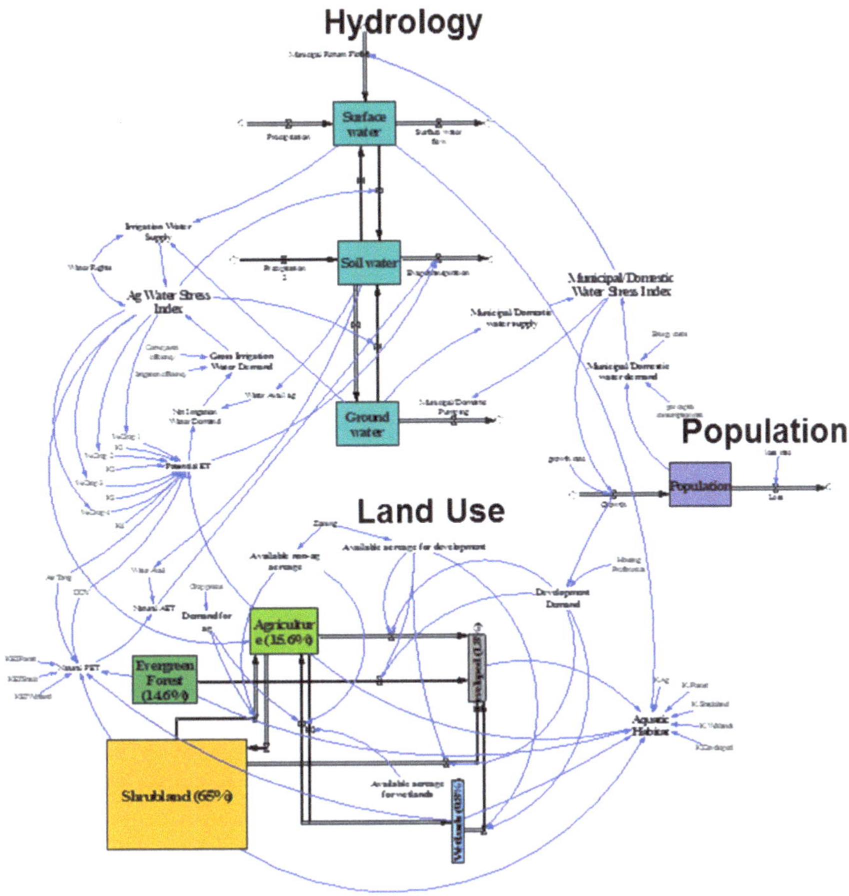

Fig. 14.5 System dynamics model framework containing coupled sub-models of hydrology, population growth, and land use dynamics

These meetings were followed by webinars with each working group to review model updates.

The technical team incorporated the working groups' revisions and presented to the entire KTAN at Workshop 2 in April 2013. Workshop discussions highlighted limitations of the system dynamics model, namely the lack of spatial representation. KTAN members' concerns related to specific locations that this model was unable to distinguish, particularly around reach level assessments and cropping systems within the basin. Although the original plan was to complete the system dynamics model before transitioning to the *Envision* model, it became clear that the system dynamics approach limited the project's salience for climate adaptation planning in the basin. Because of the enhanced capabilities of *Envision* for supporting the stakeholder engagement process and the perceived need for spatially-explicit landscape representations as the basis of the modeling effort, the KTAN made a decision to pursue development of the *Envision* application without completing the system dynamics model. While conventional modeling efforts may view this as a failure or misallocation of effort, we view the experience as a demonstration of the flexibility necessary in collaborative stakeholder modeling in order to avoid the development of unused information.

14.2.4 Alternative Scenario Development

The remainder of Workshop 2 was spent discussing alternative management scenarios the KTAN wanted to simulate within the *Envision* model. Participants were asked to provide one approach to managing the basin to mitigate impacts from future changes in climate and population, beginning their suggestion with 'What if…'; for example, 'What if we lined irrigation canals?' or 'What if we allowed unrestricted development?' We emphasized that participants not be constrained by current technical or legal limitations. The major themes that arose were based around water and land management strategies and the economic bases of the region—agriculture and tourism. Participants then broke into self-selected small groups to develop a range of future management policies. Each group was asked to characterize assumptions for two scenarios, one that emphasized a highly managed system and another for a less managed system. Additionally, the groups explored how the basin might look assuming one dominant economic base. The small groups then shared their scenarios with the larger group and integrated additional comments from the larger group into their scenarios.

From this workshop, four alternative scenarios, based on intensity of management and economic concentration, were developed for the *Envision* model. Details not captured in the workshop were supplemented by the technical team. Webinar 2 was then held in July 2013 to discuss the details of the management scenarios. KTAN members were provided information on the assumptions for each scenario in advance, and the webinar consisted of a brief presentation followed by a discussion of the validity of the assumptions and identification of any missing elements. These four scenarios would each be run with three different climate inputs (low impact, warmer/wetter, and hotter/drier) selected by Rupp et al. (2013) and downscaled from CMIP5 global climate models (Abatzoglou 2013) to provide 12 different scenario outputs (Fig. 14.6). The goal of these numerous combinations was to simulate a wide range of the many possible alternative futures to help resource users and managers prepare, plan, and/or mitigate for possible change.

14.2.5 *Envision*: A Comprehensive Systems Model to Explore Dynamics in Both Time and Space

Envision is a spatially-explicit systems modeling platform designed to assist in exploring alternative scenarios by combining sub-models of biophysical and socio-economic processes driving landscape change (Bolte 2014). An important note here

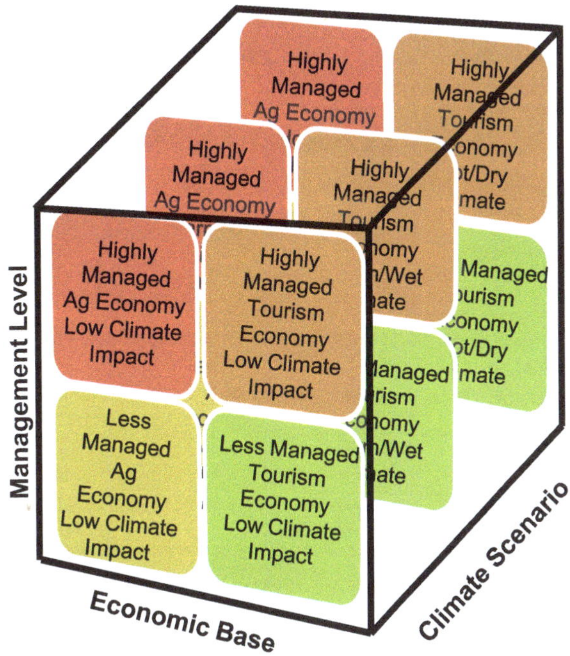

Fig. 14.6 The 12 alternative scenarios, consisting of combinations of economic base, management level, and climate scenario, developed by the KTAN and run in the *Envision* model

is that because *Envision* requires extensive technical expertise to set up and run, we moved away from the more transparent modeling approach employed earlier in the project. This was discussed during Workshop 2 and the KTAN accepted the lack of transparency since the *Envision* model would build on the same systems collaboratively explored through the conceptual and system dynamics models. No formal working group meetings were held during *Envision* model development; however, the technical team consulted KTAN members frequently via phone and email as questions arose. For example, KTAN members representing irrigation districts helped the technical team define the assumptions about reservoir operations and members from local governments provided input on urban development policies.

After a year of development, in May 2014, the *Envision* application containing coupled sub-models of hydrology, land use change, irrigation, and reservoir operations was operational and had been calibrated to historic conditions. The model was then run with the 12 alternative scenarios for the period 2010–2070. Model outputs were specified by the technical team to address the endpoints from the initial conceptual model as well as from discussions and workshops that had occurred since that time. The quantity of data was difficult to parse and the technical team recognized the need to add a visualization specialist to the team to effectively communicate the model results first to the KTAN and later to a wider public audience. A new phase of the project was implemented to create a web-based project—"Explorer"—that could guide a reader through the background and motivation of the project, model development, and the results.

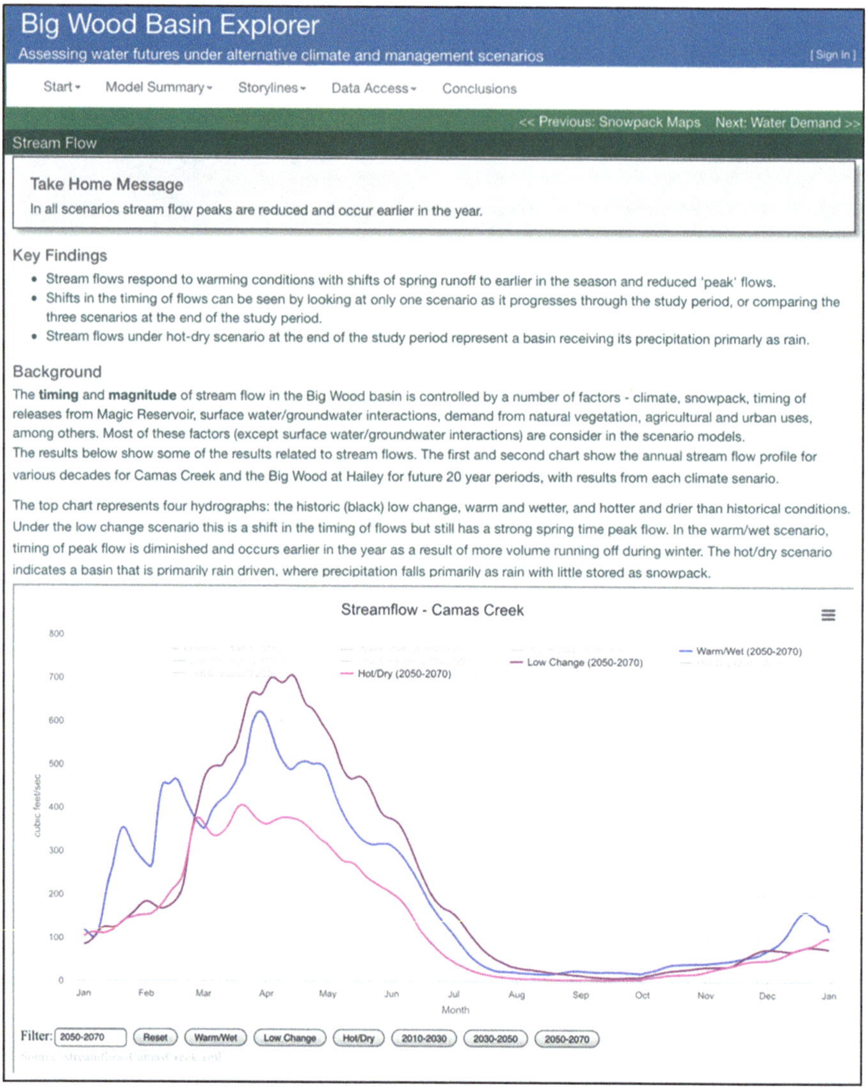

Fig. 14.7 Example of Big Wood Explorer storyline

14.2.6 *Big Wood Basin Explorer: A Tool to Foster Interaction and Exploration of a Complex Environmental Model*

The Big Wood Basin Explorer website (Fig. 14.7) is an aspect of this project that was not part of the initial scope. It was added late in the process when the technical team foresaw difficulty in guiding the KTAN through an analysis of the alternative scenarios. The large number of alternative scenarios and model outputs would make traditional charts and figures overwhelming. While the Explorer provides the type of information that would be expected on a project website—such as background and methodology—its true value comes through the presentation of results through storylines and interactive figures. Storylines refer to, in this case, ten narratives about topics of interest to the KTAN that parse and communicate the model results through a consistent format. Examples of storylines in this project were the agricultural system, municipal water demand, and snowpack. Each storyline page contains increasingly detailed information, starting with the take-home message, key findings, background information, and finally figures or maps showing model outputs. All figures are interactive, allowing readers to filter results (e.g., by a specific management or climate scenario) or turn individual data series on or off. Each user may select elements they would like to analyze or choose to simply read key findings and take-home messages developed by the technical team. Lastly, each storyline page contains a comments section where users can post feedback or questions. This format allowed us to greatly reduce the number of figures produced while still providing the breadth of results generated.

14.2.6.1 Analysis of Results

In July 2014, with preliminary model results and a working draft of the Explorer, we held Workshops 3a and 3b with the KTAN. Workshop 3a focused on issues related to water and agriculture and was held in the lower basin; while Workshop 3b focused on population growth, land use patterns, and snowpack and was held in the upper basin. Participants were invited to attend one or both workshops to review the initial model results and learn about the Explorer. Both workshops began with introductions and a brief review of the conceptual and systems dynamics models, as well the project background, including major decisions or turning points that had defined the scope and methodology of the work. Members of the technical team then highlighted the methodology of the *Envision* model. Participants asserted they understood the scope and assumptions of the model and did not identify any major concerns. Next, we presented simulated and observed data from a historic period, a format the participants were familiar with from past experience with models. Time was allowed for discussion or comments on the model, but participants vocally expressed their approval of the model's ability to capture historic trends and wanted to move on to the simulated results.

KTAN members were engaged in the results and expressed interest in exploring the Explorer in more detail. These workshops offered the first opportunity to truly explore results and we saw a notable revival in interest and participation, which had declined during the long model development phase. In fact, providing the KTAN an opportunity to interact with preliminary results led to requests for additional modeling work. Given that we were nearing the end of the project, this required a facilitated discussion between participants and the technical team to identify additional efforts that were feasible to pursue considering the time and resources remaining. The workshops concluded with a list of prioritized model additions for the technical team and all participants being tasked with detailed review of the Explorer.

14.2.7 *Future Efforts*

Remaining efforts for this project include incorporation of revisions or additions identified by the KTAN into the model and the corresponding updates to the Explorer. One final workshop inviting the KTAN as well as a wider public audience will be held to present the project results and identify tradeoffs between the management scenarios. The KTAN participants will be encouraged to help the technical team facilitate the discussion at the final workshop. Lastly, we plan to discuss the process with KTAN participants, using informal feedback, observations, surveys, and interview data collected throughout the project to analyze the utility of the approach and its transferability to other projects.

14.3 Discussion

The objective of this work was to co-generate knowledge about climate impacts with potential users. We did this by asking end-users of information to characterize the major issues in the state and in the basin before setting specific project objectives. We then initiated modeling projects focused on compiling datasets and models that addressed the initial questions of the KTAN as well as questions that emerged throughout the process. The KTAN was a venue for an exchange of information between the experts developing the model (i.e., scientists) and the experts interested in using the outputs of the model (i.e., elected officials, agency staff, private sector interests, landowners, and non-governmental organization staff). Exchange of information began with co-framing the issues of concern and continued through identifying acceptable data sources, characterizing future scenarios, validating model outputs, and iterative adjusting of model components. The goal was to keep the modeling process as transparent as possible for all to understand. The participation took place in face-to-face meetings, phone conferences, webinars, and small group meetings, all designed to elicit, share, and describe information. Each interaction was designed to be flexible, which allowed the project to adapt as necessary to meet the needs of the KTAN.

14.3.1 *Challenges*

We faced several challenges along the way and characterize them as falling into two broad categories: engagement and technical challenges. Each is addressed below.

14.3.1.1 Engagement Challenges

The decision to focus on the Big Wood Basin was made in conjunction with state agency personnel based on their knowledge of state issues. When we first talked with potential participants in the basin there was interest in the topic, but there was no sense of urgency about water scarcity. In this part of the world, water shortages are a common occurrence and infrastructure projects and individual practices have been developed to deal with water scarcity. Climate change has been characterized as a "long emergency" (Kunstler 2006), with all of the attendant difficulties in drawing people's attention away from issues perceived as more pressing (e.g., Pew Research Center 2013). Even though the project was focused on water scarcity, which might be salient locally, thinking about adapting to climate change some time in the future did not create any sense of need for new or different information.

The KTAN was structured to create opportunities for members to meet, share data sources and information and evaluate model assumptions and outcomes. However, the widely dispersed geographic location of members of the KTAN (the technical team was primarily in Oregon, while local stakeholders resided in Idaho) limited the number of in-person meetings and discussions. To address this, a new role was created to help with KTAN engagement. This engagement person and one of the graduate students working on the model were in frequent contact with the stakeholders via phone, email, and short in-person visits. While webinars were useful solutions for overcoming geographic distance, participation tended to be less than the active involvement and feedback garnered from individual phone calls and in-person visits.

The final quantitative model was highly dependent on sophisticated technical skills that few members of the KTAN possessed. Unlike the conceptual or system dynamics models, there was little for non-technical members to contribute until the model was constructed and validated. During this period, keeping all KTAN members engaged and interested became difficult. One way we kept in touch was by offering informational webinars, including one on groundwater modeling in the basin that was being conducted concurrently by state and federal agencies. Despite this drop in engagement, once model outputs became available, interest and participation quickly revived within the group.

14.3.1.2 Technical Development Challenges

As with most coupled human and natural systems, water supply and demand in the Big Wood Basin is a complex system with shared resource use. Some local particulars that made this even more challenging included a wide elevation gradient, multiple land ownerships including the federal government, different land uses including high-end resort and low-value agriculture, and cultural differences among second or third home owners and long-time residents. Wide variability in the technical abilities and understanding of KTAN members also posed challenges when framing problems and understanding output.

Technical difficulties emerged as parameters critical to understanding hydrology in the basin were revealed as the modeling proceeded. For example, modeling evapotranspiration in the basin's forests was critical and time-consuming, although it was of little interest to non-technical KTAN members. Evaluating the weaknesses in the hydrologic models in order to identify potential sources of variability was also a time-intensive process. Theoretically, the whole KTAN should provide a venue for identifying missing variables and sources of weakness in the model, but because we were unable to meet regularly due to geographic distance, this potential strength of the KTAN process was not available to help the technical team.

Once the model was running, visualizing and effectively communicating the complicated results became a non-trivial challenge. Additionally, at this point in the process the technical team was prepared to conclude modeling efforts, while seeing results led to many more questions by other KTAN members.

While both model building and stakeholder involvement present challenges unique to their practice, together they create another set of synergistic problems that neither practice has much experience managing. The use of an engagement person to act as a link between technical and non-technical KTAN members helped translate some, but not all, the difficulties that emerged through this process. In future efforts it might help to ensure there are local champions with an urgent need for new knowledge, a relatively simple set of research questions for which data can be collected and analyzed, and an adequately staffed modeling team experienced in the give-and-take of KTAN or stakeholder-directed modeling.

14.3.2 *Transferability*

In this case study, we concluded that transferability was found not in the details of the engagement process or modeling platform, but rather from the distinct approach utilized. By maintaining a focus on providing useable knowledge and allowing local end-users to provide input that directed the scope, research questions, and methodology, we were able to create a flexible and iterative process that tailored our project to our end-user's specific needs.

A similar approach is seen in a CIRC project on Oregon's northern coast, where researchers engaged a local KTAN for a project aimed at exploring sea level rise. Although details of this engagement and modeling approach differed from the Big Wood project, both shared consistencies in their underlying approach. The two KTANs possessed different levels of technical expertise and thus provided very different types of input for model development; additionally, the Oregon KTAN already existed, while the Big Wood KTAN was developed specifically for this project. However, both sought out initial and iterative end-user engagement with the objective to co-produce useable knowledge. The experiences of the participants in both projects are being studied through surveys, interviews, and observations in order to assess whether this co-production of knowledge approach leads to the generation of salient information.

14.4 Conclusions

Over a 2-year period, the Climate Impacts Research Consortium helped foster the development of a Knowledge to Action Network of water resource users and managers in Idaho's Big Wood River Basin. This network merged scientific and local expertise to co-produce usable knowledge about future climate impacts. Based on the informational needs and concerns of the KTAN, we collaboratively designed a research effort to explore questions posed by the network as well as to identify strategies to mitigate negative impacts. This included generating fully-transparent, non-technical models to develop a baseline understanding of the study system and then progressing to more technical models that allowed for comprehensive analysis of impacts of major drivers of change in the system.

We felt that in order to produce useable knowledge, the project design must be credible, salient, and legitimate for end-users. Thus, we equally emphasized the process of co-learning about the system with the actual model outputs. We held a series of workshops and webinars, with informal discussions in between. Instead of being mechanisms for the technical team to share information, these exchanges provided the project's foundation, from the scope to the assumptions to the alternative scenarios. Formal and informal feedback from the interactions was also used to modify the approach. For example, the technical modeling approach was changed at one point because the network did not feel it was adequate to address their questions. Thus, the entire network participated in the process from inception to conclusion.

We hypothesized that involvement in a KTAN would change participants' attitudes, expectations, and networks of climate information sharing. We expect to follow up in one year and then three years to see how knowledge, attitudes, and networks have changed since the project began.

Acknowledgements This work is supported by NOAA's Climate Program Office with funding from the Regional Integrated Sciences & Assessments (RISA) program.

References

Abatzoglou JT (2013) Development of gridded surface meteorological data for ecological applications and modelling. Int J Climatol 33:121–131. doi:10.1002/joc.3413

Beall A, Fiedler F, Boll J, Cosens B (2011) Sustainable water resource management and participatory system dynamics. Case study: developing the Palouse basin participatory model. Sustainability 3:720–742. doi:10.3390/su3050720

Bolte JP (2014) Envision. Oregon State University, Corvallis, OR

Bredehoeft J (2005) The conceptualization model problem—surprise. Hydrogeol J 13:37–46. doi:10.1007/s10040-004-0430-5

Buhidar B (2001) The Big Wood River Watershed Management Plan. Idaho Department of Environmental Quality, Twin Falls, Idaho

Canham CDW, Cole J, Lauenroth WK (2003) Models in ecosystem science. Princeton University Press, Princeton

Cash DW, Clark WC, Alcock F et al (2003) Science and technology for sustainable development special feature: knowledge systems for sustainable development. Proc Natl Acad Sci 100:8086–8091. doi:10.1073/pnas.1231332100

Duram LA, Brown KG (1999) Assessing public participation in US watershed planning initiatives. Soc Nat Resour 12:455–467. doi:10.1080/089419299279533

Funtowicz SO, Ravetz JR (1991) A new scientific methodology for global environmental issues. In: Costanza R (ed) Ecological economics: the science and management of sustainability. Columbia University Press, New York, pp 137–152

Holmberg J, Robert KH (2000) Backcasting - a framework for strategic planning. Int J Sustain Dev World Ecol 7:291–308. doi:10.1080/13504500009470049

Korfmacher KS (2001) The politics of participation in watershed modeling. Environ Manage 27:161–176. doi:10.1007/s002670010141

Kunstler JH (2006) The long emergency: surviving the end of oil, climate change, and other converging catastrophes of the twenty-first century. Grove Press, New York

Lach D, Sanford S (2010) Public understanding of science and technology embedded in complex institutional settings. Public Underst Sci 19:130–146. doi:10.1177/0963662508096783

Langsdale SM, Beall A, Carmichael J, Cohen SJ (2009) Exploring the implications of climate change on water resources through participatory modeling: case study of the Okanagan Basin, British Columbia. J Water Resour Plan Manage 135:373. doi:10.1061/(ASCE)0733-9496(2009)135:5(373)

Menon D, Stafinski T (2008) Engaging the public in priority-setting for health technology assessment: findings from a citizens' jury. Health Expect 11:282–293. doi:10.1111/j.1369-7625.2008.00501.x

Morgan MS (1999) Learning from models. In: Morgan MS, Morrison M (eds) Models as mediators: perspectives on natural and social science. Cambridge University Press, Cambridge, pp 347–388

Oreskes N, Shrader-Frechette K, Belitz K (1994) Verification, validation, and confirmation of numerical models in the earth sciences. Science 263:641–646

Petersen AC, Cath A, Hage M et al (2011) Post-normal science in practice at the Netherlands Environmental Assessment Agency. Sci Technol Hum Values 36:362–388. doi:10.1177/0162243910385797

Pew Research Center (2013) Deficit reduction rises on public's agenda for Obama's second term. Pew Research Center, People Press. http://www.people-press.org/2013/01/24/deficit-reduction-rises-on-publics-agenda-for-obamas-second-term/. Accessed 18 Jun 2014

Ravetz JR (2011) "Climategate" and the maturing of post-normal science. Futures 43:149–157. doi:10.1016/j.futures.2010.10.003

Rupp DE, Abatzoglou JT, Hegewisch KC, Mote PW (2013) Evaluation of CMIP5 20th century climate simulations for the Pacific Northwest USA. J Geophys Res Atmos 118:10884–10906. doi:10.1002/jgrd.50843

Sarewitz D, Pielke RA, Byerly R (2000) Prediction: science, decision making, and the future of nature. Island Press, Washington, DC

Shackley S, Young P, Parkinson S, Wynne B (1998) Uncertainty, complexity and concepts of good science in climate change modelling: are GCMs the best tools? Clim Change 38:159–205

Skinner K, Bartolino J, Tranmer A (2007) Water resource trends and comparisons between partial-development and October 2006 hydrologic conditions, Wood River Valley, South-Central Idaho. United States Geological Survey

Stave KA (2003) A system dynamics model to facilitate public understanding of water management options in Las Vegas, Nevada. J Environ Manage 67:303–313

Tidwell VC, Passell HD, Conrad SH, Thomas RP (2004) System dynamics modeling for community-based water planning: application to the Middle Rio Grande. Aquat Sci 66:357–372

University of Idaho Extension (2006) County economic base codes. In: Indicators Idaho. http://www.indicatorsnorthwest.org/. Accessed 1 Oct 2014

U.S. Census Bureau (2010) American FactFinder

Ventana Systems, Inc. Vensim PLE

Chapter 15
Science Based Modelling for Supporting Integrated Coastal Zone Management

Purwono Budi Santoso and Anthony Halog

Chapter Highlights

Approach: A systematic review of the integrated coastal zone management (ICZM) approach and all related tools for science-based modeling is undertaken.

Participant Engagement: There is an indirect participant engagement by extracting information based on previously published related researches on ICZM stakeholders' participation.

Models/Outcomes: Comparison of various science-based modelling for supporting the implementation of Integrated Coastal Zone Management (ICZM).

Challenges: Selecting and implementing an appropriate modelling tool for answering the problems in the real world, through the cyclical process of modelling from problem articulation, dynamic hypothesis, simulation, testing, policy designs and evaluation by involving diverse interests of stakeholders, is challenging, especially when applied to particular case.

P.B. Santoso
Directorate General of Marine and Fisheries Resources Surveillance, The Ministry of Marine Affairs and Fisheries of the Republic of Indonesia, Jakarta, Indonesia
e-mail: purwono.santoso@uq.net.au

A. Halog (✉)
School of Geography, Planning, and Environmental Management (GPEM), The University of Queensland, Brisbane, QLD, Australia
e-mail: a.halog@uq.edu.au

S. Gray et al. (eds.), *Environmental Modeling with Stakeholders*,
DOI 10.1007/978-3-319-25053-3_15

15.1 Introduction

15.1.1 Background

The main problem affecting coastal environments all over the world is over-population where people are attracted to overrun these narrow-productive regions which provide abundant basic human needs, functional, aesthetic and strategic resources (Cartwright 1995; Fabbri 1998). This situation will result in multiplier effects rather than just a settlement's arrangement issue in the coastal environment, such as trans-boundary or land-based marine pollution, over-exploitation, destructive fishing and conflict due to open-access regime. It reflects that the coastal environment is a very complex system with sophisticated anthropogenic pressures as shown in Fig. 15.1.

Indonesia, for example, is one of the countries in the world in which more than 70 % of its population is concentrated in the coastlines, a situation which can possibly alter the natural ecosystem rapidly (Dahuri 1998; Nurhidayah 2010). Even though Indonesia is one of the largest archipelagic countries in the world with 1.9 million km^2 land mass area and 5.8 million km^2 waters area, consisting of 17,500 islands (the exact number is still being recalculated by the joint-ministerial team for islands' toponymy) which contribute to 81,000 km length of coastline with high biological diversity, socio-cultural and economic activities (Dahuri 2011; Farhan and Lim 2010; Nurhidayah 2010; Wever et al. 2012), pressures in the forms of conflict

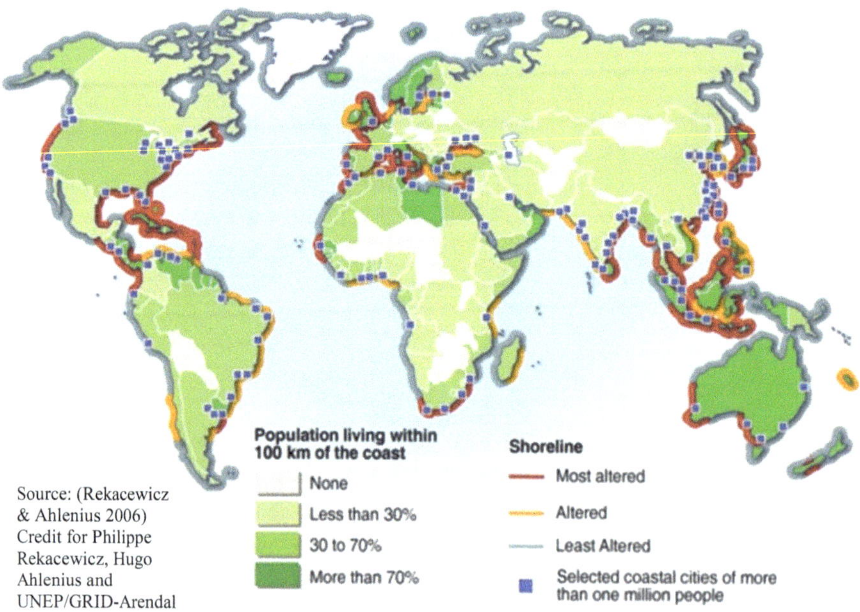

Source: (Rekacewicz & Ahlenius 2006) Credit for Philippe Rekacewicz, Hugo Ahlenius and UNEP/GRID-Arendal

Fig. 15.1 Map of the world coastal population and shoreline degradation

among users, over-exploitation, as well as degradation are still dominating the picture of Indonesian coastal environments (Nurhidayah 2010). Thus there is an urgent need of capable measures and tools to overcome those problems.

Along with the conflicts and tensions among stakeholders and sectors that contributed on magnifying the pressures in the coastal environments in every part of the world (Nurhidayah 2010), a new global trend of the integrated coastal zone management (ICZM) has emerged at the beginning of the twenty-first century with essential principles on decentralization, integration, sustainability, equity, fairness, accountability and participation (Arsana 2010; CBD 2000; Farhan and Lim 2010; Forst 2009; Post and Lundin 1996; Treves 2010; Wever et al. 2012). Hence this research project tries to examine and compares all possible science-based modeling for supporting the ICZM. Thus, the research questions are:

1. *What kinds of modeling method that have been emerged and tested to support the implementation of integrated coastal zone management?*
2. *What are the respective advantages and inherent limitations of the modelling methods evaluated?*

This research provides a significant contribution to enhancing the implementation of integrated coastal zone management in countries which are facing critical problems on managing complexities in the coastal environment.

15.1.2 Scope of Study

The scope of study is reflected through a systematic review, which is a structured literature-evaluation (Wilson 2013; Zumsteg et al. 2012), that can aid in gaining more information regarding two important things, i.e.:

(a) The concept of integrated coastal zone management (ICZM);
(b) Tools of analysis/modeling that have been developed and applied worldwide for the ICZM.

15.2 Systematic Review

15.2.1 Integrated Coastal Zone Management (ICZM)

Coastal environments are generally defined as the interface area between land and marine ecosystem which respectively alter each other (Dahuri 1998; Dutton and Hotta 1995; Haslett 2009). These environments tend to form unique ecological conditions in terms of biogeographic variations (macro scale), ecosystem patterns (mesoscale) and local physical processes (micro scale) which result in differences from one place to another (Dutton and Hotta 1995). Coastal environments are always recognized as the most productive regions on earth with abundant living and non-living resources, infrastructures and environmental services that could fulfil basic human needs, as well as functional, aesthetic and strategic facilities (Dahuri 1998; Dutton and Hotta 1995; Farhan and Lim 2013; Nurhidayah 2010; Siry 2011; Wever et al. 2012). These endowed resources have been created from dynamic interactions between terrestrial and waters ecosystem with high varieties in different places on earth, as shown in Fig. 15.2 that illustrates the tropical coastal ecosystem.

Due to their dynamics and various site specifications, it is almost impossible to find the same operational definition of coastal area from one country to another. According to Dahuri (1998), there were three criteria that have been used to determine coastal area delimitation in the world: (1) the arbitrary line that is perpendicular to the coastline; (2) legal and administrative boundaries; and (3) dynamic ecological (socio-biophysical) processes. Most countries in the world used to apply the arbitrary line and administrative boundaries to specify their coastal area delimitation due to the uniformity and applicability for general situation (Dahuri 1998; Haslett 2009; Portman et al. 2012). For instance, China has defined its coastal area by 15 km respectively to the ocean and to the inland from the coastline while Indonesia has specified 12 nautical miles (about 22.22 km) from the coastline to the ocean as coastal waters and subdistrict areas that coincide with the coastline as coastal land (Dahuri 1998; Sucofindo and DGMCSI 2009).

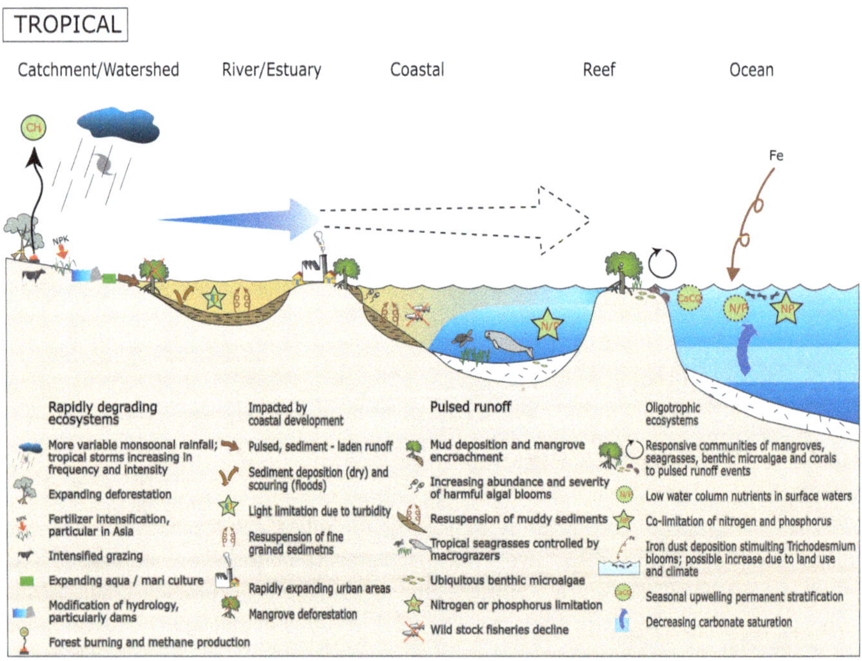

Fig. 15.2 Tropical coastal ecosystem diagram (Adrian 2005). Credit for Adrian J., Integration and Application Network, University of Maryland Center for Environmental Science (ian.umces.edu/imagelibrary/)

However in the global context, there were actually several waters-related international agreements that had existed before the establishment of the Law of the Sea (the UNCLOS 1982), e.g. The Bering Sea Fur Seals Arbitration 1893, Ramsar Convention 1971 and Marpol 1973/1978. Unfortunately those international agreements were only focused on sector-basis purposes, such as maritime jurisdiction for sealing arrangement, conservation of wetlands for protecting waterfowl (birds that are ecologically dependent on wetlands), and marine pollution from ships. The term "coastal zone management" itself was historically popularized by the US Congress when they enacted the Coastal Zone Management Act 1972 (Post and Lundin 1996) while in the same year, 1972, the first global conference on the environment was commencing in Stockholm, namely the United Nations Conference on the Human Environment which was the milestone of the emerging global awareness on balancing environmental, social and economic issues (Queffelec et al. 2009). This positive step had been followed by other countries in the late 1970s and early 1980s with various terms, e.g. coastal zone management, coastal resource management and coastal area management, but still in a single-sector approach (Post and Lundin 1996).

According to Forst (2009), in the period between 1974 and 1975, there was also an emerging ecosystem management principle on living marine resources from the World Wildlife Fund (WWF) workshops program in cooperation with the Council on Environmental Quality, the Smithsonian Institution, the Ecological Society of America and the International Union for Conservation of Nature and Natural Resources (IUCN) as an answer to the failure of the maximum sustainable yield (MSY) tools in sustaining living marine resources. Those series of workshops in conjunction with the Stockholm Conference in 1972 and initiatives from some countries on coastal management had raised the issue of conservation and ecosystem management of living marine resources into the debate in the UNCLOS 1978 among interested stakeholders, e.g. fisheries, oil and gas, tourism, maritime transportation and conservation, which were represented by governments (Forst 2009). Finally the UNCLOS that was signed on 10 December 1982 in Montego Bay and was brought into force on 16 November 1994 by 195 parties (Queffelec et al. 2009), could provide comprehensive frameworks on integrated management of coastal and marine resources which would be adopted in Chapter 17 of the Agenda 21 from the Earth Summit Rio de Janeiro 1992 (Forst 2009).

Another international agreement that had been successfully achieved by the parties in the Earth Summit Rio de Janeiro 1992 was the adoption of the Convention on Biological Diversity (CBD). However in the development of the convention, there was an expanding consciousness among the parties of the CBD of the coastal and marine conservation, resulting in the establishment of the Jakarta Mandate 1995 which had encouraged the parties to manage their coastal and marine resources in an integrated approach (CBD 2000; Queffelec et al. 2009).

However all of those international treaties, i.e. the UNCLOS 1982, Chapter 17 of the Agenda 21 Rio Summit 1992, or the Jakarta Mandate 1995 of the CBD, do not exactly specify the term "integrated coastal zone management (ICZM)". It was firstly introduced by the European Commission in 2002 when proposing a formal recommendation to the European Parliament and Council regarding the implementation of Integrated

Coastal Zone Management in Europe (Queffelec et al. 2009). Meanwhile the principles of the ICZM had been defined as a dynamic, decentralized, multi-disciplinary and iterant process to achieve sustainable coastal zone management through a framework of information collection, planning, decision making, and monitoring with the intention of gaining long term balance of environmental, economic, socio-cultural and recreational objectives within the threshold of natural dynamics and complexities.

15.2.2 Modeling in the Coastal Environment

According to Birta and Arbez (2007), a model acts as a representation of an object, system or idea with a certain form which can be static (linear) or dynamic (non-linear). The obvious differences between static and dynamic models are related to the components involved and the notion of time evolution, where a static model is only related to one aspect in a certain time and a dynamic model considers more than one component over time (Birta and Arbez 2007). In the context of science-based modeling, there are two kinds of model: simulation and analytical models. The simulation model (discrete-event or dynamic) is part of the techniques or operating models of problem entity which mimics problem behavior and its functional relationship whereas analytical model which is a set of equations or numeric algorithms that represent a problem entity or system (Swinerd and McNaught 2012).

Meanwhile in regards to their dynamic complexities, coastal environment is as a system (or multiple systems) where the agents (natural and anthropogenic agents) interrelate in biophysical and socio-economic spheres resulting in both positive and negative impacts (Fig. 15.3) (Fabbri 1998). Therefore attempts to understand coastal

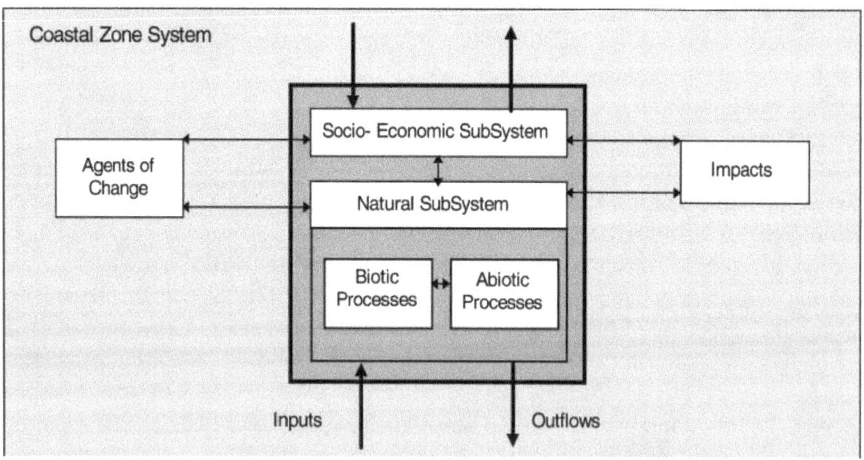

Fig. 15.3 Tropical coastal ecosystem diagram. (*Source*: Fabbri, KP 1998, 'A methodology for supporting decision making in integrated coastal zone management')

ecosystem and to address its inherent problems should be focused on finding tools of modelling that are expected to assist users in gaining insights into complex features of coastal behavior over time in simple and easy ways (Birta and Arbez 2007). Along with the development of the ICZM approach, there are tools of modeling that have been intensively researched, tested and also applied to support the ICZM approach as a response to the dynamics and complexities of coastal problems (Fabbri 1998; Schluter et al. 2012).

One of the most prominent and fascinating instruments in ICZM implementation is Geographical Information System (GIS) or Geospatial modeling. GIS has been widely acknowledged not just as a mapping tool for drawing geographical features in certain scales and projections but has also been proven as an advanced instrument for analyzing and integrating sophisticated spatial data layers and their associated non-spatial attributes to produce understandable spatial information (Gunawan 1998; Rodríguez et al. 2009). Technically, spatial data layers are constructed from two main data, i.e. geo-referenced vector data and attributes. Geo-referenced vector data is two-dimensional graphics (point, line or polygon) of a certain scaled-object on earth that have been registered into selected coordinate system projection, while attributes are substances of socio-bio-physical features that can be embedded dynamically into vector data (Gunawan 1998). At this level, GIS has proven its capability as an advanced information system which can collect, compile, retrieve, manage and visualize complicated data into informative visualization in spatial context. Furthermore, GIS also facilitates analysis, e.g. spatial analysis and geo-statistical analysis which can enhance its versatility and flexibility (Gunawan 1998; Rodríguez et al. 2009; Stephen and David 2004). Therefore this tool can tackle, at least, the ICZM problems in terms of integrating disparate datasets from many different sources, making task analysis for simulating future scenarios easier, while conventional cartography has less versatility and flexibility to deal with dynamic and complex elements in the coastal environments (David 2004).

GIS has been used widely in various fields: i.e. civil engineering, water resources management, flood management, watershed delineation, forestry and wildlife management, urban planning, soil erosion assessment, even for coastal management (Ahmad and Simonovic 2004; Fabbri 1998; Van Kouwen et al. 2008). Despite the development of GIS in addressing complex problems which can be significant for supporting ICZM implementation due to its capability in calculating spatial scenarios with geospatial modeling functionality, factually there is still evidence of limitations on diffusion process and temporal modeling, that cannot simply be denied while there are important feedbacks between time, space and pattern in every dynamic systems, including in coastal systems (Ahmad 2002; Ahmad and Simonovic 2004, 2006; Hartt 2011; Van Kouwen et al. 2008).

Fortunately, there is a tool that has strong capability in dynamic temporal modelling, called System Dynamics Modelling (SDM), which employs a system thinking method. Sterman (2000)argued that system thinking method will enhance understanding of complexities, their connectedness, interactions, feedbacks, and dynamic behavior over time. It is also useful to understand the sources of policy resistance and to assist on designing effective policies which can predict intended consequences, as well as avoid unintended consequences. A complete modelling process is iterative and should consist of, at least, problem articulation (boundary selection), dynamic hypothesis, simulation model, testing, policy formulation and evaluation. Its iteration is embedded in the whole system that constantly involves experiments and learning in both virtual and real world as shown in Fig. 15.4.

Furthermore Sterman (2000) explained that problem articulation is a step to define the problem through reference modes, time horizons, and scales whereas dynamic hypothesis is a working theory of the problem by mapping the system structure of the environment. The system structure is usually described through qualitative model (causal loop diagrams-CLD) as a conceptual model that shows polarities among key variables and feedback loops, as well as quantitative model (stock-flow diagrams-SFD) as a simulation model that shows accumulations (stocks) and movement of material and information (flows) in a system. The hypothesis can be tested by running the simulation model (SFD) to identify any contradictions or vague concepts of the conceptual model to be fixed into a confident model. Once confident model has been established through series of testing, then it can be utilized for designing policies and evaluating the system.

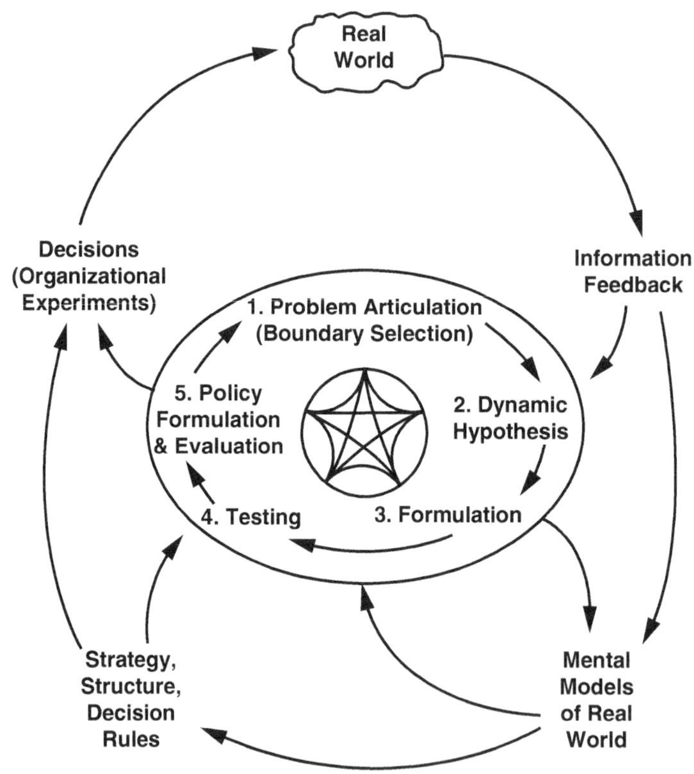

Fig. 15.4 Effective modelling process (Sterman 2000)

Thus, Ahmad and Simonovic (2004) and Hartt (2011) have tried new modeling framework which did not only capture feedbacks between time and space separately but also integrated with different capabilities on examining patterns in time and space scales. Ahmad and Simonovic (2004) proposed an integration of geographic information system (GIS) and System Dynamics Modeling (SDM) for managing water resources systems while (Hartt 2011) tried to link GIS and SDM for modeling the impacts of storm damage on coastal communities. The integration of those tools is called a Spatial System Dynamics Model.

According to (Ahmad and Simonovic 2004), this new framework aims to cover any limitations that are inherently found in each modeling tool. GIS is limited in temporal modeling and SDM does not properly represent spatial processes. Therefore there are two attempts at addressing the integration of those tools, i.e.:

(a) Implicit approach: that is acknowledging spatial dimensions into SDM where spatial features of the system are represented with aggregate stocks. Spatial dimensions will not be provided in this approach;
(b) Translating SDM equations to run in GIS tools through programming language, in other words the dynamic capability of SD models are brought into the GIS environment where the consequence is the interactive power of SD models will be lost during simulation.

In the context of modeling for water resources systems by Ahmad and Simonovic (2004), dynamic process modeling in time and space feedback interaction is achieved through dynamic data exchange (DDE) between SDM and GIS with the flow as shown in Fig. 15.5. The final result will be brought into the GIS environment so that the area of interest of the system should be initially determined and divided into cells, and then SDM is developed for each cell that interconnects with neighboring cells as shown in Fig. 15.6. The principle of those cells is similar to cellular automata (CA) models where a simple agent-based model (ABM) can be considered as a cellular automata model which offers the opportunity for spatial dimension (Swinerd and McNaught 2012).

However, the SERD (Simulation of Ecological Compatibility of Regional Development) model in Reichraming, Austria described by Gaube et al. (2009) actually has integrated three main components:

(a) Agent-based actors module that simulates decisions of significant actors, such as farmstead, the municipal authority, etc.;
(b) GIS land-use module that simulates land-use change at the level of individual parcels of land;
(c) Integrated socio-ecological stock-flow module that simulates carbon-nitrogen flows through socioeconomic and ecological stocks.

This is a comprehensive example of integrated hybrid simulation that has succeeded in answering three main characteristics of dynamic complexities, i.e. human behavior, temporal and spatial processes, all at once. Nonetheless after evaluating strong interdependencies between socioeconomic and ecological components of the system, it will need consideration of higher-level effect rather than being isolated at the local scale alone (Gaube et al. 2009).

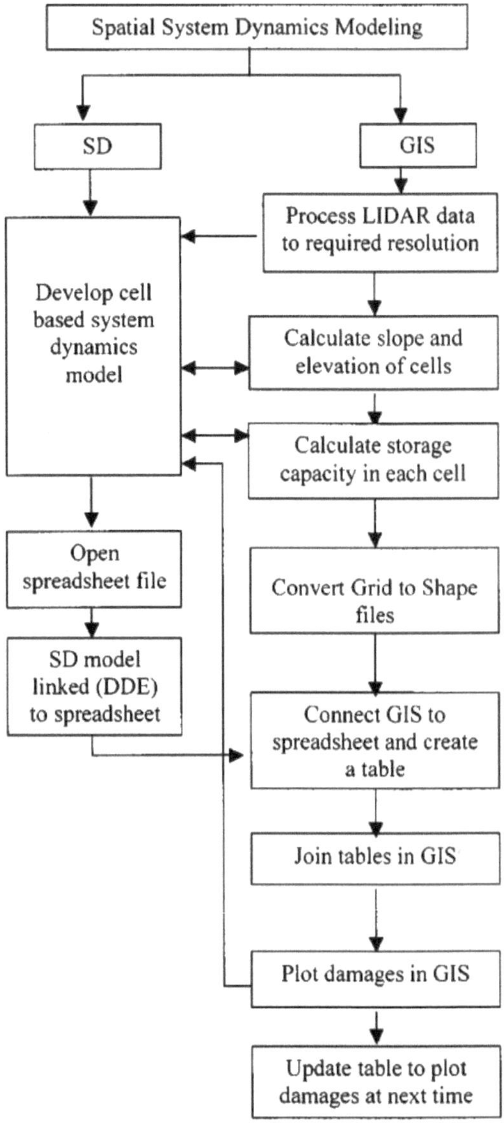

Fig. 15.5 Data flow diagram of spatial system dynamic modeling. (*Source*: Ahmad, S & Simonovic, SP 2004, 'Spatial System Dynamics: New Approach for Simulation of Water Resources Systems')

Ahmad and Simonovic (2004) concluded that the Spatial System Dynamics (SSD) is a significant achievement of hybridized simulation techniques, the strengths of which are associated with the increased speed of developing the model; the ease of model structure modification; performing sensitivity analysis and effective model results communication. The limitations related to automated results updating in GIS due to data import handling from SDM to GIS through DDE protocol, restricted portability (can only apply to GIS package for which application is made), and possible loss of interactive communication of System

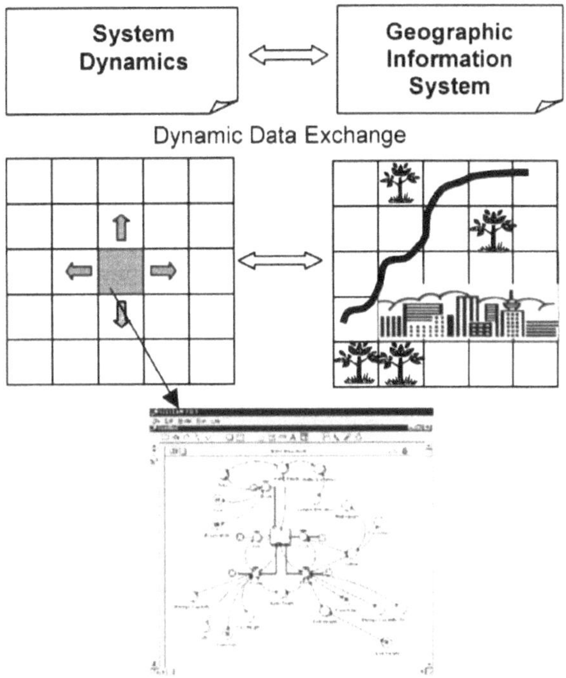

Fig. 15.6 Architecture of spatial system dynamic modeling. (*Source*: Ahmad, S & Simonovic, SP 2004, 'Spatial System Dynamics: New Approach for Simulation of Water Resources Systems')

dynamics model are the challenges that should be resolved for improving this integrated hybrid design (Ahmad and Simonovic 2004). This hybrid modeling is a generic approach that can be used for any problems such as environmental flow material studies, rainforest destruction, dissertation, wetlands management, biodiversity issues, dynamic landscape, atmospheric processes, and disaster management, however tailoring the model into specific problem domains is still required (Ahmad and Simonovic 2004).

However previous attempts of coupling several techniques to solve a problem had been practiced by Gunawan (1998) in proposing coastal resource management in the Gulf of Balikpapan, Indonesia. He coupled the GIS modelling with principal components analysis (PCA) and strength-weakness-opportunity-threat (SWOT) analysis where GIS was applied for land suitability in coastal areas, PCA was applied for analyzing socio-economic and cultural characteristics, and SWOT for arranging policy recommendations (Gunawan 1998). This was also a fair effort to utilize different tools for managing coastal resources but it seemed that the tools were actually applied separately to provide their own conclusions which were then abstracted and included into policy recommendations.

In terms of hybridized simulation with an analytical method, Shantikumar and Sargent 1983 in Swinerd and McNaught (2012) actually divided hybrid models into four classes (Fig. 15.7). According to that classification, class II is considered as the most integrated technique which contains sustained feedback (not just at one point

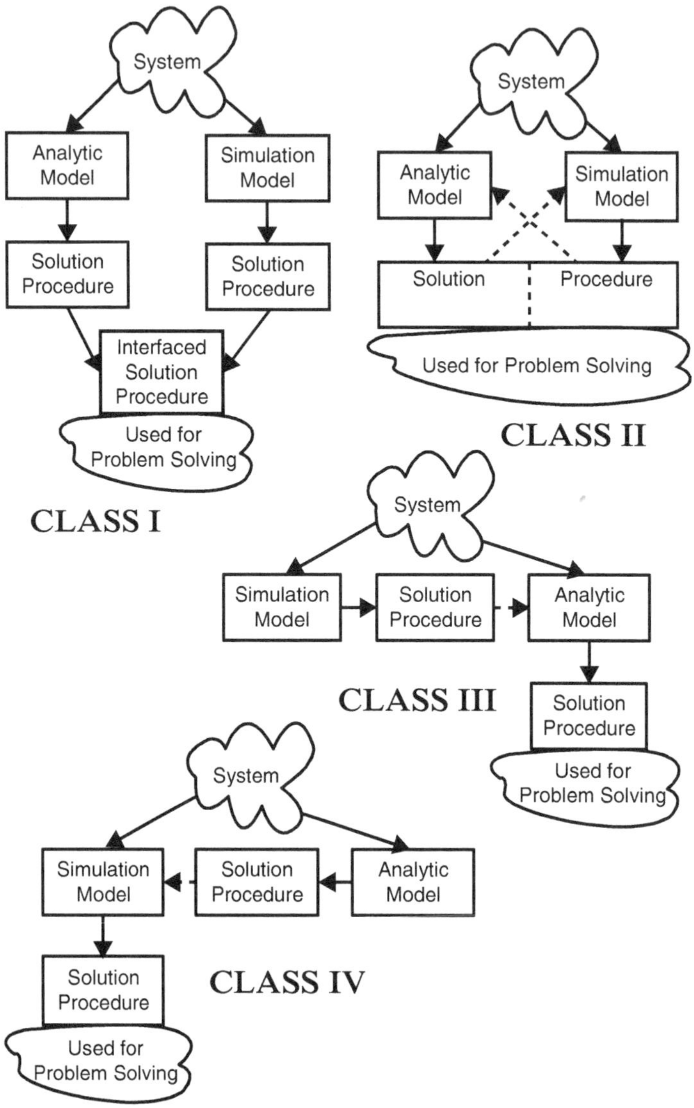

Fig. 15.7 Shanthikumar and Sargent's four classes of hybrid models. *Source*: Swinerd and McNaught (2012)

in time) between models from different paradigms. However, people are usually confused about whether the other classes (I, III & IV) are "integrated" or not; instead class I actually represents "interfaced" output between analytical and simulation models whereas classes III and IV are sequential designs (Swinerd and McNaught 2012). It seems that the spatial system dynamics (SSD) which has been introduced by Ahmad and Simonovic (2004) and expanded by Swinerd and McNaught (2012) is an example of integrated simulation model while the coupling technique by Gunawan (1998) is an interfaced or sequential approach.

Another hybridized simulation model has also been reviewed by Swinerd and McNaught (2012) by coupling the agent based modeling (ABM) and the system dynamics modeling (SDM) to representing and predicting system behavior which is dynamic and sophisticated. Whereas ABM is an inductive or bottom-up approach while SDM is a deductive or top-down approach, therefore analogically if SDM is modeling the forest then ABM is modeling the trees. Furthermore in the context of integrated coastal zone management, ABM can be used for modeling characteristics or behavior of individual or group agents in coastal environments, such as stakeholders' interests, while SDM constructs dynamic interrelationship between agents in the whole system.

With regard to the integrated approach of hybrid simulation according to Shanthikumar and Sargent's classification that requires parallel operation and interaction over time between the analytical model and the simulation model, Swinerd and McNaught (2012) proposed three options to implement ABM-SDM integrated hybrid design (Fig. 15.8), i.e.:

(a) SDM is built within agents of an ABM (agents with complex internal structure);
(b) ABM of an agent acts as stock (aggregate measure) or converter (parameter with emergent behavior) in an SDM;

Based on the options of ABM-SDM integrated hybrid design from Fig. 15.8, several combinations/examples are found from literature cited by Swinerd and McNaught (2012) which have tried to address particular complex problems, such as:

- modeling pension fund governance;
- modeling product portfolio strategies for European automotive manufacturers responding to CO_2 emission regulation;
- etc.

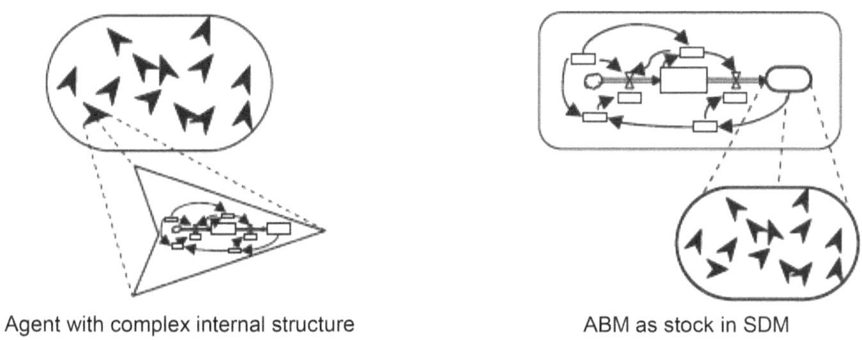

Fig. 15.8 Options of ABM-SDM integrated hybrid design. (*Source*: Swinerd, C & McNaught, KR 2012, 'Design classes for hybrid simulations involving agent-based and system dynamics models')

There are several software packages that can provide functionalities of ABM-SDM integrated hybrid design, i.e. Vensim SD package, eM-Plant, RePast AB package and AnyLogic (Swinerd and McNaught 2012). However the most capable software package that can integrate multi-method simulation, such as ABM and SDM even Discrete Event (Process-centric), is AnyLogic. Nevertheless the limitation of those software packages, including AnyLogic, is still related to the spatial processes.

Another attempt to modelling the coastal environment is the application of the Decision Support System for Integrated Coastal Zone Management (DSS-ICZM) as a computer-based information system that is designed to employ analysis in supporting ICZM decision-making in the early 2000s (Van Kouwen et al. 2008). Even though the application of DSS for coastal management is limited compared to other fields, Van Kouwen et al. (2008) found hundreds of DSSs by searching the internet and scientific journal databases and selected 13 of those to be comprehensively evaluated. Those 13 DSS tools were chosen due to two requirements: interconnectedness of biophysical and socio-economic aspects and designation to be used interactively by policymakers and managers.

In regards to evaluating those selected tools, Van Kouwen et al. (2008) specified two main parameters which were relevant to be applied in generic cases of ICZM decision-making, i.e.:

(a) Knowledge-Related Challenges and Functionalities:

- Uncertainty: quantified (measurable variability) and unquantified (lack of knowledge);
- Spatial-temporal pattern and dynamic behavioral: GIS and agent-based modeling;
- Forecasting and backcasting: policy-evaluation tools and policy-optimizing tools.

(b) Process-Related Challenges and Functionalities

- Science-management integration: link between research and policy-making;
- Stakeholders' participation: tools are designed to support multi-users through gaming techniques and simulations;
- Concept or mental model mapping: presenting complex problems into schematic and understandable information;
- Phases-oriented: appropriateness of the tools for three steps of designing process (screening, scoping, and scanning) .

The results of the evaluation show that none of the tools fulfil all requirements and functionalities completely, while the most remarkable thing is that none of them provide uncertainties visualization or apply agent-based modeling (Van Kouwen et al. 2008). Therefore it can be concluded that the ICZM-DSS tools were not perfect enough in addressing dynamic complexities of coastal environment in which improvement is still needed in several aspects.

Moreover, at the end of the research, Van Kouwen et al. (2008) gave recommendations to both coastal managers and DSS developers. Coastal managers have recommended not to apply unstructured problems to DSS tools as those tools do not have this capability, while the DSS developers have recommended to enhance DSS tools' capabilities by introducing agent-based modeling (ABM) which is effective in dealing with dynamic human behavior (Van Kouwen et al. 2008). However, while Van Kouwen et al. (2008) specified an important dynamic factor of ICZM based on human behavior which might be approached by agent-based modeling in order to improve the ICZM-DSS tools, the other experts proposed different modeling frameworks by integrating (mix and match) several significant tools such as GIS, agent based modeling and System dynamics modeling, namely hybrid simulation (Ahmad and Simonovic 2004; Crooks and Castle 2012; Hartt 2011; Swinerd and McNaught 2012).

15.3 Summary

Dynamic complexities of coastal environments, as a result of interactions among their components, have brought global attention, including in Indonesia, to the introduction of integrated coastal zone management (ICZM) principles (dynamic, decentralized, multi-disciplinary, adaptive) and framework (information collection, planning, decision/policy making, monitoring) (Nurhidayah 2010). In regards to their dynamic complexities, coastal system performance is strongly affected by the change of coastal conditions in space. Therefore the kind of modeling that can appropriately describe and represent coastal system behavior is dynamic modeling which involves components spatially in the notion of time evolution.

From this systematic review, at least the first research question can be answered through the explanation that there are remarkable tools of science-based modeling that have emerged, been researched, applied and evaluated in the world, i.e. geographical information system (GIS), system dynamics modeling (SDM), agent-based modeling (ABM), and decision support system for coastal zone management (DSS-ICZM) (Table 15.1). It has been noticed that geographical information system (GIS) with its geo-spatial modeling and simulation is the most popular tool for supporting ICZM. However due to its limitations in diffusion process and temporal simulation where dynamic characteristics of ICZM require sustained feedback between time, space and pattern, there are attempts to overcome GIS limitations by embedding other tools of modeling, especially with capabilities in temporal modeling and agents behavior. This circumstance led to the acknowledgment of hybrid simulation. Even though some hybrid simulations do not directly give examples of ICZM due to generic characteristics, those hybrid simulations can be adopted for addressing coastal problems.

Nevertheless, hybrid simulation is not just simply two or more models embedded or coupled but they should really be integrated with sustained feedbacks among them other than interfaced or even sequential relationships. Among such modeling

Table 15.1 Comparison of tools of science-based modeling for supporting the ICZM

No.	Tools of modelling	Capabilities	Limitations
1.	GIS (geospatial modelling)	Capable in calculating spatial scenarios with geospatial modeling functionality	Limited on diffusion process and temporal modeling
2.	System dynamics modelling (SDM)	Can be used for constructing dynamic interrelationship between agents (stocks, flows and converters), their connectedness, feedbacks, and dynamic behavior over time (top-down approach)	Limited on presenting spatial processes
3.	Agent based modelling (ABM)	Can be applied for modeling behavior and characteristics of individual/group actors in a system over time (bottom-up approach)	Limited on presenting spatial processes
5.	SDM-ABM modelling	Representing and predicting behavior of the system and the agents simultaneously over time	Limited on presenting spatial processes
6.	Spatial system dynamics (SSD)	Comprehensive modelling that fulfills both spatial and temporal dynamics	Limited versatility on data import handling, restricted portability (can only apply to GIS package for which application is made), and possible loss of interactive communication of system dynamics model
7.	Decision support system for coastal zone management (DSS-ICZM)	A computer-based information system that is designed to employ analysis in supporting ICZM decision-making	None of the tools provide uncertainties visualization and apply agent-based modeling

tools, there are three conventional simulation models which have capabilities to complement each other, i.e. agent based modeling (ABM), System Dynamics modeling (SDM) and geographical information system (GIS). In the context of coastal environments, ABM can be applied for modeling behavior and characteristics of individual/group actors, SDM can be used for constructing dynamic interrelationship between agents (stocks, flows and converters) over time, and finally GIS provides spatial modeling. Therefore this hybrid simulation will truly cover the main complexities in coastal ecosystems, i.e. human behavior, temporal and spatial processes all together. However technical limitation related to data import handling among models through dynamic data exchange (DDE) protocol is still the challenge for improving this hybrid simulation.

Moreover, there is also an emergence of Decision Support System for Coastal Zone Management (DSS-ICZM) in Europe that applies computer-based information system for supporting ICZM. Van Kouwen et al. (2008) evaluated 13 selected DSS tools based on several functionalities, such as uncertainty, spatial-temporal pattern and dynamic behavior, forecasting and backcasting, science-management integration, stakeholder participation, concept mapping, and phases-orientation. The result is that none of them fully employ all functionalities where there are no DSS tools which can deal with uncertainties and apply agent-based modeling.

In conclusion, every science-based modeling and simulation, stand alone or hybridized design, have their own capabilities and limitations in addressing the dynamic complexities of coastal environments. Coastal managers/policy makers should be smart in selecting the tools of science-based modeling based on the needs of management planning, while tools developers should pay attention to the technical drawbacks of the real situation.

15.4 Conclusion

There are still possibilities to enhance the ICZM implementation by applying science-based modeling that can assist policy makers in determining and forecasting priorities and policies in this complex field. GIS modeling through the method of maps overlaying is the major tool that has been recommended for doing so. This GIS modeling is also supported by several techniques of analysis but not in an "integrated" manner. Despite the advantages and disadvantages of those kinds of modeling in addressing the dynamic complexities of coastal areas, there is still opportunity in integrating those tools to assist policy makers and managers to understand dynamic environmental behavior and then set up resilient policy scenarios in sophisticated and dynamic situation.

System dynamics modeling incorporated with agent-based modeling and GIS can be powerful tools to address complexities in the coastal areas where those tools have represented participatory, integrated and dynamic aspects. System dynamic is a kind of top-down modelling and agent-based model is bottom-up modelling which involves the key players of the system that will be beneficial to minimize gaps with the real situation while GIS modelling will ease the user to understand the problems into spatial context.

Further researches to optimize the capabilities of those modeling in supporting ICZM are still needed to enhance the implementation of ICZM. However data deficiencies will be a challenge in constructing the model, therefore finding access to comprehensive sources is critical in developing system dynamics modeling incorporated with agent-based modeling is recommended for further research. A case study that utilizes one of those tools or even integration of them will be published in other peer-reviewed journal by following the cycle process of modelling: problem articulation, dynamic hypothesis, simulation, testing, policy designs, and evaluation.

References

Adrian J (2005) Tropical Ecosystem & Habitat Diagrams, IAN Science Communication Forum, viewed 21/1/2014. http://ian.umces.edu/discforum/index.php/topic,108.msg171.html?PHPSESSID=50407910eb2d53c5ec3d518fb1495369#msg171

Ahmad S (2002) An intelligent decision support system for flood management: a spatial system dynamics approach. Dissertation/Thesis, ProQuest, UMI Dissertations Publishing. http://uq.summon.serialssolutions.com/2.0.0/link/0/eLvHCXMwY2BQMDWyNElL-SjKwTARVlmmpSWlmwCyVAjoUMtXYEHIvIDxGkUpzNyEGptQ8UYYQN9cQZw9d6GUAugWWoHwArMYtga39lGTQiVPmiYZpJqDrsozN04yMkpMSTS0MzAy-SUiwTLQ3STEwTTYF2JgEr5sRE87REC7PElFRDMQYWYH86FQAJVCXp

Ahmad S, Simonovic SP (2004) Spatial system dynamics: new approach for simulation of water resources systems. J Comput Civil Eng 18(4):331–340

Ahmad S, Simonovic SP (2006) An intelligent decision support system for management of floods. Water Res Manage 20(3):391–410

Arsana IMA (2010) Ambalat dispute resolution using maritime delimitation: jurisdictional and geospatial study. Sci J Soc Polit Sci I(01):46–58

Birta LG, Arbez G (2007) Modelling and simulation: exploring dynamic system behaviour. Springer, London

Cartwright I (1995) Management of living marine resources. In: Dutton KHaIM (ed) Coastal management in the Asia-Pacific Region: issues and approaches, 1st edn. Japan International Marine Science and Technology Federation, Tokyo, Japan

CBD, Sot (2000) The Jakarta mandate - from global consensus to global action: conservation and sustainable use of marine and coastal biological diversity. Secretariat of the Convention on Biological Diversity, Montreal, Quebec

Crooks AT, Castle CJE (2012) The integration of agent-based modelling and geographical information for geospatial simulation. Springer, Netherlands, Dordrecht, pp 219–251, doi:10.1007/978-90-481-8927-4_12. http://uqtest.summon.serialssolutions.com/2.0.0/link/0/eLvHCXMwY2BQSDMGZhpQSko1NjU3T0lKSQJWJIamoJPagY2N5FSUsQ6k0txNiIEpNU-Uwc3NNcTZQxc2MRlfADl3IR7U4bE00DWxMNS1sDQy1zWJN7QAnS0Sn4RFxshQjIEF2INOBQCaayJL

Dahuri R (1998) Research needs to support the implementation of integrated coastal and marine resource management. Indones J Coast Marine Res Manage 1(2):60–76

Dahuri R (2011) Coastal and marine economics strategic development for national welfare in the global change era. The Coordinator Ministry of Economics, Jakarta, Seminar

David RG (2004) GIS an all pervasive technology in ICM. Paper presented to eco-imagine conference - GI & GIS for integrated coastal management, Seville, Spain, 13–15 May 2004

Dutton IM, Hotta K (1995) Introduction to coastal management. In: Hotta K, Dutton IM (eds) Coastal management in the Asia-pacific region: issue and approaches, 1st edn. Japan International Marine Science and Technology Federation, Tokyo, Japan, pp 3–18

Fabbri KP (1998) A methodology for supporting decision making in integrated coastal zone management. Ocean Coast Manage 39(1–2):51–62

Farhan AR, Lim S (2010) Integrated coastal zone management towards Indonesia global ocean observing system (INA-GOOS): review and recommendation. Ocean Coast Manag 53(8):421–427

Farhan AR, Lim S (2013) Improving vulnerability assessment towards Integrated Coastal Zone Management (ICZM): a case study of small islands in Indonesia. J Coast Conserv 17(3):351–367

Forst MF (2009) The convergence of integrated coastal zone management and the ecosystems approach. Ocean Coast Manag 52(6):294–306

Gaube V, Smetschka B, Wolf A, Richter A, Haberl H, Kaiser C, Wildenberg M, Adensam H, Fleissner P, Kobler J, Lutz J, Schaumberger A, Schaumberger J (2009) Combining agent-based and stock-flow modelling approaches in a participative analysis of the integrated land system in Reichraming, Austria. Landsc Ecol 24(9):1149–1165

Gunawan I (1998) Typical geographic information system applications for coastal resources management in Indonesia. Indones J Coast Marine Res Manage 1(1):7–21

Hartt MD (2011) Geographic information systems and system dynamics - modelling the impacts of storm damage on coastal communities. Dissertation/Thesis, ProQuest, UMI Dissertations Publishing. http://uq.summon.serialssolutions.com/2.0.0/link/0/eLvHCXMwY2BQMDWyN-ElLSjKwTARVlmmpSWlmwCyVAjoUMtXYEHIvIDxGkUpzNyEGptQ8UQY5N9cQZw9d-WNEYn5KTE28E7BuDlhACs6YYAwuwZ5wKAM38GEw

Haslett SK (2009) Coastal systems. Routledge, New York

Nurhidayah L (2010) Integrated coastal zone management in Indonesia: the implementation and its challenges. Available at SSRN: http://ssrn.com/abstract=1666807 or doi:10.2139/ssrn.1666807

Portman ME, Esteves LS, Le XQ, Khan AZ (2012) Improving integration for integrated coastal zone management: an eight country study. Sci Total Environ 439:194

Post JC, Lundin CG (1996) Guidelines for integrated coastal zone management. Environmentally sustainable development studies and monographs, vol 9. The World Bank, Washington

Queffelec B, Cummins V, Bailly D (2009) Integrated management of marine biodiversity in Europe: perspectives from ICZM and the evolving EU maritime policy framework. Mar Policy 33(6):871–877

Rekacewicz P, Ahlenius H (2006) Coastal population and altered land cover in coastal zones (100 km of coastline) GRID-Arendal, 17 Feb 2012, jpg. http://www.grida.no/graphicslib/detail/coastal-population-and-altered-land-cover-in-coastal-zones-100-km-of-coastline_7706

Rodríguez I, Montoya I, Sánchez MJ, Carreño F (2009) Geographic information systems applied to integrated coastal zone management. Geomorphology 107(1):100–105

Schluter M, Quaas M, StoVen M, McAllister RRJ, Arlinghaus R, Bunnefeld N, Eisenack K, Holker F, Milner-Gulland EJ, Muller B, Nicholson E (2012) New horizons for managing the environment: a review of coupled socio-ecological systems modeling. Nat Res Model 25(1):219–272

Siry HY (2011) In search of appropriate approaches to coastal zone management in Indonesia. Ocean Coast Manag 54(6):469–477

Stephen DK, David RG (2004) Applying the geospatial technologies to estuary environments. In: GIS for coastal zone management. CRC Press, viewed 2013/12/17. doi:10.1201/9781420023428.ch18

Sterman JD (2000) Business dynamics: systems thinking and modeling for a complex world. McGraw-Hill, Irwin

Sucofindo, DGMCSI (2009) Final Project Report: The Initiation of Coastal Zoning Plan of Ternate, Directorate General of Marine Coastal and Small Islands (DG of MCSI), Ministry of Marine Affairs and Fisheries (MMAF) of the Republic of Indonesia, Jakarta

Swinerd C, McNaught KR (2012) Design classes for hybrid simulations involving agent-based and system dynamics models. Simul Model Pract Theory 25:118

Treves T (2010) The development of the law of the sea since the adoption of the UN convention on the law of the sea: achievements and challenges for the future. In: Vidas D (ed) Law, technology and science for oceans in globalisation - IUU fishing, oil pollution, bioprospecting, outer continental shelf. Martinus Nijhoff, Leiden – Boston, pp 41–58

Van Kouwen F, Dieperink C, Schot P, Wassen M (2008) Applicability of decision support systems for integrated coastal zone management. Coast Manage 36(1):19–34

Wever L, Glaser M, Gorris P, Ferrol-Schulte D (2012) Decentralization and participation in integrated coastal management: policy lessons from Brazil and Indonesia. Ocean Coast Manag 66:63

Wilson V (2013) Research methods: systematic reviews. Evid Based Libr Inform Pract 8(3):83–84

Zumsteg JM, Cooper JS, Noon MS (2012) Systematic review checklist. J Ind Ecol 16:S12–S21

Chapter 16
Assessing Flood Impacts, Wetland Changes and Climate Adaptation in Europe: The CLIMSAVE Approach

M. Mokrech, A.S. Kebede, and R.J. Nicholls

Chapter Highlights

Approach: A meta-modeling approach is used to develop a continental (European) scale integrated assessment methodology. The method allows dynamic and cross-sectoral simulations of flood impacts and wetlands change/loss to be developed under varying conditions including climate and socio-economic changes.

Participant Engagement: A series of six professionally facilitated, participatory scenario development workshops involving stakeholders and scientists were carried out during the project period in order to: (1) develop plausible socio-economic scenarios, and (2) test and provide feedback on the design and functionality of the CLIMSAVE Integrated Assessment Platform (IAP) for Europe.

Models/Outcomes: A user-friendly and interactive web-based tool was developed to allow stakeholders to assess climate change impacts, adaptations, and vulnerabilities for a range of sectors including coastal and fluvial flooding, and wetlands.

Challenges: The dynamic link and feedback of adaptation plans over a long time span is challenging to simulate. Incorporating adaptation responses into scenarios is planned for future work and is an important frontier for participatory modeling.

M. Mokrech (✉)
School of Science and Computer Engineering, University of Houston Clear Lake, 2700 Bay Area Blvd., Box 540, Houston, TX 77058, USA
e-mail: mokrech@uhcl.edu

A.S. Kebede • R.J. Nicholls
Faculty of Engineering and the Environmental, Tyndall Centre for Climate Change Research, University of Southampton, Highfield, Southampton SO17 1BJ, UK

S. Gray et al. (eds.), *Environmental Modeling with Stakeholders*,
DOI 10.1007/978-3-319-25053-3_16

16.1 Introduction

Floods have already had significant socio-economic impacts in Europe (EEA 2010). These impacts are expected to be exacerbated due to future changes in climate and sea-level rise (IPCC 2013). Several studies have already developed methods for assessing flood impacts and analyzed the implications of future climate and socio-economic conditions at global (Jongman et al. 2014; Hinkel et al. 2014; Hallegatte et al. 2013; Hirabayashi et al. 2013), continental (Mokrech et al. 2015; Rojas et al. 2013; Meyer et al. 2013; Jongman et al. 2012; Feyen et al. 2012; Hinkel et al. 2010), and national/sub-national scales (Dawson et al. 2009; Mokrech et al. 2008; Richards et al. 2008; Evans et al. 2004a; Evans et al. 2004b). For example, at the European scale, Feyen et al. (2012) investigated the implications of future climate on fluvial flooding under current socio-economic conditions; and Rojas et al. (2013) accounted for future climate and population changes and investigated the effect of adaptation by increasing protection levels. The Dynamic Interactive Vulnerability Assessment (DIVA) integrated model of coastal systems, developed by the DINAS-COAST (Dynamic and interactive assessment of national, regional and global vulnerability of coastal zones to climate change and sea-level rise) project, is capable of assessing biophysical and socio-economic impacts of sea-level rise and socio-economic development as well as analyzing selected adaptation options. The DIVA model has been used (Hinkel et al. 2010) to investigate flood impacts and adaptation in Europe due to sea-level rise and storm surges for the A2 and B1 IPPC SRES scenarios (IPCC 2007). The analysis highlights the role of socio-economic pressures (population and GDP) in driving coastal flood impacts during the first half of the current century; while in the second half, the consequences of sea-level rise become more significant. No studies have yet assessed the combined impacts of coastal and inland flooding at regional-to-continental scale in view of climate and socio-economic changes, except Mokrech et al. (2015). In this chapter, we present the overall methodology that was developed within the CLIMSAVE project for assessing the socio-economic impacts of flooding as well as the environmental impacts on wetland habitats under future conditions.

Projections of future climate and socio-economic conditions can be uncertain (Berkhout et al. 2014). Investigating flood impacts under these uncertain conditions can be ineffective for planning adaptation. The use of a dynamic and interactive model allows the assessment of various scenarios under user-defined socio-economic and climate conditions leading to a better understating of the implications of future conditions. Many studies to date investigate only a limited number of predefined scenarios, often for multiple climate realizations under a given scenario. Thus, any changes in climate, socio-economic conditions, and/or adaptation options require a major research effort to analyze possible changing and interacting impacts. Holman et al. (2008) have suggested integrated assessment methodologies to address this limitation, where an integrated platform can be developed and the concept of meta-modeling can be used to overcome modeling complexity and to allow dynamic links between sectors and user interactions. Achieving this, however, requires that the modeling process involves scientists as well as stakeholders to inform the development of alternative socio-economic scenarios and to design and assess the usability of the integrated platform. This approach has already been implemented at the regional scale by developing the Regional Impact Simulator (RegIS) for assessing the socio-economic and environmental impacts under future climate and socio-economic conditions in the UK (Holman et al. 2008) and more recently has been extended to the European continental scale in the CLIMSAVE (Climate Change Integrated Assessment Methodology for Cross-Sectoral Adaptation and Vulnerability in Europe) project (Harrison et al. 2013, 2015). The aim of this project is to develop a broad-scale model that combines coastal and fluvial flood impact assessment models for Europe that can be integrated into the CLIMSAVE IAP (Integrated Assessment Platform) with rapid simulations that allow users to interactively examine flood impacts under varying climates, socio-economic conditions, and adaptation options.

As part of this effort, this chapter presents the CFFloodmeta-model and describes briefly the participatory modeling activities that have been conducted to design socio-economic scenarios for Europe, to quantify socio-economic parameters, to identify relevant adaptation choices, and to design and assess the CLIMSAVE IAP. The chapter is structured to present the methodology in the CLIMSAVE approach section including the integrated assessment framework, datasets, scenario development, flood meta-model, and adaptation options. The flood impact assessment section discusses illustrative model results for Europe. Finally, key findings and future work are included that relate not only to our work, but also the field of participatory modeling at large.

16.2 The CLIMSAVE Approach

Five main design principles can be used to characterize the modeling approach adopted in the overall CLIMSAVE project including the CFFlood meta-model:

1) Interdisciplinary: climate change impacts, adaptation, and vulnerability for a range of sectors including agriculture, forests, biodiversity, coasts, water resources, and urban growth can be assessed;
2) Integration: the Driver-Pressure-State-Impact-Response (DPSIR) framework (Harrison et al. 2013; Holman et al. 2008) is used to integrate sectoral assessment models, establish the needed links, and facilitate interactions between models;
3) Interactivity: users of the CLIMSAVE IAP can interact with the platform to explore impacts attributed to climate and socio-economic changes as well as examine predefined scenarios and exploratory scenarios that they can create to reflect the uncertainty associated with climate and socio-economic projections;
4) Effective user engagement: achieved via simulations and visualizations on the CLIMSAVE IAP based on rapid, but credible meta-models. The use of computationally simpler but efficient modeling techniques (e.g., simplified

process-based models, multiple regression on model outputs, artificial neural network models), so called “meta-models” can be effective in allowing much greater complexity of model linkages and feedbacks (Holman et al. 2008; Carmichael et al. 2004).

5) Participatory: stakeholder involvement is needed to develop alternative socio-economic futures, quantify key models’ parameters (e.g., population, GDP), provide guidelines on feasible adaptation choices under different futures, and assess the CLIMSAVE IAP for constructive feedback on its design, functionality, and overall social validity. The following sections explain the methodologies used with a focus on the CFFloodmeta-model for the European case study.

16.2.1 Model Data

The data inputs for the CFFloodmeta-model are acquired mainly from available European datasets, such as the “coordination of information on the environment”(i.e. CORINE) land cover data, but also global datasets such as the enhanced Shuttle Radar Topography Mission (SRTM) topographical dataset. Table 16.1 describes the key datasets and the necessary processing for the model.

Table 16.1 Key datasets used in the development of the CFFlood model

Name	Description
Fluvial flood maps	Derived from LISFLOOD (i.e. a two-dimensional hydrodynamic model designed to simulate floodplain inundation) simulations for 2-, 5-, 10-, 20-, 50-, 100-, 250- and 500-year flood events at 100 m spatial resolution (Feyen et al. 2012). Data is gridded at 10′ spatial resolution
CORINE land cover	2006 dataset—version 12/2009 at 100 m spatial resolution. Data is tabulated in the fluvial flood zones and at 25 cm elevation bands in coastal areas and then gridded at 10′ spatial resolution
The ESRI enhanced global SRTM elevation data	SRTM dataset at 90 m spatial resolution; Void filled SRTM elevation data using the United States Geological Survey (USGS) Global 30 Arc-Second Elevation (GTOPO30) dataset. Data is tabulated at 25 cm elevation interval and gridded at 10′ spatial resolution
Population density, GDP of the Nomenclature of Territorial Units for Statistics (NUTS) dataset	The NUTS data of EUROSTAT at the NUTS3 statistical regions (~1–10^6 km^2). Data is statistically summarized at the 25 cm elevation bands and at 10′ spatial resolution.
Extreme sea-levels (astronomical tides and storm surges)	Water depth for 1-, 10-, 100- and 1000-year events (DIVA database, Vafeidis et al. 2008) with an average segment length of 70 km. Data is gridded at 10′ spatial resolution
Land uplift/subsidence	Annual rate of change (DIVA database, Vafeidis et al. 2008) with an average segment length of 70 km. Data is gridded at 10′ spatial resolution

16.2.1.1 Indicative Flood Protection Data for Europe

Flood protection levels are essential for assessing flood impacts. Currently in Europe there is little information about these protection levels in coastal and river flood zones. Studies, such as those cited in Feyen et al. (2012) and Jongman et al. (2014) used economic indicators and modeling approaches to estimate flood protection standards. In the CFFlood model, an indicative flood protection dataset at the European level is constructed following the UK (Department for Environment Food and Rural Affairs) DEFRA methodology (MAFF 1999), where ranges of Standard of Protection (SoP) of coastal and fluvial flood defenses are determined based on land use/cover classes and the economic value of the land. Table 16.2 shows the minimum and maximum indicative standards of protection that are implemented for six land use categories in fluvial and coastal flood zones based on the CORINE land use/cover dataset. The resulting flood protection dataset has been revised using published data on flood protection in individual regions/nations including Belgium, the Netherlands, Northern Germany, and London (Mokrech et al. 2015). For example, up to 10,000- and 1000-year flood protection levels are included for the Netherlands and London (UK), respectively. This method is seen to be consistent for establishing a European baseline dataset on flood protection for exploratory purposes.

Table 16.2 Ranges of indicative standards of protection associated with land use classes (from CORINE), (following MAFF 1999)

			Indicative standard of protection	
			Fluvial	Coastal
Land use band	Description	Land use (CORINE classes—third level)	Return period (years)	Return period (years)
A	Intensively developed urban areas	111	50–200	100–300
B	Less intensive urban areas with some high grade agricultural land and/or environmental assets	112, 121, 122, 123, 124, 131, 141, 142, 211, 212, 213,221, 222, 223	25–100	50–200
C	Large areas of high-grade agricultural land and/or environmental assets with some properties	132, 133	5–50	10–100
D	Mixed agricultural land with occasional properties at risk of flooding	241, 242, 243, 244	1.25–10	2.5–20
E	Low-grade agricultural land (often grass) or seasonally occupied properties at risk	31, 311, 312, 313, 321, 322, 323, 324, 333	0–2.5	0–5
F		All other classes	0	0

16.2.2 CLIMSAVE IAP

The development of the CLIMSAVE IAP has been iterative; it has undergone modifications throughout the duration of the project in response to progressive stakeholder feedback over three sets of workshops held over the course of 2 years. The IAP includes a collection of mathematical models that together provide quantitative estimates of sectoral impacts due to climate and socio-economic changes in the form of maps and graphs. The effect of a range of adaptation options on reducing impacts and vulnerability can be simulated, as well as estimates of their cost-effectiveness. The fundamental concept underpinning the specification of the IAP is to deliver rapid interactivity for the user utilizing the World Wide Web (available at www.CLIMSAVE.eu). This approach is designed to broaden accessibility and participation of stakeholders, thereby increasing impact in research communities.

16.2.3 Scenarios

16.2.3.1 Climate Change Scenarios

An ensemble of climate change scenarios were developed to represent alternative emissions and climate sensitivity that are consistent with the Intergovernmental Panel on Climate Change (IPCC 2007) climate scenarios. Thus, climate pressures (e.g., temperature, precipitation, sea-level rise) are available for four emissions scenarios (i.e., A1B, A2, B1, or B2) at three climate sensitivity levels (low, medium, or high) as well as for credible designed ranges. For example, the CFFloodmeta-model allows the exploration of up to 2 m of sea-level rise by 2100 (following current guidance by Nicholls et al. 2014).

16.2.3.2 Socio-Economic Scenarios

Four qualitative socio-economic scenarios are developed using participatory methods; mainly based on a "story-and-simulation" approach in an iterative procedure (Gramberger et al. 2015). Scientific methods and stakeholder knowledge are brought together to develop storylines that cover a range of aspects including social and economic developments, but also cultural, institutional, and political aspects in a set of integrated future outlooks (Linham et al. 2010). Stories are developed during a set of three stakeholder workshops. Additional to the stories, flow-charts, graphs depicting temporal developments, and quantitative estimates of main drivers of future changes (e.g., population and GDP) are produced as inputs for the mathematical models within the CLIMSAVE IAP. The four socio-economic scenarios for Europe are developed with extensive stakeholder input at two specific future time periods: the 2020s and the 2050s. Collectively, they show both population and GDP changes (increases and decreases) under the following conditions:

1. **We Are the World (WAW) Scenario**: Effective government change with a focus from GDP to welfare; less inequality and global cooperation. GDP change is +26 % by 2020s and +94 % by 2050s, while population change is +1 % by 2020s and +5 % by 2050s.
2. **Should I Stay or should I Go (SISOG) Scenario**: Failure to address economic crisis leads to increased gaps between rich and poor, political instability and conflicts, people live in an insecure and instable world. No change in GDP by 2020s and then −36 % by 2050s, while population changes +5 % by 2020s and +23 % by 2050s.
3. **Icarus Scenario**: Short-term policy planning and a stagnating economy lead to disintegration of social fabric and shortage of goods and services. No change in GDP, while population change is +5 % by 2020s and the −9 % by 2050s.
4. **Riders on the Storm (ROS) Scenario:** Strong economic recessions but successively countered with renewable and green technologies. No change in GDP by 2020s and then +54 % by 2050s, while population change is +5 % by 2020s and 16 % by 2050s.

In regards to the CFFloodmeta-model, change in GDP is used to reflect the change in economic conditions and how flood damage is influenced by such changes. Change in population density is used to estimate the number of people in flood zones.

16.2.4 Adaptation

Based on an extensive literature review, a list of adaptation options were generated and discussed by stakeholders and CLIMSAVE experts (Mokrech et al. 2015). The designed options, partly based on stakeholder opinions, were associated with the socio-economic scenarios and made as default choices. However, non-default options can also be explored. In regard to flooding, the following adaptation options can be examined in the CLIMSAVE IAP:

a) Flood protection upgrade by 50, 100, 500, and 1000 %: this is applied directly to the present indicative protection and uniformly across Europe.
b) Resilience measures: new properties are not affected by flooding due to the resilience measure applied (e.g., raising them above ground levels) up to a pre-defined threshold of flood event (e.g., 100-year event), while old properties continue to suffer from flood damage.
c) Mixed response: this provides a more realistic adaptation approach, where a plausible combination of flood protection improvement (i.e., 100 % upgrade) and realignment of flood defenses are implemented.
d) Retreat from rural areas in floodplains with the aim to generate accommodation space for creating habitats with the aim of maintaining or doubling stocks.

16.3 Flood Modeling

The CFFloodmeta-model is a two-dimensional simplified process-based model that consists of three main components: (1) coastal flood impacts, (2) fluvial flood impacts, and (3) wetland change/loss. These three components are coupled with a range of adaptation measures for reducing adverse flood impacts (socio-economic as well as environmental) under future conditions. Figure 16.1 shows the main modeling steps with data inputs and outputs at the 2010 baseline year as well as at the 2020s and 2050s time slices. Other models within the IAP that provide inputs into the flood model include the RUG (Regional Urban Growth) model and the WaterGAP meta-model. The modeling is nested at multiple spatial scales, where input data is resampled from high resolution data sets (e.g., 100 m resolution CORINE land use data and 100 m fluvial flood maps) and the results are communicated to the IAP at 10′ resolution.

The notion of the meta-model is to determine the flood zones and the land cover classes within these zones in each 10′ grid cell, and then use them to build detailed databases so this information can be retrieved quickly for computational algorithms within an integrated assessment framework. This approach allows interactions with other models and provides rapid dynamic assessments of flood impacts, adaptation to impacts, and sensitivity analysis. The meta-model is developed around the Driver-Pressure-State-Impact-Response (DPSIR) integrated assessment framework (Holman et al. 2005a, b; Rapport and Friend 1979) in order to establish dynamic links between the various models in the CLIMSAVE IAP (Harrison et al. 2013, 2015) as well as to build a consistent structure for the modeling elements.

The concept of "overlay analysis" is used to outline coastal flood zones by examining the regional extreme sea level relative to topography (Mokrech et al. 2015). Future regional extreme sea levels are obtained by combining present-day extreme sea levels and future relative sea-level rise (i.e., absolute rise in sea level and varying vertical land movement around the European coastline), as appropriate. Thus, flood zones are calculated and estimates of the people living in these zones are calculated using local population density. The method uses the Standard of Protection (SoP) parameter for analyzing the effect of relative sea-level rise on the protection level provided by flood defenses. It assumes that SoP decreases and flood frequency increases with a rise of extreme sea level (Lowe et al. 2001; Mokrech et al. 2015, 2008). This effect will vary along the European coastline as the vertical land movement and the slope of the exceedance curve varies spatially.

The fluvial flood component uses fluvial flood maps for Europe that are produced at 100 m resolution with a similar planar approximation approach based on LISFLOOD extreme river water level simulations (Feyen et al. 2012). The flood maps represent fluvial catchments across Europe including the extent and water depth at 2-, 5-, 10-, 20-, 50-, 100-, 250-, and 500-year return periods, assuming no flood defenses. These maps have been used to define the fluvial flood zones in the CLIMSAVE project. They are analyzed in conjunction with the CORINE land use and the socio-economic data (i.e., population and GDP) from the NUTS3 statistical datasets. The estimated SoP parameter is used to analyze the effect of change in peak river flows on flood

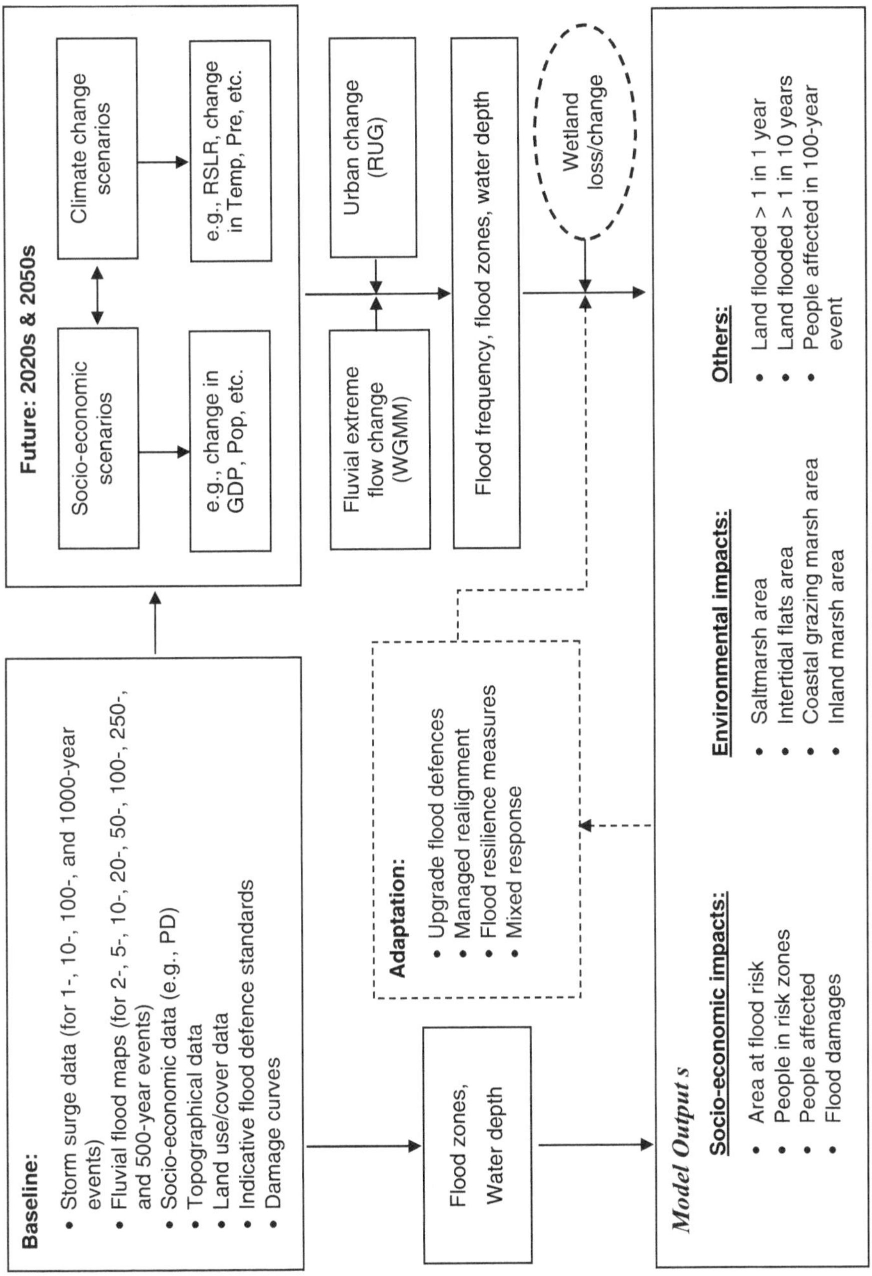

Fig. 16.1 Overview of the main steps, inputs, and outputs of the CFFLOOD meta-model. *PD* population density, *GDP* gross domestic product, *Temp* temperature, *Pre* precipitation, *RSLR* relative sea-level rise

protection following Mokrech et al. (2008). The changes in the peak river flow are derived from the WaterGAP meta-model (WGMM) (Wimmer et al. 2015), which emulates the performance of the WaterGAP3 model (Alcamo et al. 2003; Döll et al. 2003; Verzano 2009) on hydrology and water use. To reduce model runtime and input data requirements, the spatial resolution of WaterGAP3 (5 arc minute) has been aggregated to 92 European river basins greater than 10,000 km^2. Each river basin represents either a large natural river catchment or a cluster of several smaller catchments with similar hydro-geographic conditions. The climate change impacts on peak river flow is represented by the changes in the median of the annual maximum river discharge (Q_{med}), where the latter are derived from catchment-specific response surfaces that relate changes in Q_{med} with changes in temperature and precipitation. Response surfaces were derived from pre-run WaterGAP3 simulations for the period 1971–2000, in which spatio-temporal patterns in the baseline climate dataset were incrementally modified with respect to temperature ([0,0.5,…,6 °C]) and precipitation ([−50, −45,… ,+50 %]) (Mitchell and Jones 2005). When WGMM is run with scenario input data of gridded mean annual air temperature and mean annual precipitation, it first computes the relative change in temperature and precipitation compared to the baseline in each river basin. In a second step, scenario Q_{med} is interpolated by inverse distance weighting of Q_{med} at the four neighboring grid points in the response surface. Finally, the relative change in Q_{med} compared to the baseline value is computed and passed to CFFLOOD as an estimate of changes in peak river discharge (see S6 for model performance). Thus, protection levels of flood defenses are degraded with the increases in peak river flows (e.g., a 10 % increase in peak river flow may degrade the 100-year flood defense to the 60-year level).

By comparing the investigated flood event with the degraded level of flood protection due to relative sea-level rise and/or change in peak river flow, the meta-model determines whether or not the flood zones are flooded. Thus, the number of people affected by flooding is estimated using local population density and urban land use classes. Considering the 10′ cell size and the meta-modeling approach, the failing mechanisms of flood defenses (e.g., breaching and overtopping) are not investigated—the assumption here is that the flood risk zones will be flooded if the flood defense's SoP is exceeded.

Structural and content flood damages are calculated for residential and non-residential properties based on the broad assessment methodology outlined by Linham et al. (2010). The method uses the notion that the value of physical losses from a flood is no more than the value of the assets exposed to this hazard. For developed economies such as in Europe, the net capital asset is approximated to be three times the GDP. The proportions of structural assets are considered at 36 and 42 % for residential and non-residential properties respectively. Only a proportion of those assets located in a risk area are considered to be exposed to flooding; in densely populated urban areas a significant proportion of buildings are multi-storied and a large part of the assets are above any conceivable flood level. Hence, classes of population density are used to determine the proportions of assets at risk of flooding. Then, the Dutch Depth-Damage curve (Linham et al. 2010) is used to estimate structural and content losses caused by flooding.

Flood constraints on agricultural production are also calculated and provided to the agricultural model in the CLIMSAVE IAP. These include: (1) land flooded more than once a year is considered not suitable for any type of farming; (2) land flooded more than once each 10 years is considered not suitable for arable farming (Mokrech et al. 2008). In addition, the number of people affected by flooding in a 100-year event is calculated as a vulnerability indicator and communicated to the IAP.

Wetland changes and losses in floodplain are assessed following the broad scale model of McFadden et al.(2007). The investigated habitats comprise "saltmarsh," "intertidal flats," and what we term as "coastal grazing marsh" in coastal floodplains and "inland marshes" in fluvial floodplains. Saltmarsh and intertidal flats exist seaward of coastal defenses and are subjected to tides, while coastal grazing marshes are largely artificial habitats that exist landward of coastal defenses in areas that would otherwise be intertidal habitats. The wetland change/loss component accounts for both habitat loss and change using a standardized index, where the three influencing factors of *accommodation space*, *sediment supply*, and *rate of relative sea-level rise* are considered. Consequently, habitats such as saltmarsh, coastal grazing marsh, and intertidal flat can be either lost under high forcing conditions or can experience transition under the low-to-moderate forcing conditions as shown in Fig. 16.2. The direct effects of sea-level rise and the effects of defense abandonment due to managed realignment are also included. In river valleys, change in inland marshes is a function of change in river flows where existing marshes can increase or decrease as a function of change in floodplains.

The CORINE land cover data is used to establish the baseline of the intertidal habitats: saltmarsh and intertidal flats, and fluvial habitats (inland marshes). However, the designated habitats landward of coastal flood defenses are not defined in the CORINE

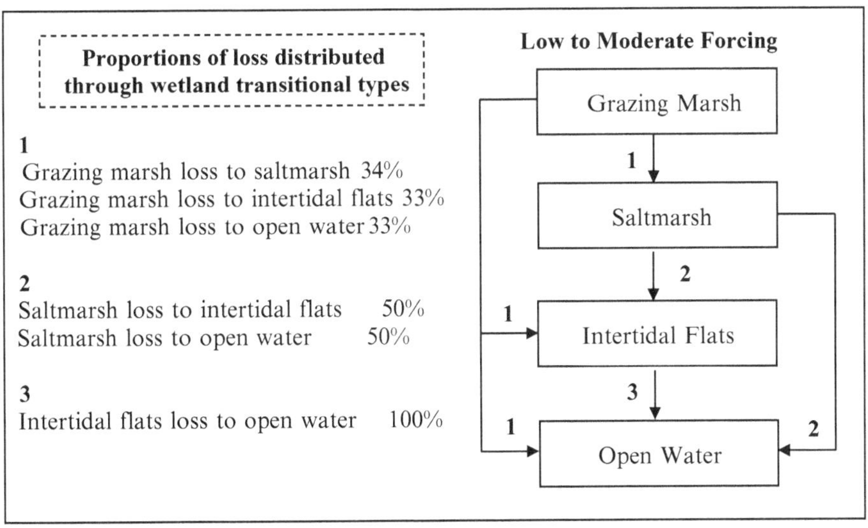

Fig. 16.2 Example of modeling wetlands loss/change for coastal areas. Adapted from McFadden et al. (2007)

land cover dataset. There is no standard European nomenclature for these areas and they are variously termed as: "coastal grazing marsh" (in the UK), or "summer polders" (in the Netherlands/Germany), to give two examples. They are also widely designated under the Habitats Directive for their environmental value. To develop a generic European methodology, pasture areas located within the coastal floodplain are assumed to be potential areas for "coastal grazing marsh" and this term is used for all such habitats in CLIMSAVE. If defenses are abandoned or realigned, the new intertidal land experiences a transition to saltmarsh and intertidal flats.

16.4 Impact Assessment

The coastal and fluvial flood impact analysis without flood protection indicates that almost 28 million people (i.e., 6 % of the total population of the European Union) live within a 100-year flood inundation area. The economic damage is estimated to be €236 billion if this area is flooded. If flood protection is considered at the two designed levels (see Table 16.2), the number of people impacted will range from 0.24 to 17.4 million, while economic loss is estimated at €0.6 to €79 billion for the maximum and minimum protection standards respectively. These numbers demonstrate the benefits of flood protection, especially for the maximum protection level in reducing impacts without accounting for climate change and sea-level rise. In essence, the analysis suggests that Europe has adapted to a large degree to current flood risks, but these risks will grow with climate change and sea-level rise, and with economic growth in the floodplain if this occurs.

The CFFlood model within the IAP is capable of exploring impacts under predefined climate and socio-economic scenarios as well as under a wide range of exploratory scenario combinations that online users can define by varying climate, sea-level rise, socio-economic parameters, flood protection, and adaptation options. The people affected under the A1B climate scenario and the four pre-defined socio-economic scenarios for four events (i.e., 10-, 50-, 100-, and 200-year) are summarized in Fig. 16.3. The general trend of impact reflects the differences in population density under the four socio-economic futures. For example, the number of people affected under the minimum protection level (i.e., the default option on the IAP) for the very extreme flood event of 200 years is the highest under the SISOG scenario with the highest population change of +23 % from baseline, while it is the least under the Icarus scenario with population change of −9 % from baseline. Similarly, the economic impact of flooding is affected by change in economic conditions (i.e., change in GDP).

By isolating coastal flooding from fluvial flooding and exploring a range of conditions and flood events, we found that increases in socio-economic impacts of flooding can be mainly attributed to the effects of sea-level rise and to changes in future socio-economic conditions; the effect of change in temperature and precipitation on fluvial flood impact is small by comparison. When isolating the climate factor from the social factor, the quantitative and spatial distribution of the fluvial flooding shows

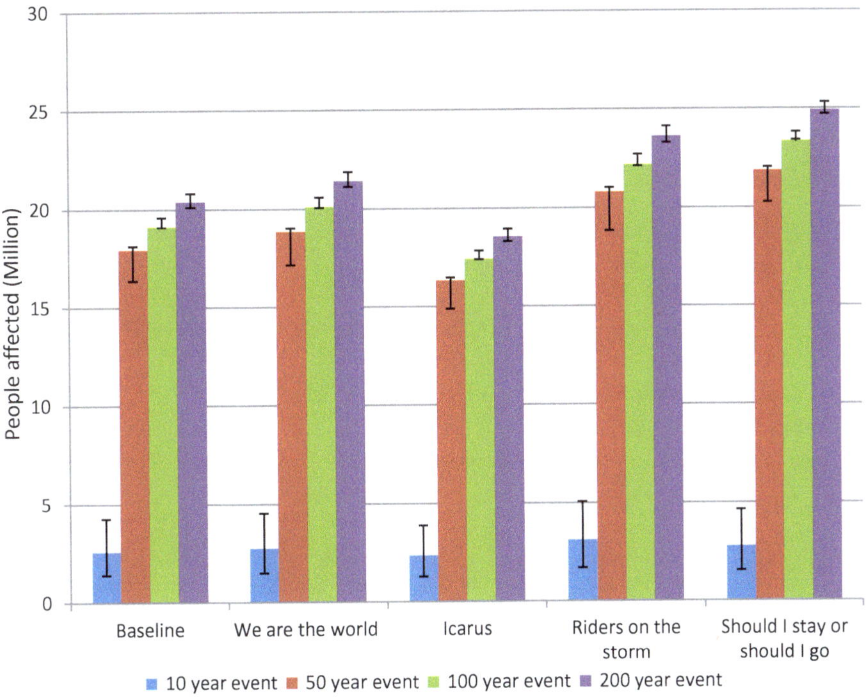

Fig. 16.3 People affected under different flood events in 2050s (i.e., 10-, 50-, 100-, and 200-year events) for future climate and socio-economic scenarios at the minimum level of flood protection at the middle climate variables; as well as sensitivity ranges that correspond to low to high climate variables

a general reduction in people at risk under both the WAW and ROS socio-economic scenarios by the 2050s. Under the Icarus socio-economic scenario, the number of people flooded reduces over almost the whole of Europe except in some areas in western and northern European regions. Under the SISOG scenario, there is a considerable spatial variation in people affected with some areas in Western Europe showing a reduction in people flooded while other areas show a clear opposite trend, for example Eastern regions of Europe. This can also be consistent with the fact that the increase in social pressure (e.g., +23 % change in population by 2050s under the SISOG scenario) leads to larger flood impacts while a decrease in social pressure leads to a decrease in flood impacts. In this context, there is no significant difference in the number of people flooded in the 2020s under the low, medium, and high sensitivities of the investigated A1B emission scenario as well as across the socio-economic scenarios as minimal climate and social variations are expected by the 2020s under those scenarios. The economic damage follows a different pattern across the investigated scenarios as GDP is the primary parameter that influences damages in the implemented methodology. For example, the economic damage is the largest under the WAW scenario as the GDP increase is the highest (+94 %).

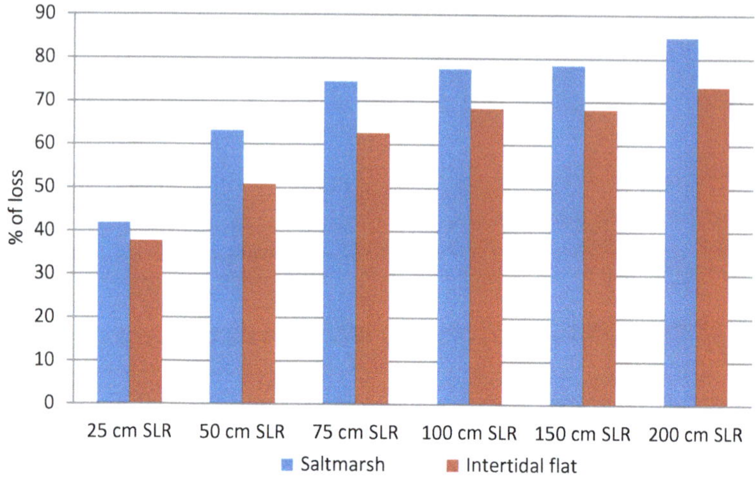

Fig. 16.4 Loss pattern of coastal wetlands under exploratory value of sea-level rise

The model indicates that environmental impacts due to sea-level rise can be significant. Figure 16.4 shows the systematic trend of loss in saltmarsh and intertidal flats. One meter of sea-level rise may lead to a loss of almost 80 % of saltmarsh and 70 % of intertidal flats in Europe. The area of coastal grazing marsh is mainly an indicative estimate of the potential of this habitat. It is mostly managed habitat and it will change into saltmarsh due to a change in salinity. On the other hand, in river valleys, change in inland marshes is a function of change in river flows where existing marshes can increase or decrease as a function of change in floodplains and management. The retreat options of maintaining habitats and doubling habitats have been analyzed using the CLIMSAVE IAP. It was found that habitats (e.g., saltmarsh) can be created but it will be very hard to maintain them at the baseline level under high-end climate scenarios.

To explore the potential benefits of the designed adaptation options, we examined an extreme climate and socio-economic scenario: one meter of sea-level rise, 25 % increase in winter and summer precipitation, 3 °C increase in temperature, 25 % increase in population, and 25 % increase in GDP; and evaluate flood consequences for the adaptation options. We found that the number of people at risk of flooding increases from almost 28 million (at the baseline) to 41 million people (i.e., +46 %). The minimum level of flood protection reduces the impact to almost 17 million at the baseline conditions and to 37 million under the extreme scenario (i.e., +32 %). Thus, while the performance of the current defense systems under current conditions can be effective, it is not effective under the investigated extreme scenario—more aggressive policies (e.g., 500 % or more upgrade of flood protection) are needed in order to reduce impacts of such extreme scenarios. These policies should also consider managed realignment of defenses in order to create the accommodation space needed for habitat creation. On the other hand, the resilience measures (e.g., elevated buildings) at the minimum level of flood protection may perform well, but they are not enough on their own to reduce flood impacts to the baseline level. The economic damages under the investigated extreme scenario demonstrate a similar pattern as in the number of people flooded with the exception that even aggressive adaptation options such as upgrading defense by 500 % or more will not be effective in reducing economic damages to the baseline level, which can be mainly attributed to the increase in GDP. This shows that to maintain present risk levels under high-end scenarios, defenses will have to be raised even more than these upgrades imply; or, alternative measures will have to be considered such as landward realignment of defenses. A mixed response depending on land use and economic and environmental value looks most likely.

A sensitivity analysis of the sectoral and cross-sectoral effects of climate and socio-economic drivers on flood impacts has been conducted. Figure 16.5 shows the sensitivity of people flooded due to different sectoral (or direct) and/or cross-sectoral (or indirect) climate and socio-economic drivers. Out of the six drivers considered, the climatic drivers (temperature and precipitation) are identified as indirect drivers, while the socio-economic driver (population change) is identified as a combined (i.e., both direct and indirect) driver. Sea-level rise and change in flood protection are direct drivers; due to which the number of people flooded shows the highest sensitivity with a range greater than 17 million people.

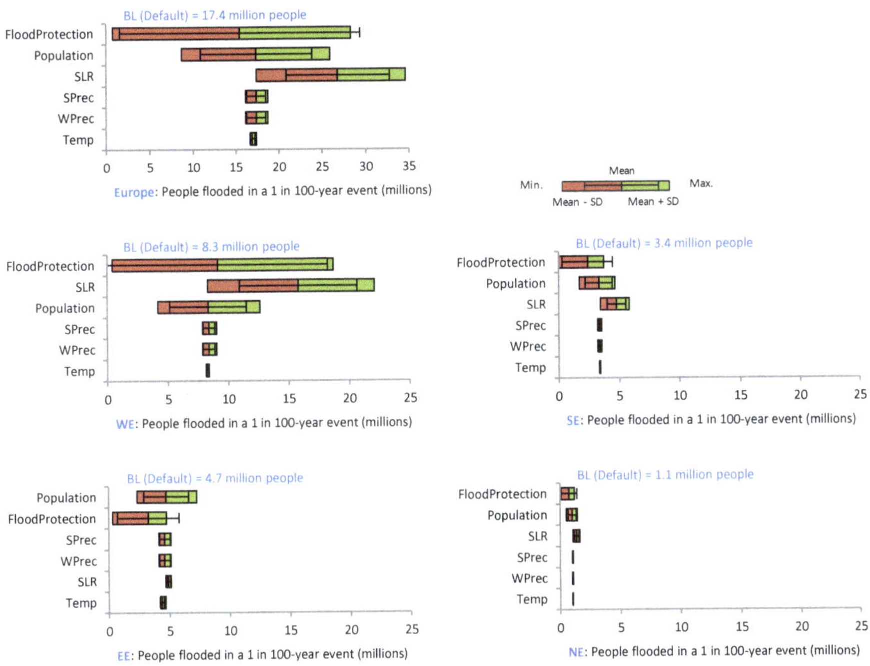

Fig. 16.5 Summary statistics (i.e., minimum, mean, maximum and standard deviation) of the sensitivity of people flooded in a 1 in 100 year flood event due to different climate and socio-economic change drivers for Europe and its four river-basin regions. *Note*: The drivers are sorted on the vertical axes based on the sensitivity range value. *BL* baseline, *WE* Western Europe, *EE* Eastern Europe, *SE*: Southern Europe, *NE* Northern Europe

16.5 Findings and Future Work

Participatory modeling led by scientists and stakeholders has been conducted to design the CLIMSAVE IAP and to develop alternative socio-economic futures where variations in economic development and innovation are realized in the qualitative storylines and the quantitative drivers of change (e.g., population and GDP growth). A range of adaptation options are also integrated within the socio-economic scenarios and the CLIMSAVE IAP.

The CFFlood flood meta-model within the CLIMSAVE IAP is capable of estimating coastal and fluvial flood impacts due to changes in climate and socio-economic conditions at the baseline year (i.e., 2010) and two time periods (i.e., the 2020s and the 2050s). The analysis of an illustrative set of results indicates that flood protection is very effective under baseline conditions. In the future, climate change may challenge this situation, especially under high-end scenarios in all coastal areas, and to a lesser extent in Northern European fluvial flood plains. Hence, there is a potentially large, but highly uncertain, need for adaptation. In terms of socio-economic changes, future socio-economic conditions have a major influence on the level of economic damage. Under socio-economic scenarios such as Icarus and SISOG with economic decline, the reduction in economic impact is significant and it is due to this cause. The highest economic damages are likely under the WAW scenario, reflecting the large economic growth. It is worth noting that economic growth gives a greater capacity to adapt, and vice versa, so the overall implications of these results need to be carefully considered. In addition, the impacts on coastal wetlands due to sea-level rise can also be assessed in the CFFlood meta-model. Incremental losses of saltmarsh and intertidal flats (70–80 % from baseline) associated with high-end scenarios are simulated highlighting the need for corresponding adaptation efforts.

Although the CFFloodmeta-model offers a unique opportunity to quantify the socio-economic impacts of coastal and fluvial flooding across Europe for current as well as future conditions, there are a number of improvements that can be considered in future research. These include: (1) improving the flood protection dataset; (2) increasing flexibility for investigating time slices (ideally 10-year time steps until 2100); and (3) developing dynamic analysis of adaptation and feedbacks that may vary temporally and spatially in Europe reflecting regional and national policies for managing flood risk. Participatory modeling is seen as an appropriate approach for developing a multi-criteria indicator that feeds into the decision-making process with regards to adaptation.

References

Alcamo JM, Döll P, Henrichs T, Kaspar F, Lehner B, Rösch T, Siebert S (2003) Development and testing of the WaterGAP2 global model of water use and availability. Hydrol Sci J 48:317–337

Berkhout F, van den Hurk B, Bessembinder J, de Boer J, Bregman B, van Drunen M (2014) Framing climate uncertainty: socio-economic and climate scenarios in vulnerability and adaptation assessments. Reg Environ Chang 14:879–893

Carmichael J, Tansey J, Robinson J (2004) An integrated assessment modeling tool. Glob Environ Chang 14:171–183

Dawson RJ, Dickson ME, Nicholls RJ, Hall JW, Walkden MJA, Stansby P, Mokrech M, Richards J, Zhou J, Milligan J, Jordan A, Pearson S, Rees J, Bates P, Koukoulas S, Watkinson A (2009) Integrated analysis of risks of coastal flooding and cliff erosion under scenarios of long term change. Clim Chang 95:249–288

Döll P, Kaspar F, Lehner B (2003) A global hydrological model for deriving water availability indicators: model tuning and validation. J Hydrol 270:105–134

EEA (2010) Mapping the impacts of natural hazards and technological accidents in Europe – an overview of the last decade. EEA Technical Report. European Environment Agency, Copenhagen, Denmark, p 144. doi:10.2800/62638. ISSN 1725-2237

Evans E, Ashley R, Hall J, Penning-Rowsell E, Saul A, Sayers P, Thorne C, Watkinson A (2004a) Foresight future flooding scientific summary: volume I—future risks and their drivers. Office of Science and Technology, London

Evans E, Ashley R, Hall J, Penning-Rowsell E, Saul A, Sayers P, Thorne C, Watkinson A (2004b) Foresight future flooding scientific summary: volume II—managing future risks. Office of Science and Technology, London

Feyen L, Dankers R, Bódis K, Salamon P, Barredo JI (2012) Fluvial flood risk in Europe in present and future climates. Clim Change. doi:10.1007/s10584-011-0339-7

Gramberger M, Zellmer K, Kok K, Metzger M (2015) Stakeholder Integrated Research (STIR): a new approach tested in climate change adaptation research. Clim Chang 128(3–4):201–214

Hallegatte S, Green C, Nicholls RJ, Corfee-Morlot J (2013) Future flood losses in major coastal cities. Nat Clim Chang 3(9):802–806

Harrison PA, Holman IP, Cojocaru G, Kok K, Kontogianni A, Metzger M, Gramberger M (2013) Combining qualitative and quantitative understanding for exploring cross-sectoral climate change impacts, adaptation and vulnerability in Europe. Reg Environ Change 13(4):761–780

Harrison PA, Holman IP, Berry PM (2015) Assessing cross-sectoral climate change impacts, vulnerability and adaptation: an introduction to the CLIMSAVE project. Clim Chang 128(3–4):153–167

Hinkel J, Nicholls RJ, Vafeidis A, Tol RSJ, Avagianou T (2010) Assessing risk of and adaptation to sea-level rise in the European Union: an application of DIVA. Mitig Adapt Strateg Glob Chang 5(7):1–17

Hinkel J, Lincke D, Vafeidis AT, Perrette M, Nicholls RJ, Tol RSJ, Marzeion B, Fettweis X, Ionescu C, Levermann A (2014) Coastal flood damage and adaptation costs under 21st century sea-level rise. Proc Natl Acad Sci. doi:10.1073/pnas.1222469111

Hirabayashi Y, Mahendran R, Koirala S, Konoshima L, Yamazaki D, Watanabe S, Kim H, Kanae S (2013) Global flood risk under climate change. Nat Clim Change 3:816–821. doi:10.1038/nclimate1911

Holman IP, Rounsevell MDA, Shackley S, Harrison PA, Nicholls RJ, Berry PM, Audsley E (2005a) A regional, multi-sectoral and integrated assessment of the impacts of climate and socioeconomic change in the UK. Part I. Methodology. Clim Chang 71:9–41

Holman IP, Nicholls RJ, Berry PM, Harrison PA, Audsley E, Shackley S, Rounsevell MDA (2005b) A regional, multi-sectoral and integrated assessment of the impacts of climate and socioeconomic change in the UK, Part II, results. Clim Chang 71:43–73

Holman IP, Rounsevell MDA, Cojacaru G, Shackley S, McLachlan C, Audsley E, Berry PM, Fontaine C, Harrison PA, Henriques C, Mokrech M, Nicholls RJ, Pearn KR, Richards JA (2008) The concepts and development of a participatory regional integrated assessment tool. Clim Chang 90:5–30

IPCC: Climate Change (2007) The physical science basis. In: Solomon S, Qin D, Manning M, Chen Z, Marquis M, Averyt KB, Tignor M, Miller HL (eds) Contribution of Working Group I to the Fourth Assessment Report of the Intergovernmental Panel on Climate Change. Cambridge University Press, Cambridge, UK/New York, NY, USA

IPCC: Climate Change (2013) The physical science basis. In: Stocker TF et al (ed) Working Group 1 (WG1) Contribution to the Intergovernmental Panel on Climate Change (IPCC) 5th Assessment Report (AR5). Cambridge University Press

Jongman B, Ward PJ, Aerts JCJH (2012) Global exposure to river and coastal flooding: long term trends and changes. Glob Environ Chang 22:823–835
Jongman B, Hochrainer-Stigler S, Feyen L, Aerts JCJH, Mechler R, Botzen WJW, Bouwer LM, Pflug G, Rojas R, Ward PJ (2014) Increasing stress on disaster risk finance due to large floods. Nat Clim Chang 4:264–268. doi:10.1038/nclimate2124
Linham M, Green C, Nicholls RJ (2010) Costs of adaptation to the effects of climate change in the world's large port cities. AVOID - Avoiding dangerous climate change report AV/WS1/D1/02. Department of Energy and Climate Change (DECC) and Department for Environment Food and Rural Affairs (DEFRA), London
Lowe JA, Gregory JM, Flather RA (2001) Changes in the occurrence of storm surges around the United Kingdom under a future climate scenario using a dynamic storm surge model driven by the Hadley Centre climate models. Clim Dyn 18:179–188
MAFF (1999) Flood and coastal defense project appraisal guidance (FCDPAG3), vol 3. Economic appraisal. Defra publications ref. PB 4650, Defra, London
McFadden L, Spencer T, Nicholls RJ (2007) Broad-scale modeling of coastal wetlands: what is required? Hydrobiolgica 577:5–15
Meyer V et al (2013) Review article: assessing the costs of natural hazards-state of the art and knowledge gaps. Natl Hazards Earth Syst Sci 13(5):1351–1373
Mitchell TD, Jones PD (2005) An improved method of constructing a database of monthly climate observations and associated high-resolution grids. Int J Climatol 25:693–712
Mokrech M, Nicholls RJ, Richards JA, Henriques C, Holman IP, Shackley S (2008) Regional impact assessment of flooding under future climate and socio-economic scenarios for East Anglia and North West England. Clim Chang 90:31–55
Mokrech M, Kebede AS, Nicholls RJ, Wimmer F, Feyen L (2015) An integrated approach for assessing flood impacts due to future climate and socio-economic conditions and the scope of adaptation in Europe. Clim Chang 128(3–4):245–260
Nicholls RJ, Hanson S, Lowe JA, Warrick RA, Lu X, Long AJ (2014) Sea-level scenarios for evaluating coastal impacts. Wiley Interdiscip Rev Clim Chang 5(1):129–150
Rapport D, Friend A (1979) Towards a comprehensive framework for environmental statistics: a stress-response approach. Statistics Canada, Ottawa
Richards JA, Mokrech M, Berry PM, Nicholls RJ (2008) Climate change and floodplain ecosystems: regional assessment and adaptation potential. Clim Chang 90:141–167
Rojas R, Feyen L, Watkiss P (2013) Climate change and river floods in the European Union: socio-economic consequences and the costs and benefits of adaptation. Glob Environ Chang 23:1737–1751
Vafeidis AT, Nicholls RJ, McFadden L, Tol RSJ, Spencer T, Grashoff PS, Boot G, Klein RJT (2008) A new global coastal database for impact and vulnerability analysis to sea-level rise. J Coast Res 24(4):917–924
Verzano K (2009) Climate change impacts on flood related hydrological processes: further development and application of a global scale hydrological model. Reports on Earth System Science, 71-2009, Max Planck Institute for Meteorology, Hamburg
Wimmer F, Audsley E, Savin CM, Malsy M, Dunford R, Harrison PA, Schaldach R, Flörke M (2015) Modeling the effects of cross-sectoral water allocation schemes in Europe. Clim Chang 128(3–4):153–167

Chapter 17
Linking Participatory, Bayesian, and Agent-Based Modeling Techniques to Simulate Coupled Natural-Human System: A Case Study with Ranchers in Sonora, Mexico

Aloah Pope and Randy Gimblett

Chapter Highlights

Approach: We used Bayesian cognitive mapping to calculate the probability a rancher would make important decisions under a variety of environmental conditions; then demonstrated how to apply this approach in an agent-based model to explore uncertainty in decision-making.

Participant Engagement: Stakeholders and scientists engaged in two modeling workshops carried out in Rayon, Sonora, Mexico to develop a Bayesian cognitive map and populate it with data.

Models/Outcomes: The results of the Bayesian cognitive map and agent-based model were compared under potential normal and drought conditions.

Challenges: Developing Bayesian cognitive maps of human decision-making is a time-intensive endeavor and is a relatively untested approach to modeling decision-making. The workshops did not entice sufficient numbers of ranchers living outside of Rayon to attend, so the spatial extent of the study had to be limited.

17.1 Introduction

Competition between human use and ecological function for water from semi-arid rivers has caused increased conflicts in recent years. Riparian corridors in semi-arid regions are oases of biological diversity; however, their appealing features also

A. Pope (✉) • R. Gimblett
School of Natural Resources and the Environment, University of Arizona, Tucson, AZ, USA
e-mail: aloahpope@email.arizona.edu; gimblett@cals.arizona.edu

S. Gray et al. (eds.), *Environmental Modeling with Stakeholders*,
DOI 10.1007/978-3-319-25053-3_17

attract human settlements. River diversions and groundwater pumping have significantly reduced riparian ecosystem function, which has, in turn, impacted local human processes. Degradation of coupled natural-human systems within a semi-arid watershed produces drought, flooding, land cover vegetation change, altered agricultural and urban behaviors, as well as diminished ecosystem services.

Computer modeling provides an excellent conduit in which to study coupled natural-human systems (Schlüter et al. 2012). Such models are also capable of addressing issues that make traditional approaches to studying coupled natural-human systems difficult, such as non-linearity and uncertainty. To understand how future climatic conditions may influence social systems, we developed a model of ranchers' livelihoods in Sonora, Mexico, where conflict over water use is a debated topic. It is currently unknown how ranchers will react and adapt to challenges associated with impending climate change in the region.

17.1.1 The Rio Sonora Watershed and Its Ranchers

The Rio Sonora is a semi-arid watershed with north–south drainage among transverse mountains. Precipitation varies widely both annually and seasonally, averaging around 500mm annually (Lizarrago-Celaya et al. 2010). Most of the precipitation occurs during the monsoon season in the form of thunderstorms in July and August, while the remaining rain is dropped during winter as small events. Between the winter and monsoon rain events lies a dry, hot summer period. The basin experiences large variability in climate, with periods of drought and pluvial that can last several years to decades (Sheppard et al. 2002). While general circulation models predict a more arid climate in the future (Seager et al. 2007; Diffenbaugh et al. 2008), a dynamically downscaled model predicts increased precipitation (Robles-Morua et al. 2014), highlighting the uncertainties surrounding the effects of climate change for the Rio Sonora Basin.

Ranching was first introduced to the region in the early eighteenth century (Perramond 2010). By the end of the nineteenth century, much of the ranchland was privatized (Baroni 1991). Now, the Rio Sonora Watershed produces one of the greatest volumes of livestock, primarily cattle, in Mexico. Within the river's riparian corridor lie the lushest vegetation in the region, predominately cottonwoods and fodder crops. Outside the riparian corridor, the landscape is marginal (Liverman 1990), including cacti, desert shrubs, and mesquite trees. In a harsh semi-arid climate with limited water resources, ranching remains one of the few remaining options to support livelihoods (West 1993).

Not all ranches are created equal; ranches vary in size, from less than 500 hectares to over 10,000 hectares. The geography of the land varies from desert shrubland to grassy plains. The carrying capacity, or number of cattle per hectare a rancher can successfully produce on his land, depends on the composition of its vegetation. Lush riparian vegetation can sustain more cattle per hectare than desert shrubs, while variation in precipitation can significantly alter the amount of natural fodder produced within a land cover type. In a drought year, reduced rainfall results in reduced carrying capacity. In such a case, a rancher must make decisions to prevent cattle loss, such as selling off cattle early or purchasing additional feed.

Water resources are an important component to ranching operations. Water is a limited and costly, but necessary, resource for the production of cattle (Wilder and Whiteford 2006). One of the most common strategies for increasing water resources is to build stock ponds. Stock ponds are hand-dug depressions in the land that can store water from a precipitation event for up to several weeks. In the past 30 years, groundwater has become vital, producing water when rainwater is limited. During that time, overexploitation has reduced the groundwater so much that many of the wells have gone dry. Annual precipitation can refill a dry well, similar to the way it fills a stock pond, from which ranchers can then extract water.

The ranching lifestyle stresses individuality and is a valued subculture in Mexico (Jordan 1993). Some ranchers rotate their herds around the pasture to maximize productivity of the land after a rainfall event, while others spread their herd out to minimize labor costs. If water is in short supply and forage is minimal on the range, a rancher may be forced to sell part of his herd. Under more favorable water and forage conditions, a rancher can wait to sell his cattle until market prices peak. The method a rancher adopts to manage his herd is one of many decisions heavily influenced by both water and money. In an uncertain political and environmental climate, concern rises about the future of the rancher industry, rancher families, and their way of life.

As a case study for the Rio Sonora Watershed, a series of workshops were held in the city of Rayon, Sonora, Mexico. Founded in 1638, Rayon is one of the largest towns in the San Miguel River basin (Fig. 17.1), a sub-basin of the Rio Sonora, with a population of approximately 1500 people (INEGI 2010). For the city of Rayon, ranching is the most important component of the local economy. It is the purpose of this study to identify important decisions of ranchers in the study area, and attempt to predict the likelihood of decisions being made under various environmental conditions using Bayesian cognitive mapping for characterizing decision-making.

17.2 Bayesian Cognitive Mapping

Drawing from the fields of Bayesian probability theory and cognitive mapping, Bayesian cognitive mapping attempts to create a probability of the likelihood a certain decision will be made (Sedki and de Beaufort 2012). Cognitive mapping, or developing a network-based representation of an expert's cognition, produces a qualitative and static representation of decisions (Eden et al. 1992). Bayesian networks were developed based on probability theory to represent expert knowledge in situations in which that knowledge is ambiguous or incomplete (Spiegelhalter et al. 1993). By embedding cognitive maps with conditional probabilities, researchers can create a qualitative acyclic graph of the decision-making process, while within each decision relationships between components are expressed as conditional

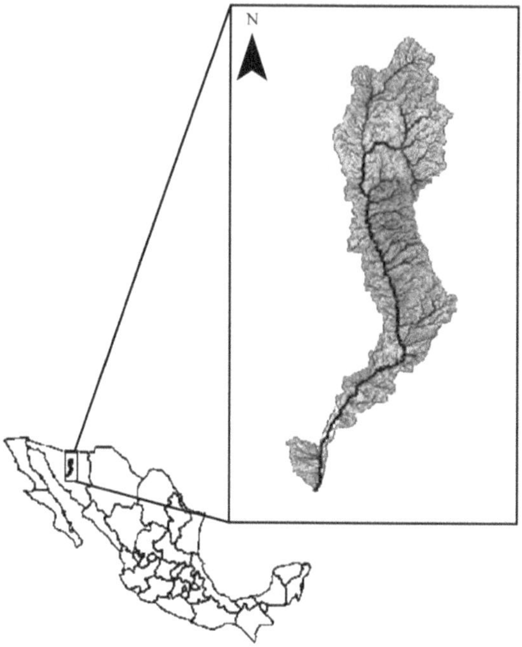

Fig. 17.1 The Rio Sonora Watershed lies within the Mexican state of Sonora

probabilities. Research in the fields of perceptual psychology and computational neuroscience provides increasing support that predictions of human behavior can be modeled with Bayesian networks (Griffiths and Tenenbaum 2006; Keshvari et al. 2012; Knill and Pouget 2004; Trommershauser et al. 2012; Kording 2014). A Bayesian approach to model decision-making has gained popularity in the fields of artificial intelligence and venture capitalism (Mishra et al. 2001; Kemmerer et al. 2002; Korb and Nicholson 2004). However, it is a relatively unknown approach in natural resource management that is typically incorporated as an intermediary node within a system-level integrated water resource model (Bromely et al. 2002; Giordano et al. 2010). Using this approach, researchers can incorporate inference and uncertainty in the complex decision-making process.

Bayesian causal networks are composed of three features: nodes, links, and probabilities. Nodes represent system variables and can be either discrete or continuous. For example, a discrete amount of rainfall could be below average, average, or above average whereas a continuous amount of rainfall can be binned across a full range of rainfall values. Links represent causal relationships between two nodes. A link, for example, from rainfall to crop yield would describe the effect of crop yield caused by the amount of rainfall. In Bayesian networks, the "effect" node (i.e., crop yield) is described as a child of the parent "cause" node of rainfall. Within each node lies a set of probabilities that defines the relationships between parents and their children. A unique probability specifies the likelihood a child will be in a certain state for each combination of parent states. The set of probabilities are called conditional probability tables (CPTs).

In our previous example of crop yield, let us say that both rainfall and crop choice are its parents (Fig. 17.2). In order to populate this Bayesian causal network with data, we would need to the know the probability of high, normal, or low crop yield given each combination of below average, average, or above average rainfall and corn, wheat, or soybean crop choice (Table 17.1). The combination of nodes and links create a directed acyclic graph, the structure of a Bayesian network. When developing the structure of a Bayesian network, it is important to minimize the number to nodes to the least number needed to accurately describe the system. This is to ensure the network is comprehensible to others, especially stakeholders who will be asked to fill in the conditional probability tables. The incorporation of conditional probabilities tables transforms the structure into a functioning Bayesian network. At this point, a Bayesian network can be analyzed. By changing the state of a parent node, one can observe its effects through-

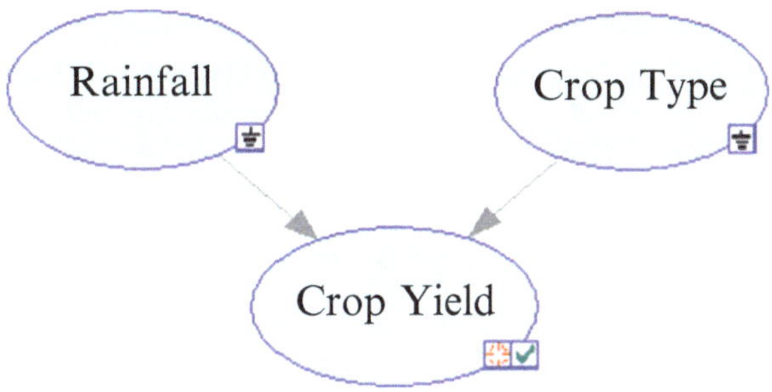

Fig. 17.2 Sample acyclic graph defining the causal relationship between rainfall, crop type, and crop yield

Table 17.1 Template conditional probability table for developing probabilities in Bayesian causal maps, wherein P(H) equals the probability of high yield and P(N) equals the probability of normal yield under all combinations of parent node (rainfall and crop choice) conditions

Rainfall	Below average			Average			Above average		
Crop choice	Corn	Wheat	Soybean	Corn	Wheat	Soybean	Corn	Wheat	Soybean
High yield	*P(H)*	*P(H)*	*P(H)*	*P(H)*	*P(H)*	*P(H)*	*P(H)*	*P(H)*	*P(H)*
Normal yield	*P(N)*	*P(N)*	*P(N)*	*P(N)*	*P(N)*	*P(N)*	*P(N)*	*P(N)*	*P(N)*
Low yield	*1-P(H)-P(N)*	*1-P(H)-P(N)*	*1-P(H)-P(N)*	*1-P(H)-P(N)*	*1-P(H)-P(N)*	*1-P(H)-P(N)*	*1-P(H)-P(N)*	*1-P(H)-P(N)*	*1-P(H)-P(N)*

out the network. Because perfect knowledge of a parent's state is rare, probability distributions across a parent's state can be used as input in Bayesian networks, incorporating uncertainty.

17.2.1 *Acyclic Graph Creation*

The construction of a Bayesian cognitive map requires stakeholder involvement in two steps: development of an acyclic graph structure and data collection (Cain 2001). A Bayesian cognitive map was developed through a series of stakeholder workshops held in Rayon, Sonora, Mexico. By the best estimates of the Secretariat of Agriculture, Livestock, Rural Development, Fisheries and Food, and the National Institute of Statistics and Geography, we surveyed between 3 and 5 % of the agriculturist (ranchers and farmers) population in Rayon. Development of the acyclic graph starts by choosing key variables. Among ranchers in Sonora, Mexico, the two big issues are livelihood sustainability and water security. Specifically, making enough money each year to support their families and receiving enough water to support their herds of cattle. Next, states must be chosen for key variables. In this model, we chose below normal, normal, and above normal states for nodes "Annual Profit" and "Water Availability" to encompass the entire range of conditions a stakeholder may perceive each year. The Bayesian network can now be developed both forward—to identify the important decisions of ranchers—and backward—to define when a rancher considers environmental conditions to be low, normal, and high.

17.2.1.1 Forward Development

In the first stakeholder workshop, held January 14, 2013, a survey was given to a group of ranchers to help us understand in detail some of the decisions they make based on alternative scenarios. We asked ranchers, "What options would you consider if you were unsatisfied with your annual profit?" Twenty-seven percent of ranchers stated they would consider changing careers, 18 % would consider obtaining an additional job, and 9 % would consider requesting government support or decreasing activity. Since changing careers was the most common response, we chose it as a decision node and followed up by asking ranchers, "What other variables would lead you to the decision to change careers?" Eighty-one percent of ranchers listed water scarcity, 27 % annual profit, 18 % extreme climate, and 9 % decreased natural forage or livestock sickness as contributing variables. Since annual profit and water availability were the top two responses, we added links from the child node "Change Careers" to parent nodes "Annual Profit" and "Water Availability."

The survey also asked, "What options would you consider if you were low on water?" Thirty-six percent of ranchers stated they would consider reducing herd size, 27 % would consider changing water consumption, and 9 % would consider requesting government support. Since both reducing herd size and changing water consumption were popular responses, we chose to create a node for each and asked follow-up questions: "What variables would go into your decision to change herd size?" and "What do you consider important decisions that may alter water consumption?" In the decision to change herd size, 45 % of ranchers listed water availability, 27 % listed annual profit, and 9 % listed livestock theft. Since annual profit and water availability were the top two responses, we added links from the child node "Change Herd Size" to parent nodes "Annual Profit" and "Water Availability." Of decisions that may alter water consumption, 36 % of ranchers considered increasing water efficiency to be important, 18 % considered increasing water from wells via digging a new well or reducing herd size, and 9 % considered using floodwater to support forage production or rotating cattle on pasture. As per popular answers, nodes "Increase Efficiency" and "Dig New Well" were added to the Bayesian network. Follow-up discussions with stakeholders elucidated that although a rancher may decide to dig a new well, it does not necessarily mean that a rancher will be given permission from the government to build one. Therefore, an additional node "New Well Approved" was added and linked from "Dig New Well."

17.2.1.2 Backward Development

In the second stakeholder workshop, held January 30, 2014, a survey was given to a group of ranchers asking them to define when they considered annual profit and water availability to be below normal, normal, or above normal. Ranchers defined below normal annual profit to be when forage production was low, and they therefore *must* invest in the purchase of additional feed for the cattle. In contrast, ranchers defined above normal annual profit to be when there is high productivity of forage, and they therefore *can* invest in the purchase of additional feed for the cattle. Ranchers defined normal annual profit to be when there is average productivity. Water availability was considered below normal when there was a drought and when the water table was below the depth of their well. Based on stakeholder responses, a parent node "Invest" was created and linked to "Annual Profit." The states within "Invest" were defined as must, indifferent, and can. Since the purchase of additional feed was defined as a response to forage conditions and only done in years of either high or low profit, a node "Purchase Additional Feed" was added to the model with links to both "Annual Profit" and "Water Availability."

Water availability was considered normal if there was average rainfall. Water availability was considered above normal if there was above average rainfall, or if the water table was above the depth of their well. Based on stakeholder responses, parent nodes of "Rainfall" and "Well water" were created and linked to "Water Availability." The states of "Rainfall" were defined as below normal, normal, and above normal. The states of "Well water" were defined as below water table or above water table. With this information, we have the completed acyclic graph structure of Sonoran Ranchers decisions (Fig. 17.3).

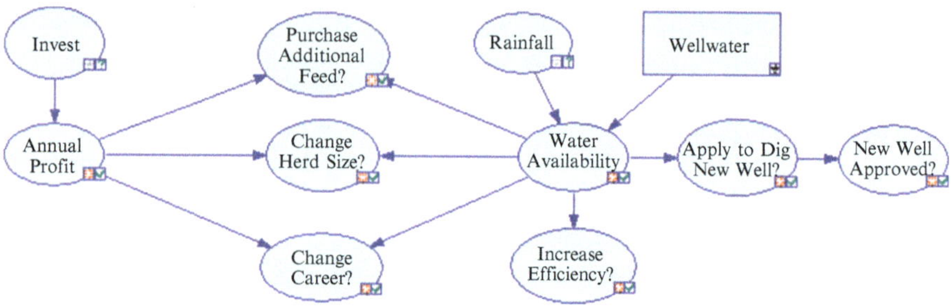

Fig. 17.3 Acyclic graph of important rancher decisions and contributing variables in Rayon, Sonora, Mexico

17.2.2 Conditional Probabilities

The goal of developing the Bayesian cognitive map is to create a unique probability of a decision being made for each combination of contributing variables as described in the sample conditional probability table in Table 17.1. In order to populate the conditional probability table, a questionnaire to ranchers was distributed at the second stakeholder workshop. For each iteration of contributing variables (below normal/normal/above normal), stakeholders were asked to respond with the likelihood they would make a positive decision on a Likert scale, from "Not At All Likely" to "Completely Likely." Each response on the Likert scale then translated to a probability from 0 to 100 %. The average probability within each iteration of contributing variables was used to calculate the conditional probability table for decisions "Purchase Additional Feed?," "Change Herd Size?," and "Change Career?" (see Table 17.2). The average probability a rancher would choose to dig a new well if water availability was below normal is 77 %, if water availability was normal is 43 %, and if water availability was above normal is 67 %. If digging a new well was

Table 17.2 Average probability of making a positive decision under all iterations of low, normal, and high values of annual profit and water availability

Water availability	*Low*	*Normal*	*High*	*Low*	*Normal*	*High*	*Low*	*Normal*	*High*
Annual profit	*Low*	*Low*	*Low*	*Normal*	*Normal*	*Normal*	*High*	*High*	*High*
Purchase additional feed?	59	55	53	64	58	68	55	68	71
Change herd size?	79	57	84	52	34	50	54	41	42
Change career?	55	50	68	55	44	54	34	41	44

chosen, ranchers believed they would have a 61 % chance of gaining approval from the government to dig. Additionally, the average probability a rancher would choose to increase water efficiency if water availability is below normal is 85 %, if water availability was normal is 72 %, and if water availability was above normal is 66 %.

The "Annual Profit" node was defined so that it is low if "Invest" is set to must, but high if "Invest" is set to can. The "Water Availability" node is defined as above normal regardless of how "Rainfall" is set if "Well water" is set to below the water table, meaning the well always has access to water. If "Well water" is set to above the water table, "Water Availability" depends on how "Rainfall" is set. The "Water Availability" node is defined so that it is below normal if "Rainfall" is set to below normal, but above normal if "Rainfall" is set to above normal. The Bayesian cognitive model was developed in GeNIe, a Bayesian network software tool (http://genie.sis.pitt.edu/).

17.2.3 Bayesian Cognitive Map of Sonoran Ranchers

Once compiled, the Bayesian cognitive map automatically calculates the probability a positive decision will be made based on user-input on parent nodes "Invest," "Rainfall," and "Well water." To illustrate the utility of the Bayesian cognitive map in predicting rancher behavior, imagine a rancher with an uncertain annual profit, a well that is too shallow to obtain rainwater, and a year with variable, but normal potential rainfall. To create this scenario in the Bayesian cognitive map, we input probabilities in "Invest" to be 33.3 % likely to be can, indifferent, and must. For rainfall, we input 25 % below normal, 50 % normal, and 25 % above normal rainfall. Then, for "Well water," we input above well water. The scenario of unknown annual profit and normal potential rainfall results in conditional probabilities for each decision (Fig. 17.4). Under this scenario, there is a 64 % chance the rancher will purchase additional feed, a 43 % he will change herd size, a 45 % chance he will change careers, a 74 % chance he will attempt to increase water efficiency, and a 35 %

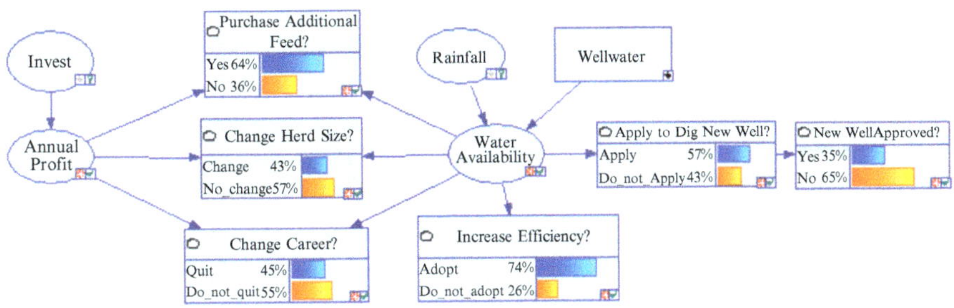

Fig. 17.4 Bayesian cognitive map of important decisions when an individual rancher's annual profit is unknown, his well is dry, and potential rainfall is under normal conditions

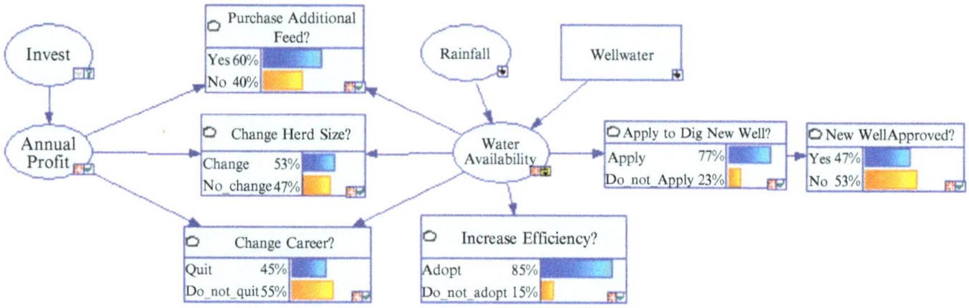

Fig. 17.5 Bayesian cognitive map of important decisions when an individual rancher's annual profit is unknown, his well is dry, and potential rainfall is under drought conditions

chance he will dig a new well. Compare these probabilities of making a positive decision when it is a potential drought year instead (Fig. 17.5). Under a drought (Rainfall: below normal = 70 %, normal = 20 %, and above normal = 10 %), a rancher is less likely to purchase feed (60 %), more likely to change herd size (53 %), just as likely to change careers (45 %), more likely to increase water efficiency (85 %), and more likely to dig a new well (47 %).

17.3 Temporal Dynamics

Bayesian cognitive maps represent a snapshot in time—in this case an agricultural year. If we are concerned about the future of a coupled natural-human system—in this case semi-arid ranchland—we must find a way to incorporate temporal dynamics. One way to do so is to apply the same Bayesian cognitive map to consecutive years, but allow the variables at the top of the network to change. This could be natural change in a coupled natural-human systems, such as fluctuating rainfall patterns, or changing node states based on the results of the Bayesian cognitive map from the previous time step. To demonstrate the application of Bayesian cognitive mapping in a temporal landscape, we integrated our Bayesian cognitive map of Sonoran ranchers into an agent-based model.

Agent-based models (ABM) are simulations of autonomous entities (agents) that respond heterogeneously to their environment (patches). Each agent has a defined set of simple rules from which to behave. Patches host a set of variables that can update during each time step that subsequently changes the environment. Since agent-based modeling is a bottom-up approach, researchers can explore how intricacies in micro-scale behavior can influence macro-scale patterns (Schlüter et al. 2012). Linking Bayesian cognitive maps with agent-based modeling techniques is a new approach in modeling coupled natural-human systems. A primary concern in the coupled modeling approach is that uncertainty in social systems is not being addressed (Schlüter et al. 2012). Using this Bayesian approach with ABM, researchers can incorporate inference and uncertainty in the complex decision-making process. The ABM was developed in Netlogo, a multi-agent programmable modeling environment, because it is freely available and user-friendly for non-modelers (http://ccl.northwestern.edu/netlogo/).

17.3.1 Decision Rules

Since our Bayesian cognitive map was developed from participation with individual ranchers, each agent in our agent-based model will also represent an individual rancher. Our Bayesian cognitive map also tells us that each rancher's value of water availability, annual profit, and herd size is independent of other ranchers. These variables are defined as rancher attributes. Each of these rancher attributes can be in three states: below normal, normal, and above normal. Value of water availability is determined by a combination of rainfall and well conditions. The conditions of the "Rainfall" node as well as the "Well water" node are used as inputs into the Bayesian cognitive map. User input of potential rainfall conditions are used to test various future climate scenarios. Normal rainfall conditions are defined as 25 % below normal, 50 % normal, and 25 % above normal rainfall. Drought conditions for potential rainfall are defined as 70 % below normal, 20 % normal, and 10 % above normal rainfall. If a rancher decides to dig a new well, the node "Well water" switches to below water table and the resulting new probabilities are calculated. The value of annual profit is determined by the output of the previous time step. If water availability is considered below normal and a rancher agent chooses to purchase feed, annual profit is set to below normal. If water availability is considered above normal or normal and a rancher agent chooses to purchase feed, annual profit is set to above normal. The new value of profit is used as input into the next year's Bayesian cognitive map. Herd size is determined by the output of the Bayesian cognitive map. If a rancher decides to change herd size and water availability is below normal, herd size drops to below normal. If a rancher decides to change herd size and water availability is above normal, herd size improves to above normal. Each time an agent state changes, the Bayesian cognitive map updates to calculate new probabilities. All attributes are set to normal at initialization of the model.

The output of the Bayesian cognitive maps are used as the probability of approving the following decisions in the model: purchasing additional feed, changing herd size, and digging a new well. In order to incorporate uncertainty via Bayesian probabilities, all decisions use the following equations:

$$\text{IF } P_R < P_D, \text{THEN } D = 1;$$

$$\text{IF } P_R > P_D, \text{THEN } D = 0$$

wherein P_R equals a random probability between 0 and 100 and P_D equals the calculated probability from the Bayesian cognitive map for making decision D. If the model calculates $D=1$, a positive decision has been made.

17.3.1.1 Scenarios

Since the future of the local climate, specifically water security, is one of the greatest concerns for Sonoran Ranchers, the agent-based model can be used to test how a potential extended drought may influence herd size and ranchers' annual profits. Using Netlogo's BehaviorSpace Tool, 30 repetitions were simulated for two treatments: normal and drought potential rainfall conditions for 10 consecutive years. The percentage of ranchers with below normal, normal, and above normal states of herd size and annual profits were compared between treatments. A Two-Way ANOVA was used to test whether there was any statistical difference between the two treatments.

17.3.2 Agent-Based Model Results

At the end of 10 years, 61 % percent of ranchers in the normal potential rainfall scenario had normal herd sizes, 24 % of ranchers had greater than normal herd sizes, and 15 % had below normal herd sizes. Annual profits of 34 % of ranchers in the normal potential rainfall scenario were normal, 51 % were above normal, and 15 % were below normal. In the drought scenario, 32 % of ranchers experienced below normal herd sizes, 54 % normal herd sizes, and 14 % above normal herd sizes. Forty-two percent of ranchers in the drought scenario experienced below normal annual profits, 35 % normal annual profits, and 23 % above normal profits. Herd sizes and annual profits among the normal and drought scenarios were significantly different ($p<0.01$, Fig. 17.6). In comparison to a predicted normal potential rainfall condition, drought conditions shifted ranchers into smaller herd sizes and reduced annual profits. A final step would be presenting the completed model to stakeholders for model exploration and validation.

17.4 Conclusions and Further Work

Developing Bayesian cognitive maps of human decision-making is a time-intensive endeavor; however, the approach is holistic, easy to use, encourages stakeholder participation, and values individual variation. By comparing potential normal and drought rainfall conditions, the Bayesian cognitive map elucidated the changing likelihoods of important water decisions being made by individual ranchers. Applying these likelihoods into an agent-based model

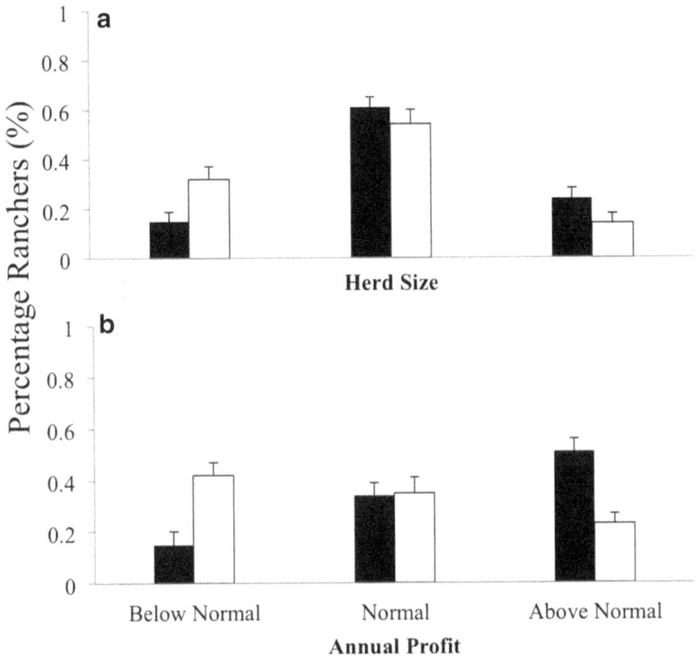

Fig. 17.6 Percentage of ranchers who observed below normal, normal, or above normal states of (**a**) herd size and (**b**) annual profits for a decade of predicted normal (*black bars*) and drought (*open bars*) conditions

allowed us to observe individual variation in ranchers' decision-making from the effects of drought on herd size and profit. These results would not have been created using traditional approaches to decision-making, such as deterministic modeling.

This chapter provides an introduction into how Bayesian cognitive maps can be created via participatory approaches and translated into an agent-based model to help explore the uncertainty of human decision-making. For our purposes, we wanted to understand how ranchers may be impacted by changes in future precipitation patterns. Increased sophistication of the model is currently underway to include real-time rainfall data, water availability based on depths of each rancher's well, estimates of herd size for each rancher, and water demand based on herd size. The purpose of increasing sophistication is to calculate any possible disjunctions between water availability and water demand among ranchers. Using this approach, the only change needed in the Bayesian cognitive map is how water availability is defined. Instead of below normal, normal, and above normal rainfall, we will be using deficit, balance, or surplus of water availability versus cattle demand. Increasing detail within the model is not the only way to expand it; recent work also explores how human water consumption effects with riparian vegetation, including changes in the quality of its ecosystems services (Pope and Gimblett 2015). By using a Bayesian cognitive mapping approach, we have a sophisticated base upon which we can explore the interdependencies of this complex system.

By creating a holistic decision-making model of ranchers, researchers and stakeholders alike can examine how micro-scale changes in semi-arid ranchlands can influence macro-scale patterns. Stakeholders will now have the ability to interact with both the Bayesian cognitive map and the agent-based model to explore the dynamics of their semi-arid ranchlands, highlight the most influential components, and even help identify what parts are not properly understood. Stakeholders also gain the ability to run scenarios—such as changes in climate—to get an idea of how ranchers in their area may react. Facing a future of uncertainty and increasing human influence, these types of models can be extremely useful tools for decision-makers, increasing the ability to make better-informed decisions to improve resilience of ranchers in the Sonoran Desert. Since external stressors and internal conflicts are likely to worsen if current trends of increasing water demand and decreasing water supply in semi-arid regions continue, this work may improve research and assessment in other threatened coupled natural-human systems.

Acknowledgements We thank the staff, students, and PIs of The University of Arizona, Universidad de Sonora, Colegio de Sonora, and University of Copenhagen working on the "Strengthening the Resilience of Arid Region Riparian Corridors" project. We particularly thank Alan Navarro and America Lutz for help with stakeholder workshop coordination and execution. This research was sponsored by a NSF Dynamics of Coupled Human-Natural Systems grant (DEB-1010495).

References

Baroni A (1991) Agricultura, ganaderia y sociedad en la Cuenca media del Rio Sonora de 1900 a 1950. In: Camou Healy E (ed) Potreros, Vegas y Mahuechis. Gobierno del Estado de Sonora, Hermosillo, pp 61–119

Bromely J, Jackson N, Giacomello A-M, Bradford RJ (2002) Participatory development of Bayesian networks as an aid to integrated water resource planning. In: Proceedings of the international conference on policies and tools for sustainable water management in the European Union, Venice, November 2002

Cain J (2001) Planning improvements in natural resources management; guidelines for using Bayesian networks to manage development projects. Institute of Hydrology, Wallingford, UK

Diffenbaugh NS, Giorgi F, Pal JS (2008) Geophys Res Lett 35, L16709

Eden C, Ackermann F, Cropper S (1992) The analysis of cause maps. J Manag Stud 29:309–323

Giordano R, Preziosi E, Romano E (2010) An integration between cognitive map and Bayesian belief network for conflicts analysis in drought management. In: Proceedings of the international congress on environmental modeling and software, Ottawa, July 2010

Griffiths T, Tenenbaum J (2006) Optimal predictions in everyday cognition. Psychol Sci 17:767–773

Jordan T (1993) North American Cattle Ranching Frontiers: origins, diffusion and differentiation. University of New Mexico Press, Albuquerque

Kemmerer B, Mishra S, Shenoy P (2002) Bayesian causal maps as decision aids in venture capital decision making: methods and applications. In: Proceedings of the academy of management conference, August 2002

Keshvari S, van den Berg R, Ma W (2012) Probabilistic computation in human perception under variability in encoding precision. PLoS ONE 7:1–9

Knill D, Pouget A (2004) The Bayesian brain: the role of uncertainty in neural coding and computation. TRENDS Neurosci 27:712–719

Korb K, Nicholson A (2004) Bayesian artificial intelligence. Chapman and Hall, Boca Raton

Kording K (2014) Bayesian statistics: relevant for the brain? Curr Opin Neurobiol 25:130–133

Liverman D (1990) Drought impacts in Mexico: ritual, capital, and environmental degradation in the Cordillera of Northern Luzon, 1900–1986. University of California Press, Berkley

Lizarrago-Celaya C, Watts C, Rodrigues J, Garatuza-Payan J, Scott R, Saiz-Hernandez J (2010) Spatio-temporal variations in surface characteristics over the North American Monsoon region. J Arid Environ 74:540–548

Mishra S, Kemmerer B, Shenoy P (2001) Managing venture capital investment decisions: a knowledge-based approach. Working Paper, School of Business, University of Kansas

Perramond E (2010) Political ecologies of cattle ranching in Northern Mexico. The University of Arizona Press, Tucson

Pope A, Gimblett HR (2015) Linking bayesian and agent-based models to simulate complex social-ecological systems in Sonora, Mexico. Frontiers in Environmental Science

Robles-Morua A, Che D, Mayer AS, Vivoni ER (2014) Hydrologic assessment of proposed reservoirs in the Sonora River Basin, Mexico, under historical and future climate scenarios. Hydrol Sci J. doi:10.1080/02626667.2013.878462

Schlüter M, McAllister R, Arlinghaus R, Bunnefeld N, Eisenack K, Hölker F, Milner-Gulland E, Müller B (2012) New horizons for managing the environment: a review of coupled social-ecological systems modeling. Nat Resour Model 25:219–272

Seager R, Ting M, Held I, Kishnir Y, Lu J, Vecchi G, Huang H-P, Harnik N, Leetmaa A, Lau N-C, Li C, Velez J, Naik N (2007) Model predictions of an imminent transition to a more arid climate in southwestern North America. Science 316:1181–1184

Sedki K, de Beaufort L (2012) Cognitive maps and Bayesian networks for knowledge representation and reasoning. In: 24th international conference on tools with artificial intelligence, Greece.

Sheppard PR, Comrie AC, Packin GD, Angersbach K, Hughes MK (2002) The climate of the US Southwest. Clim Res 21:219–238

Spiegelhalter D, Dawid A, Lauritzen S, Cowell R (1993) Bayesian analysis in expert systems. Stat Sci 8:219–247

Trommershauser J, Kording K, Landy M (2012) Sensory cue integration. Oxford University Press, New York

West R (1993) Sonora: its geographical personality. University of Texas Press, Austin

Wilder M, Whiteford S (2006) Flowing uphill toward money: groundwater management and ejidal producers in Mexico's free trade environment. In: Randall L (ed) Changing structure of Mexico: political social and economic prospects. M. E. Sharpe, New York, pp 341–358

Index

S. Gray et al. (eds.), *Environmental Modeling with Stakeholders*,
DOI 10.1007/978-3-319-25053-3

C

D

Kolibri 360 GmbH
Ferdinand-Jühlke-Straße 7
99095 Erfurt, Deutschland
produktsicherheit@kolibri360.de